# ENGINEERING SCIENCE IN SI UNITS

Other works by the same authors

by Edward Hughes

*Fundamentals of Electrical Engineering*
*Electrical Technology*

By Arthur Morley and Edward Hughes

*Basic Engineering Science*
*Elementary Engineering Science*
*Mechanical Engineering Science*
*Electrical Engineering Science*
*Principles of Electricity in SI Units*

# Engineering Science in SI Units

EDWARD HUGHES

D.Sc.(Eng.), Ph.D., C.Eng., F.I.E.E.

*Fellow of the Heriot-Watt College, Edinburgh*
*Formerly Vice-Principal and Head of the Engineering Department*
*Brighton College of Technology*

and

CHRISTOPHER HUGHES
B.Sc.(Eng.), C.Eng., M.I.Mech.E.

*Senior Lecturer in Mechanical Engineering*
*North Gloucestershire College of Technology,*
*Cheltenham*

WITH DIAGRAMS

Longman

LONGMAN GROUP LIMITED
London

*Associated companies, branches and representatives throughout the world*



*First published* 1970
*Second Impression* 1971
*Third Impression* † 1972

ISBN 0 582 42543 3

Printed in Singapore by
New Art Printing Co., (Pte.) Ltd.

# Preface

In 1960 the General Conference of Weights and Measures recommended that the International System of Units should be universally adopted. This system is an extension and refinement of the traditional metric system. It embodies features which make it logically superior to any other system as well as being more convenient in practice: it is rational, coherent and comprehensive.

The idea of a decimal system of units was conceived by Simon Stevin (1548–1620) and the metric system was legally adopted in France in 1795, at the end of the Revolution. This metric system was based on the metre as the unit of length and the gramme as the unit of mass. The metre was intended to be one ten-millionth part of the distance from the north pole to the equator at sea level through Paris, and the gramme was to be the mass of one cubic centimetre of water at 0°C.

In 1873 the British Association for the Advancement of Science selected the centimetre, the gramme and the second as basic units of length, mass and time for scientific purposes—hence the c.g.s. system. About 1900 practical measurements began to be based on the metre, the kilogramme and the second; and in 1950 the International Electrotechnical Commission recommended that this MKS system of mechanics should be linked with the ampere as a fourth basic unit, giving the MKSA system.

Since 1875 all international matters concerning the metric system have been the responsibility of the Conférence Générale des Poids et Mesures (CGPM). In 1954 the CGPM adopted a rationalized and coherent system of units based on the four MKSA units, together with the kelvin as the unit of temperature and the candela as the unit of luminous intensity, and in 1960 gave it the title of 'Système International d'Unités' for which the abbreviation is SI in all languages.

About thirty countries have already decided to make SI the only legally accepted system and it is only a matter of time for it to become the universal currency of science and commerce. SI should be legally in force in this country by 1971 and most students entering engineering and science courses in 1969 and after will be examined in SI units. The adoption of SI will simplify engineering calculations considerably and avoid the necessity of memorizing relationships such as 1 hp = 550 ft lbf/s = 746 W.

This textbook covers, in terms of SI units only, the syllabus in Engineering Science of the G1 and G2 General Course in Engineering and also the syllabuses of a number of Technician Courses. It deals with the basic principles of Mechanics, Heat and Electricity. The text includes 119 worked examples and 384 problems. Some of the latter have been taken from examination papers; and for permission to reproduce these questions we are grateful to the East Midland Educational Union, the Northern Counties Technical Examinations Council, the Union of Educational Institutions and the Union of Lancashire & Cheshire Institutes.

We wish to express our thanks to Mr Alan Jackson, B.Sc.(Eng.), for reading through the manuscript and making a number of valuable suggestions.

E.H.
C.H.

*Notes:* (a) In the third impression of this book, the spelling 'kilogram' has been adopted in accordance with B.S. 3763: 1970.

(b) At the 1971 General Conference of Weights and Measures (CGPM), the *pascal* (= 1 N/m$^2$) and the *siemens* were approved as the units of pressure (and stress) and of electric conductance respectively.

# Contents

# Symbols and abbreviations

Based on British Standard 3763: 1970, and *The International System of Units*, 1970 (HMSO).

*Notes on the use of symbols and abbreviations*

1. A unit symbol is the same for the singular and the plural: for example, 10 kg, 5 V.

2. Full point should be omitted after a unit symbol.

3. Full point should be used in a multi-word abbreviation: for example, s.t.p., e.m.f.

4. In a compound-unit symbol, the product of two units is preferably indicated by a dot, especially in manuscript. The dot may be dispensed with when there is no risk of confusion with another unit symbol: for example, N·m or N m, but not mN.
A solidus (/) denotes division: for example, m/s, J/kg.

5. A unit symbol should be used only after a numerical value: for example, *m* kilograms, 5 kg; *I* amperes, 10 A.

6. A hyphen is inserted between the numerical value and the unit when the combination is used adjectivally: for example, a 2-metre (or 2-m) rod; a 240-volt (or 240-V) supply.

7. The abbreviations 'a.c.' and 'd.c.' should only be used adjectivally: for example, a.c. circuit, d.c. motor.

## 1. GENERAL

| Quantity | Quantity symbol | Unit | Unit symbol |
|---|---|---|---|
| Acceleration, linear | $a, f$ | metre per second squared | $m/s^2$ |
| Acceleration, local gravitational | $g$ | metre per second squared | $m/s^2$ |
| Acceleration, angular | $\alpha$ | radian per second squared | $rad/s^2$ |

| Quantity | Quantity symbol | Unit | Unit symbol |
|---|---|---|---|
| Angle, plane | $\alpha$, $\theta$ | radian | rad |
| | | degree | ° |
| Density | $\rho$ | kilogram per cubic metre | kg/m³ |
| Density, relative | $d$ | – | – |
| Distance along path | $s$ | metre | m |
| Efficiency | $\eta$ | – | – |
| Diameter | $d$ | metre | m |
| Energy | $W$ | joule | J |
| | | kilojoule | kJ |
| | | megajoule | MJ |
| | | watt hour | W h |
| | | kilowatt hour | kW h |
| Force | $F$ | newton | N |
| | | kilonewton | kN |
| Length | $l$ | metre | m |
| | | kilometre | km |
| Mass | $m$ | kilogram | kg |
| | | tonne (= 1000 kg) | t |
| Power | $P$ | watt | W |
| | | kilowatt | kW |
| Pressure | $p$ | newton per square metre | $N/m^2$ |
| | | pascal (= 1 $N/m^2$) | Pa |
| Time | $t$ | second | s |
| | | minute | min |
| | | hour | h |
| Torque | $T$ | newton metre | N m |
| Velocity, linear | $u$, $v$ | metre per second | m/s |
| | | kilometre per hour | km/h |
| Velocity, angular | $\omega$ | radian per second | rad/s |
| Volume | $V$ | cubic metre | $m^3$ |
| | | litre (= 0·001 $m^3$) | l |
| Weight | $W$ | newton | N |
| Work | $W$ | joule | J |

## 2. MECHANICS AND THERMODYNAMICS

| Quantity | Quantity symbol | Unit | Unit symbol |
|---|---|---|---|
| Coefficient of friction | $\mu$ | – | – |
| Energy, kinetic | $T$ | joule | J |
| Energy, potential | $V$ | joule | J |
| Momentum | $p$ | kilogram metre per second | kg m/s |
| Young's modulus | $E$ | newton per square metre | $N/m^2$ |
| | | or pascal | Pa |
| | | giganewton per square | $GN/m^2$ |
| | | metre or gigapascal | GPa |
| Coefficient of linear expansion | $\alpha$ | per degree Celsius | /°C |

| Quantity | Quantity symbol | Unit | Unit symbol |
|---|---|---|---|
| Coefficient of cubic expansion | $\gamma$ | per degree Celsius | /°C |
| Heat, latent | $L$ | joule | J |
| | | kilojoule | kJ |
| Heat, specific latent | $l$ | joule per kilogram | J/kg |
| | | kilojoule per kilogram | kJ/kg |
| Heat, quantity of | $Q$ | joule | J |
| Specific heat capacity | $c$ | joule per kilogram degree Celsius | J/kg °C |
| Temperature, thermodynamic | $T$ | kelvin | K |
| Temperature, Celsius | $\theta$, $t$* | degree Celsius | °C |
| Temperature interval | – | kelvin, degree Celsius | K, °C |

## 3. ELECTROMAGNETISM

| Quantity | Quantity symbol | Unit | Unit symbol |
|---|---|---|---|
| Charge or Quantity of electricity | $Q$ | coulomb | C |
| Conductance | $G$ | siemens | S |
| Conductivity | $\sigma$ | siemens per metre | S/m |
| Current: | | | |
| Steady value | $I$ | ampere | A |
| | | milliampere | mA |
| | | microampere | μA |
| Instantaneous value | $i$ | | |
| Current density | $J$ | ampere per square metre | A/m² |
| Difference of potential: | | | |
| Steady value | $V$ | volt | V |
| Instantaneous value | $v$ | | |
| Electromotive force: | | | |
| Steady value | $E$ | volt | V |
| Instantaneous value | $e$ | | |
| Frequency | $f$ | hertz | Hz |
| Inductance, self | $L$ | henry (plural, henrys) | H |
| Inductance, mutual | $M$ | henry (plural, henrys) | H |
| Magnetic flux | $\Phi$ | weber | Wb |
| Magnetic flux density | $B$ | tesla | T |
| Resistance | $R$ | ohm | Ω |
| | | microhm | μΩ |
| | | megohm | MΩ |
| Resistivity | $\rho$ | ohm metre | Ω m |
| | | microhm metre | μΩ m |

* In B.S. 1991: Part 1, 1967, $\theta$ and $t$ are given as alternative symbols for temperature, with $\theta$ as the preferred symbol. Most standard textbooks on Thermodynamics use $t$; consequently, $t$ has been adopted in this book.

Temperature *difference* in degrees Celsius is the same as in kelvins, i.e. change of temperature of 1°C = change of temperature of 1 K.

## 4. PREFIXES DENOTING DECIMAL MULTIPLES or SUB-MULTIPLES

| Value | Prefix | Symbol |
|---|---|---|
| $10^{12}$ | tera | T |
| $10^{9}$ | giga* | G |
| $10^{6}$ | mega | M |
| $10^{3}$ | kilo | k |
| $10^{2}$† | hecto | h |
| $10^{1}$† | deca | da |
| $10^{-1}$† | deci | d |
| $10^{-2}$† | centi | c |
| $10^{-3}$ | milli | m |
| $10^{-6}$ | micro | μ |
| $10^{-9}$ | nano | n |
| $10^{-12}$ | pico | p |

There should be no space or hyphen between the prefix and the name of the unit which it qualifies. Similarly, there should be no space or hyphen between the symbols for the prefix and the unit. For example: kilogram, kg; microsecond, us. A double prefix such as kilokilogram should never be used.

## 5. GREEK LETTERS USED AS SYMBOLS IN THIS BOOK

| Letter | Capital | Lower case |
|---|---|---|
| Alpha | – | $\alpha$ (angle, coefficient of linear expansion, temperature coefficient of resistance) |
| Beta | – | $\beta$ (coefficient of superficial expansion) |
| Gamma | – | $\gamma$ (coefficient of cubic expansion) |
| Eta | – | $\eta$ (efficiency) |
| Theta | – | $\theta$ (angle, temperature) |
| Mu | – | $\mu$ (micro, coefficient of friction) |
| Pi | – | $\pi$ (circumference/diameter) |
| Rho | – | $\rho$ (density, resistivity) |
| Sigma | – | $\sigma$ (conductivity) |
| Phi | $\Phi$ (magnetic flux) | $\varphi$ (angle) |
| Omega | Ω (ohm) | $\omega$ (angular velocity) |

* *Giga* is derived from a Greek word meaning giant and dictionaries state that its pronunciation should be the same as 'giga' in gigantic.

† Powers which are a multiple of 3 are generally preferred, but because of the large value that may result in the case of some derived units, such as volume, these smaller multiples are permissible, but their use should be limited as far as possible.

CHAPTER 1

# Units

## 1.1 The International System of Units (SI)

The International System of Units, known as SI in every language, derives all the units used in the various technologies from the following *seven* base units:

| *Quantity* | *Unit* | *Symbol* |
|---|---|---|
| length | metre | m |
| mass | kilogram | kg |
| time | second | s |
| electric current | ampere | A |
| temperature | kelvin | K |
| luminous intensity | candela | cd |
| amount of substance | mole | mol |

The candela and the mole are not dealt with in this book and will therefore not be referred to again.

The SI* is a *coherent* system of units, i.e. the product or quotient of any two unit quantities in the system is the unit of the resultant quantity. For example, unit area (= 1 square metre) results when unit length (= 1 metre) is multiplied by unit length,

or 1 square metre = 1 metre × 1 metre.

Further examples illustrating the coherent nature of SI are:

(*a*) unit velocity (= 1 m/s) results when unit length (= 1 m) is divided by unit time (= 1 s);

(*b*) unit force (= 1 newton) results when unit mass (= 1 kg) is multiplied by unit acceleration (= 1 $m/s^2$).

* It is incorrect to speak of 'SI system'.

In practice, it is often convenient to use a multiple or submultiple of the SI base unit, but the choice should, in general, be confined to powers of ten* which are a multiple of $\pm$ 3, thereby effecting a considerable reduction in the number of multiples and submultiples,

e.g.

$$1 \text{ kilometre} = 10^3 \text{ metres},$$
$$1 \text{ millimetre} = 10^{-3} \text{ metre},$$
$$\left.\begin{array}{r}1 \text{ micrometre}\\ \text{or micron}\end{array}\right\} = 10^{-6} \text{ metre}.$$

For *communication* purposes it is usually convenient to express a quantity in terms of a multiple or submultiple of the SI base unit, thereby avoiding the use of very large or very small numbers. For example, it is convenient to express the calorific value of coal as, say, 30 megajoules per kilogram rather than 30 000 000 joules per kilogram. On the other hand, for *calculation* purposes and especially in equations, it is advisable to express the quantities in terms of the SI base units and to insert the unit symbol in brackets† after the numerical value. For example, when clockwise and anticlockwise moments are being equated, the values should be expressed in the following form:

$$1000\,[\text{N}] \times 0{\cdot}008\,[\text{m}] = 16\,[\text{N}] \times 0{\cdot}5\,[\text{m}].$$

This precaution ensures that the two sides of the equation are dimensionally the same.

## 1.2 SI Base Units of Length, Mass and Time

(*a*) *Metre.* The metre is the SI unit of length and is defined as the length equal to 1 650 763·73 wavelengths of the orange line in the spectrum of an internationally-specified krypton discharge lamp.

* It is important that students should be familiar with the use of indices and with the multiplication and division of numbers expressed in the form $10^n$. For example, 2 000 000 can more conveniently be written as $2 \times 10^6$, and an error of a nought is far more likely to be made in the extended than in the shorter form.

When multiplying, say, $10^6$ by $10^3$, we add the indices; and when dividing $10^6$ by $10^3$, we subtract the indices: thus,

$$10^6 \times 10^3 = 10^9 \quad \text{and} \quad 10^6 \div 10^3 = 10^3.$$

In general, $10^a \times 10^b = 10^{(a+b)}$ and $10^a \div 10^b = 10^{(a-b)}$

† Brackets, [ ], are more satisfactory than parenthesis, ( ), as they indicate more definitely that, say, 10[m] represents 10 metres and not 10 × m.

This definition reproduces the metre with an accuracy of one part in a hundred million ($10^8$). The metre was formerly defined as the distance between two lines on a certain platinum-iridium bar at 0°C. This bar is kept at the International Bureau of Weights and Measures at Sèvres, near Paris, and is still used as a reference standard. Its exact length is periodically determined in terms of the wavelengths of radiation from the krypton lamp.

(*b*) *Kilogram.* The kilogram is the SI unit of mass and is defined as the mass of a platinum-iridium cylinder kept at Sèvres. The mass of another body can be compared with that of this standard cylinder and can be determined with an accuracy of one part in a hundred million ($10^8$) by means of a specially-constructed balance.

(*c*) *Second.* The second is the SI unit of time and is defined as the interval occupied by 9 192 631 770 cycles of radiation corresponding to the transition of the caesium-133 atom. This definition enables the fantastic precision of one part in ten thousand million ($10^{10}$) to be achieved. The second is approximately 1/86 400 of the mean solar day.

1 minute [min] = 60 seconds [s]

and 1 hour [h] = 3600 seconds.

The minute and the hour are not decimal multiples of the second and are therefore non-SI units, but they are so firmly established in practice that their use is likely to continue indefinitely.

The SI base units of electric current and of temperature respectively are discussed in the chapters relating to their applications (chapters 14 and 11 respectively).

The degree of precision in the determination of the fundamental units of length, mass and time is really beyond human comprehension and is possible only with the elaborate apparatus available at national laboratories such as the National Physical Laboratory in this country. No student should be expected to memorise the figures given above.

For normal practical purposes, sub-standards of *length* and *mass* are used. The exact value of these sub-standards are periodically checked against those of the standards kept at the national laboratories. Sub standards of *time*, such as clocks and watches, can easily be checked against time signals transmitted from observatories or from broadcasting sources such as the B.B.C.

## 1.3 SI derived* units of area and volume

If the floor of a room is 6 m long and 4 m wide,

$$\text{area of floor} = 6\ [\text{m}] \times 4\ [\text{m}] = 24\ \text{m}^2.$$

It will be noted that the unit symbol for square metre is '$\text{m}^2$' and not 'sq. m.'.

If a room has a floor area of 24 $\text{m}^2$ and a height of 5 m,

$$\text{volume of room} = 24\ [\text{m}^2] \times 5\ [\text{m}] = 120\ \text{m}^3.$$

An alternative unit of volume, often used in the metric system, is the *litre* (symbol, l). In 1964, the General Conference of Weights and Measures decided that the litre should be used as a special name for 1000 $\text{cm}^3$ (or 0·001 $\text{m}^3$) and not as the volume of 1 kg of pure water at maximum density, as had previously been the practice. Precise measurement has shown that 1 kg of pure water at maximum density and under atmospheric pressure is 1·000 028 litres; but for most practical purposes, we can assume that 1 litre of water has a mass of 1 kg.

## 1.4 Density and relative density

The *density* of a body is defined as the *mass per unit volume* and is represented by the Greek letter $\rho$ (rho). Thus, if a body has a mass of $m$ kilograms and a volume of $V$ cubic metres,

$$\text{density} = \rho = m/V \text{ kilograms/cubic metre.}$$

It was mentioned in section 1.3 that the mass of 1 litre of water is practically 1 kg.

$$\begin{aligned} \text{Since } 1\ \text{m}^3 &= 1000 \text{ litres} \\ \therefore \quad \text{mass of } 1\ \text{m}^3 \text{ of water} &= 1000\ \text{kg} \\ \text{and density of water} &= 1000\ \text{kg/m}^3. \end{aligned}$$

The *relative density* (symbol, $d$) of a material is defined as the ratio:

$$\frac{\text{density of the material}}{\text{density of water}}$$

For example, if the density of copper is 8900 $\text{kg/m}^3$,

$$\text{relative density of copper} = \frac{9800\ [\text{kg/m}^3]}{1000\ [\text{kg/m}^3]} = 8{\cdot}9.$$

* Derived units are obtained from the multiplication or division of base units, e.g. $\text{kg/m}^3$ for density.

Since the relative density is a ratio of two quantities expressed in the same units, it is purely a number and therefore has no units.

**Example 1.1** *A block of steel,* 200 *mm* × 100 *mm* × 60 *mm, has a mass of* 9·42 *kg. Calculate (a) the density and (b) the relative density of the steel.*

(*a*) $$\text{Volume of block} = 0{\cdot}2\ [\text{m}] \times 0{\cdot}1\ [\text{m}] \times 0{\cdot}06\ [\text{m}] = 0{\cdot}0012\ \text{m}^3,$$

$$\therefore \quad \text{density of steel} = \frac{9{\cdot}42\ [\text{kg}]}{0{\cdot}0012\ [\text{m}^3]} = 7850\ \text{kg/m}^3.$$

(*b*) $$\text{Since the density of water} = 1000\ \text{kg/m}^3$$

$$\therefore \quad \text{relative density of steel} = \frac{7850\ [\text{kg/m}^3]}{1000\ [\text{kg/m}^3]} = 7{\cdot}85.$$

## 1.5 Determination of density and relative density

(*a*) *Solids.* If the solid is in the form of a rectangular or cylindrical block, its dimensions can be measured and its volume calculated. The mass of the block can be determined by means of a beam balance such as an ordinary chemical balance, and the values of the density and of the relative density can then be calculated as in Example 1.1.

In the case of a solid having an irregular shape, a simple, though not very accurate, method of determining its *volume* is to fill vessel

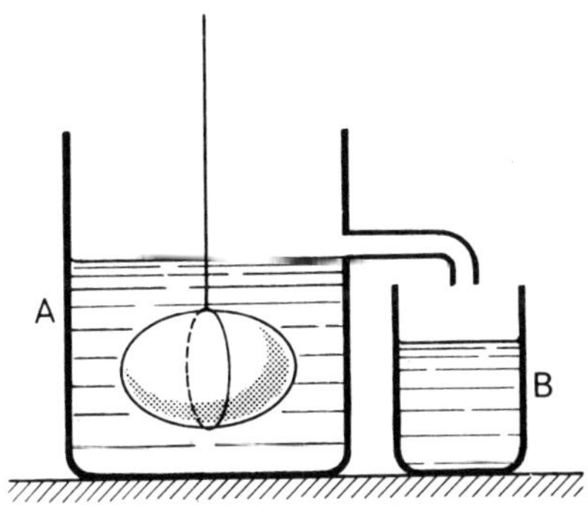


Fig. 1.1 Determination of the density of a solid.

A, shown in fig. 1.1, with water until it overflows. The object is then immersed in the water and the amount of water displaced is collected in vessel B. The mass of B is determined before and after the water is collected, and the difference in the readings gives the mass of water displaced. The volume of the solid can then be calculated from the fact that the volume of 1 kg of water is $0{\cdot}001\ m^3$.

(*b*) *Liquids.* If a vessel of known volume is available, the mass of that volume of liquid is the difference between the mass of the vessel (i) empty and (ii) filled with the liquid. Hence the density of the liquid can be calculated.

The relative density of a liquid can be accurately determined by means of a special glass bottle known as a 'relative density bottle'. The glass stopper has a bore of small diameter centrally through it and fits accurately into the opening, as shown in fig. 1.2. Consequently, when the bottle is filled with a liquid and the stopper inserted, the surplus liquid flows out and the bottle will then contain a definite volume of the liquid.

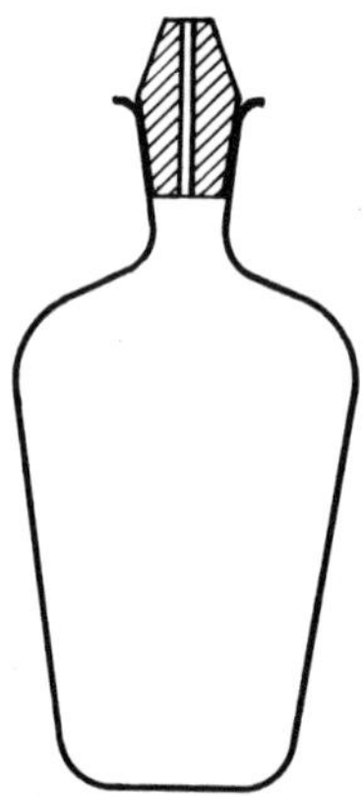

Fig. 1.2 A relative density bottle.

The procedure is to weigh the bottle after it has been dried internally and externally. It is then filled with water, the external surface is dried and the bottle is again weighed. The test is repeated with the liquid whose relative density is required.

Then,

$$
\begin{aligned}
\text{if } x &= \text{mass of bottle empty,}\\
y &= \text{mass of bottle filled with water}\\
\text{and} \quad z &= \text{mass of bottle filled with the other liquid,}\\
\text{mass of water} &= y - x\\
\text{and} \quad \text{mass of other liquid} &= z - x\\
\therefore \text{relative density of the other liquid} &= \frac{z - x}{y - x}.
\end{aligned}
$$

## 1.6 Force

Mechanics is largely concerned with force and the effects of force. Everyone who has exerted muscular force in lifting an object, pushing a load or overcoming any kind of resistance, has some conception of force.

Force is generally recognized and measured by its effects. Force exerted on a body tends to change its motion, and unless resisted by another force, will do so; for example, the pushing at a door by a hand, the pulling of a train by a locomotive and the acceleration of a falling body by the force of gravity.

Another effect which always occurs when a force is exerted on a body is that the material of the body is deformed, i.e. its shape is altered. For example, it may be stretched, compressed, bent or twisted. Frequently, this deformation is so small as to be imperceptible to the unaided eye.

## 1.7 Mass and weight

The *mass* of a body is usually defined as the quantity of matter in the body, but the nature of mass cannot be fully appreciated without reference to its quality of *inertia* or reluctance to change of velocity. For instance, a cricket ball, thrown with a given force in a given direction, will travel a certain distance; but if a ball of the same size, made of lead, was thrown in the same direction with the same force, its acceleration and maximum velocity would be much less than those of the cricket ball, and it would not travel as far. This is due to the lead ball having a greater inertia or reluctance to acceleration than the cricket ball, and so we say that the mass of the lead ball is greater than that of the cricket ball.

A great English scientist, Sir Isaac Newton (1642-1727), showed that the earth attracts every body to itself with a force proportional to the mass of the body, and that the value of this force decreases as the body moves further away from the surface of the earth. The force with which a body is attracted towards the earth is termed the *weight* of that body, and its direction is always towards the centre of the earth.

Since the earth is not a perfect sphere, its radius being less at the poles than at the equator, the pull of the earth on a given mass is slightly different at different places on the earth's surface. For instance, if the scale of a *spring* balance is calibrated in London to read exactly 1 kilogram when a mass of 1 kg is suspended from the balance, then, if the balance and the same mass of 1 kg were taken to the north or south pole, the scale reading would be about 1·002 kg, whereas at the equator the reading would be about 0·997 kg.

Had the mass of 1 kg been placed in one pan of a beam balance, such as an ordinary chemical balance, and pieces of metal placed on the other pan until the beam was exactly balanced, then this equilibrium would hold whether the balance were used in London, at the equator, on the moon or at any point where the masses were subject to a gravitational force. This is due to the gravitational force at a given place being the same for the two pans and their contents. Hence it will be seen that the beam balance measures the *mass* of a body in comparison with the known mass of another body, and the result is independent of the value of the gravitational force at the place where the measurement is being made. A spring balance, on the other hand, measures the *weight* of the body; and its reading, for a given suspended mass, varies from place to place on the earth's surface and would be considerably less on the moon's surface.

## 1.8 Unit of force

At this stage we can only deal with the subject in a brief and simple manner—full discussion will be found in Chapter 7. All we need say here is that the SI unit of force is termed the *newton* (symbol, N), to commemorate the name of Sir Isaac Newton, and is defined as *the force which, when applied to a mass of* 1 *kilogram gives it an acceleration of* 1 *metre per second every second* (*or* 1 $m/s^2$).

When a mass of 1 kg rests on, say, a table as shown in fig. 1.3, it exerts a downward force on that table, but the magnitude of that force is different at different points on the earth's surface, as already mentioned in section 1.7. This means that if a body is allowed to fall freely (i.e. if the air resistance is negligible), the acceleration varies at different places. For instance, at sea level in

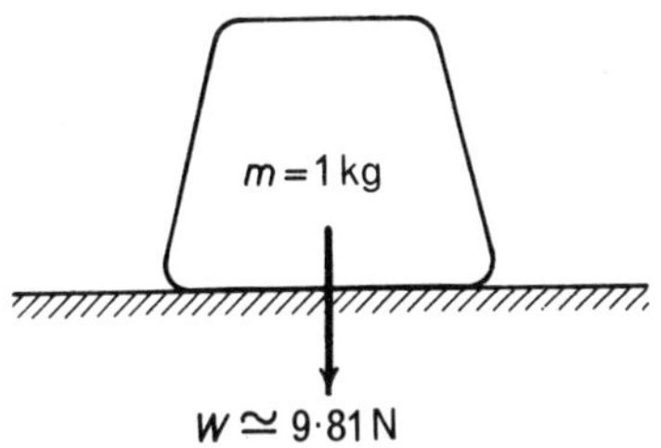


Fig. 1.3 Mass and weight.

the vicinity of London, gravitational acceleration is almost exactly 9·81 m/s², whereas at the equator the acceleration is about 9·780 m/s² and at each pole it is about 9·832 m/s². From the definition of the newton, it follows that the gravitational force at sea level on a mass of 1 kg (i.e. the weight of 1 kg) is almost exactly 9·81 newtons in London, 9·78 newtons at the equator and 9·832 newtons at each pole, as shown in fig. 1.4.

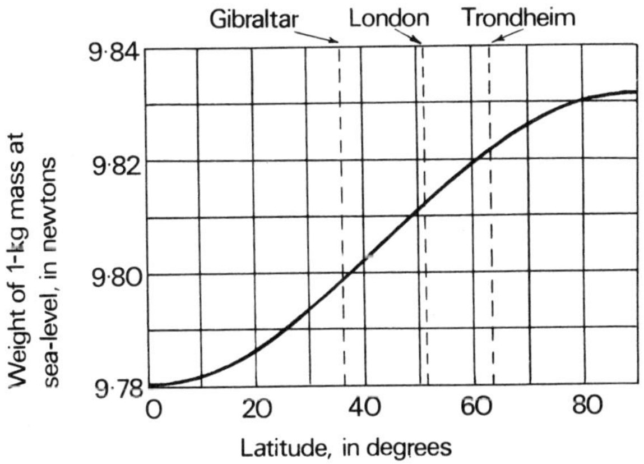


Fig. 1.4 Variation of weight with latitude.

It will be seen from fig. 1.4 that the weight of a 1-kg mass at sea level is within the range 9·81 N $\pm$ 0·1 per cent for latitudes extending from 38° to 61°, i.e. from the north coast of Africa to the central region of Norway, in the northern hemisphere, and for the corresponding region in the southern hemisphere. It follows that within this range of latitude, we can say that for most practical purposes:

$$\text{weight of a body} \simeq 9{\cdot}81\, m \text{ newtons},$$

where $m$ is the mass of the body in kilograms. The sign $\simeq$ will be used in this relationship, wherever it appears in this book, to indicate that '9·81' may be an approximate value and that the precise value depends upon the latitude at which the figure is used. Thus the force exerted on the table by the mass of 1 kg in fig. 1.3 is practically 9·81 N.

For very rough estimates, we could assume the weight of a body having a mass of $m$ kilograms to be 10 $m$ newtons.

**Example 1.2.** *A steel block having a mass of* 80 *kg rests on a table, the area of contact with the table being* 2000 *mm². Calculate* (a) *the downward force on the table and* (b) *the average pressure on the table in kilonewtons per square metre.*

(*a*) Downward force

$$\text{on table} = \text{weight of block}$$
$$\simeq 80\,[\text{kg}] \times 9{\cdot}81\,[\text{N/kg}]^{*} = 784{\cdot}8\text{ N}.$$

(*b*)

$$\text{Area of contact} = 2000\text{ mm}^2$$
$$= \frac{2000\,[\text{mm}^2]}{1\,000\,000\,[\text{mm}^2/\text{m}^2]} = 0{\cdot}002\text{ m}^2.$$

$\therefore$ average pressure

$$\text{on table} = \text{force per unit area}$$
$$= \frac{\text{downward force}}{\text{area of contact}}$$
$$= \frac{784{\cdot}8\,[\text{N}]}{0{\cdot}002\,[\text{m}^2]} = 392\,400\text{ N/m}^2$$
$$= 392{\cdot}4\text{ kN/m}^2.$$

* This expression can alternatively be stated thus;

$$\text{weight of block} \simeq 80\,[\text{kg}] \times 9{\cdot}81\,[\text{m/s}^2] = 784{\cdot}8\text{ N}.$$

When the body under consideration is *stationary*, [N/kg] is better since it emphasizes the relationship between mass and weight, namely that a mass of 1 kg has a weight of approximately 9·81 N.

## 1.9 Work

When a force is exerted against some form of resistance through a distance in the direction of the force, *work* is said to be done. For example, work is done if a body is lifted vertically upward against the gravitational pull of the earth; or if a spring is stretched, compressed or bent against the elastic resistance to deformation; or if a body is accelerated by a force applied to it.

The SI unit of work is the *joule** (symbol, J) to commemorate the English physicist James P. Joule (1818–89), famous for his experiments on the relationship between mechanical and thermal energies. The *joule* is defined as *the work done when a force of* 1 *newton is exerted through a distance of* 1 *metre in the direction of the force.* Hence, if a force $F$, in newtons, is exerted through a distance $s$, in metres, in the direction of the force,

$$\text{work done, in joules} = F\,[\text{newtons}] \times s\,[\text{metres}]$$

$$= Fs \qquad (1.1)$$

$$1 \text{ kilojoule [kJ]} = 1000 \text{ J},$$

$$1 \text{ megajoule [MJ]} = 1\,000\,000 \text{ J}.$$

**Example 1.3.** *The work done in moving a body through a distance of* 30 *m is* 600 *J. Assuming the force to act in the direction of motion, calculate the average value of the force.*

From the expression (1.1), we have:

$$600\,[\text{J}] = F \times 30\,[\text{m}]$$

$$\therefore \qquad F = 20 \text{ N}.$$

**Example 1.4** *A body having a mass of* 20 *kg is lifted through a distance of* 15 *m. Calculate the work done.*

$$\text{Weight of the body} \simeq 20\,[\text{kg}] \times 9{\cdot}81\,[\text{N/kg}] = 196{\cdot}2 \text{ N},$$

$$\therefore \qquad \text{work done} = 196{\cdot}2\,[\text{N}] \times 15\,[\text{m}] = 2943 \text{ J}$$

$$= 2{\cdot}943 \text{ kJ}.$$

## 1.10 Power

*Power* is defined as *the rate of doing work* and is a measure of the work done per second. The SI unit of *power* is the *watt* (symbol, W), named after the famous Scottish engineer James Watt (1736-1819). *The watt is equal to* 1 *joule per second.*

* *Joule* is pronounced 'jool', rhyming with 'tool'.

$$1 \text{ kilowatt [kW]} = 1000 \text{ W},$$
$$1 \text{ megawatt [MW]} = 1\,000\,000 \text{ W}.$$

**Example 1.5** *A horizontal force of* 60 *N is applied to a body to move it at a uniform velocity through a distance of* 20 *m in* 8 *s in the direction of the force. Calculate the value of the power.*

$$\text{Work done} = 60 \text{ [N]} \times 20 \text{ [m]} = 1200 \text{ J},$$
$$\therefore \quad \text{power} = \text{work done per second}$$
$$= \frac{1200 \text{ [J]}}{8 \text{ [s]}} = 150 \text{ W}.$$

**Example 1.6** *Calculate the power required to lift a mass of* 300 *kg at a constant speed through a vertical height of* 200 *m in* 4 *min.*

$$\text{Force required to lift load} \simeq 300 \text{[kg]} \times 9{\cdot}81 \text{[N/kg]} = 2943 \text{ N}.$$
$$\text{Work done in 4 min} = 2943 \text{ [N]} \times 200 \text{ [m]}$$
$$= 588\,600 \text{ J}.$$
$$\therefore \quad \text{power} = \frac{588\,600 \text{ [J]}}{(4 \times 60) \text{ [s]}} = 2453 \text{ W}$$
$$= 2{\cdot}453 \text{ kW}.$$

*Summary of Chapter* 1

The SI units of length, mass and time are the metre, the kilogram and the second respectively.

Density is the mass per unit volume and the relative density of a material is the ratio of the density of that material to the density of water.

The SI unit of force is the newton, namely the force required to give a mass of 1 kg an acceleration of 1 $m/s^2$.

The SI unit of work is the joule. It is the work done when a force of 1 N is exerted through a distance of 1 m in the direction of the force.

$$\text{Work done, in joules} = F\,s \tag{1.1}$$

Power is the rate of doing work and the SI unit of power is the watt which is 1 joule/second.

## EXAMPLES 1

1. Calculate the area, in square metres, of: (*a*) a square of side 600 mm, (*b*) a triangle of base 400 mm and height 500 mm and (*c*) a circle of diameter 1·5 m.
2. Calculate the volume, in cubic metres, of: (*a*) a room, 7 m × 4 m × 3 m and (*b*) a sphere having a diameter of 1·5 m.
3. Calculate the length of each side of a cube having a volume of 3 $m^3$.
4. Calculate the diameter of a sphere having a volume of 0·8 $m^3$.
5. A steel plate measures 500 mm × 200 mm × 10 mm. If the relative density of the steel is 7·8, calculate the mass of the plate.
6. If the density of aluminium is 2700 kg/$m^3$, what is the volume, in cubic centimetres, of 5 kg of this material?
7. A block of wood, measuring 300 mm × 150 mm × 60 mm, has a mass of 2·3 kg. Calculate the density of the wood.
8. If the mass of 7500 $mm^3$ of lead is 85 g, calculate the relative density of the lead.
9. A 2-m length of iron pipe has a mass of 137 kg. The external and internal diameters of the pipe are 160 mm and 120 mm respectively. Calculate (*a*) the density and (*b*) the relative density of the iron.
10. Hydrogen has a density of 89·9 g/$m^3$ at a certain temperature and a certain pressure. Calculate the volume, in cubic metres, of hydrogen having a mass of 0·1 kg at the same temperature and pressure.
11. A 10-kg mass is suspended at the end of a cord. Calculate the approximate value of the force, in newtons, exerted by the cord.
12. A body is suspended from a spring balance. The reading on the balance is 4 N. Assuming the local gravitational force to be 9·81 N/kg, calculate the mass of the body in grams.
13. A vertical column supports a body having a mass of 6 Mg (or 6 tonnes). Calculate the approximate value of the upward force, in kilonewtons, exerted by the column.
14. A cylindrical vessel has an internal diameter of 100 mm and an internal length of 300 mm. It is stood on end and filled with oil having a relative density of 0·8. Calculate (*a*) the total thrust on the bottom of the cylinder and (*b*) the pressure on the bottom of the cylinder, in kilonewtons per square metre.
15. A force of 30 N acts through a distance of 4 m in the direction of the force. What is the work done?
16. A force of 700 N does 2500 J of work. Through what distance does the force move, assuming that it acts in the direction of motion?
17. Calculate the force, in kilonewtons, required to do 3 MJ of work when it moves through a distance of 200 m in the direction of the force.
18. A horizontal force of 20 N is applied to a body to move it at a uniform velocity through a distance of 80 m in 6 s in the direction of the force. Calculate (*a*) the work done and (*b*) the power.
19. The work done by a force in moving a body at a uniform speed through a distance of 20 m, in the direction of the force, is 3 kJ. If this work is done in 10 s, calculate (*a*) the force and (*b*) the power.

20. The power required to lift a certain body through a distance of 30 m in 6 s is 200 W. Calculate (*a*) the work done and (*b*) the mass of the body.
21. If 40 MJ of work are done in lifting 5 Mg (or 5 t) of coal, through what height is it lifted? If the time taken to lift the 5 Mg of coal is 120 s, what is the power required? Neglect any losses.
22. If 60 $m^3$ of water are pumped per hour to a height of 80 m, calculate (*a*) the power required, in kilowatts, and (*b*) the work done, in megajoules, in 10 min. Neglect any losses and assume that 1 $m^3$ of water has a mass of 1000 kg.

## ANSWERS TO EXAMPLES 1

1. 0·36 $m^2$, 0·1 $m^2$, 1·765 $m^2$.
2. 84 $m^3$, 1·77 $m^3$.
3. 1·44 m.
4. 1·15 m.
5. 7·8 kg.
6. 1850 $cm^3$.
7. 852 $kg/m^3$.
8. 11·3.
9. 7800 $kg/m^3$, 7·8.
10. 1·112 $m^3$.
11. 98·1 N.
12. 408 g.
13. 58·86 kN.
14. 18·5 N, 2·354 $kN/m^2$.
15. 120 J.
16. 3·57 m.
17. 15 kN.
18. 1600 J, 267 W.
19. 150 N, 300 W.
20. 1200 J, 4·08 kg.
21. 815 m, 333 kW.
22. 13·08 kW, 7·848 MJ.

CHAPTER 2

# Triangle and polygon of forces

## 2.1 Forces and reactions

One of the most important facts to be appreciated in mechanics is that to every force there is an equal and opposite resistance or reaction. This is easily stated but not so readily fully understood, but consideration of simple examples will help.

Suppose a piece of metal to be suspended by a string as shown in fig. 2.1. The *weight* of the metal exerts a *downward* pull on the string, and the string exerts an *upward* force on the piece of metal.

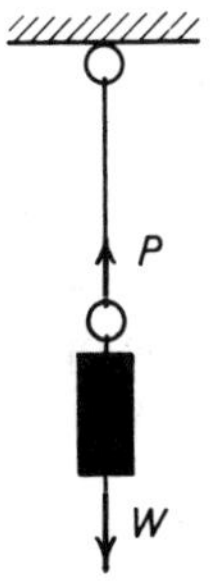

Fig. 2.1 Force exerted on a body.

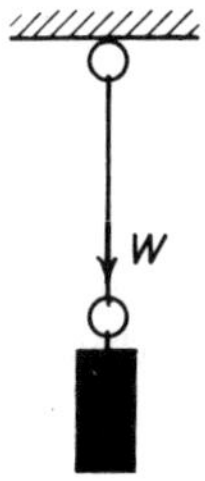

Fig. 2.2 Force exerted by a body

Since the body is stationary, it is in a state of *equilibrium*, and force $P$ is therefore exactly equal in magnitude and opposite in direction to force $W$. A body cannot remain in equilibrium under the action of a *single* force. For instance, if the string in fig. 2.1 were cut, the only force acting on the metal would be its weight, and the metal would fall to the ground, its speed increasing as it fell.

The forces shown in fig. 2.1 are those exerted *on* the piece of metal. The force of gravity $W$ is acting downward; and force $P$ is therefore acting upward and is the *reaction* to the downward force $W$ exerted *by* the piece of metal on the string, as shown in fig. 2.2.

Let us next consider a rigid* bar AB, fig. 2.3(a). Suppose the bar to be pulled by two forces *P* applied, say, by helical springs attached to the loops at its ends. Force *P* is transmitted by the material throughout the whole length of the bar. If we consider any short length CD of the bar, it is pulled at C by a force *P*, acting towards A, exerted by the material in the bar lying to the left of C. It is also pulled at D by an equal force *P*, acting towards B, exerted by the material to the right of D. Also, piece CD, when forming part of the whole bar, as in fig. 2.3(a), exerts a rightward pull *P* at C on the portion AC of the bar, and a leftward pull *P* at D on portion DB of the bar, as shown in fig. 2.3(b).

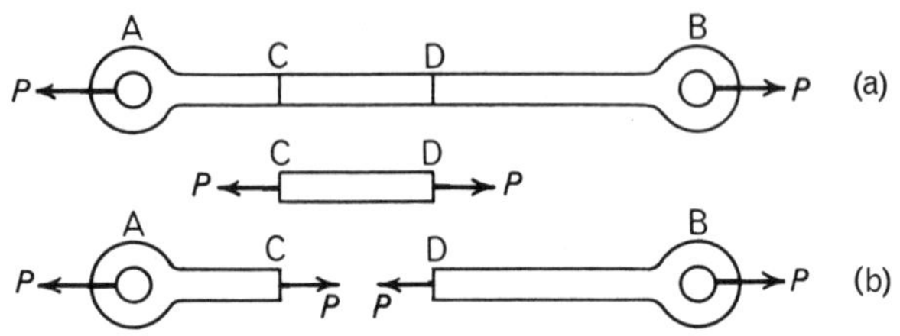


Fig. 2.3 Tension in a bar.

It will be seen that at any cross-section of the bar there are equal and opposite forces *P*. Each of the two pieces, into which any cross-section divides the bar, pulls the piece on the opposite side of the section towards itself. It is this dual action which constitutes *tension*.

The forces *P* shown in fig. 2.3(a) are those exerted outwards *on* the bar *by* the helical springs at A and B. The forces exerted *by* the bar *on* the springs are equal to *P* but are inwards. When arrowheads are used to indicate forces in such a bar, it is necessary to state whether we are indicating the forces exerted *on* the bar or those exerted *by* the bar on the springs or other means used for exerting the tension.

## 2.2 Representation of a force by a vector

A force has the following characteristics:

(*a*) magnitude,

(*b*) direction (line of action and sense),

(*c*) point of application.

* A rigid body is a body whose size and shape are unaffected by forces acting upon it. No solid is perfectly rigid, but we assume it is so unless otherwise stated.

For example, in fig. 2.1, the line of action of the force $P$ exerted by the string on the suspended metal is vertical and its sense is upwards. The magnitude of this force may be, say, 30 N and is the same at all points along the length of the string.

A quantity which has magnitude and direction is referred to as a *vector quantity,* whereas a quantity which has magnitude only is known as a *scalar quantity*. Quantities such as mass and time possess magnitude only and therefore are scalar quantities. Force, on the other hand, possesses both magnitude and direction and is consequently a vector quantity.

In the case of a body suspended from a string (fig. 2.1), it is not sufficient to state that the line of action of force $P$ is vertical; we must also state or indicate that this force is acting upwards, i.e. the *sense* of the force must be specified.

A vector quantity can be represented by a straight line whose length represents to a known scale the magnitude of the quantity, and whose direction is parallel to the line of action of that quantity. Such a line is termed a *vector*. For example, if the *weight*, $W$, of the suspended mass in fig. 2.1 is 30 N, then, using a scale of, say 1 mm to 1 N, we can represent this weight by a vertical line *ab*, 30 mm long, as shown in fig. 2.4(a). The direction of this force is indicated by the arrowhead. Also, when we say that the force is represented by vector *ab*, we mean that the direction of the force is downward from *a* to *b*, not upward from *b* to *a*.

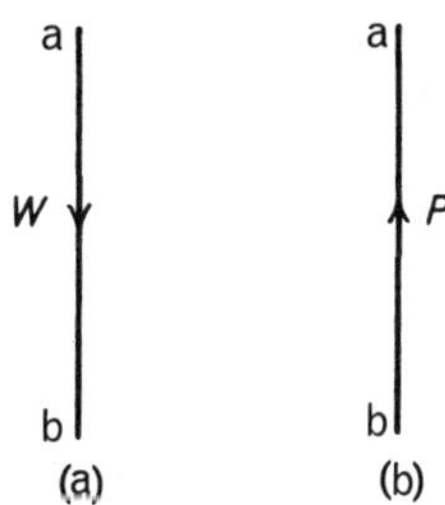


Fig. 2.4 Representation of a force by a vector.

The force $P$ exerted *on* the mass *by* the string is equal and exactly opposite to force $W$ and is therefore represented to the same scale by vector *ba* in fig. 2.4(b); i.e. vector *ba* represents a force of 30 N acting vertically upward.

## 2.3 Equilibrium

When two or more forces act upon a body and are so arranged that the body remains at rest or moves at a constant speed in a straight line, the forces are said to be in *equilibrium*.

It was mentioned in section 2.1 that a body cannot remain in equilibrium under the action of a single force. The effect of a single force on a body is to accelerate the body in the direction of the force.

The simplest case of equilibrium is that of two equal and opposite forces in the same straight line, such as the weight $W$ of the piece of metal acting vertically downward in fig. 2.1, and the equal upward reaction or supporting force $P$ exerted by the string. In this case, the body remains at rest.

The forces may be two equal and opposite *pulls*, as in figs. 2.1 and 2.3, in which case the body is said to be in *tension*. On the other hand, the forces may be two equal and opposite *pushes*, in which case the body is said to be in *compression*. For instance, if a helical spring, anchored at one end, has a push applied at the other end, the spring is compressed; and the push, together with the equal and opposite reaction exerted by the anchorage at the other end, are compressive forces.

Another example of the equilibrium of two equal and opposite forces is that of a locomotive pulling a train at a *constant* speed along a straight level track. The forward pull exerted by the locomotive on the train is equal to the backward drag exerted by the train on the locomotive. This backward drag is due to friction at the axles and the rails together with some air resistance. If the engine were to exert a larger pull than the drag due to friction and air resistance, the forces would not be in equilibrium and the train would accelerate.

In this chapter, we shall confine our attention to the equilibrium of forces acting on a body which remains stationary.

We shall next consider the condition for the equilibrium of a body subjected to three inclined co-planar forces, i.e. three non-parallel forces acting in the same plane.

## 2.4 Three inclined forces: Triangle of Forces

Fig. 2.5 shows three strings in the same horizontal plane, knotted together or attached to a small ring. Spring balances P, Q, and R, calibrated in newtons, are attached to the respective strings.

Suppose the strings attached to balances P, Q and R to be subjected to tensions of 5 N, 4 N and 6 N respectively. The directions of the three strings are marked on a sheet of paper pinned to a board behind the strings. The paper is then removed and a triangle *cab* is constructed by drawing side *ca* parallel to the string attached to balance R, and making *ca* 120 mm long to represent the force of 6 N at a scale of 1 mm to 0·05 N. The triangle is then completed by drawing lines from *a* and *c* respectively parallel to the directions of the other two strings to meet at *b*. On measuring these two sides, it is found that *ab* represents 5 N and *bc* represents 4 N to the same scale as that used for *ca*. The degree of accuracy in this experiment will depend upon the care exercised and the quality and condition of the spring balances.

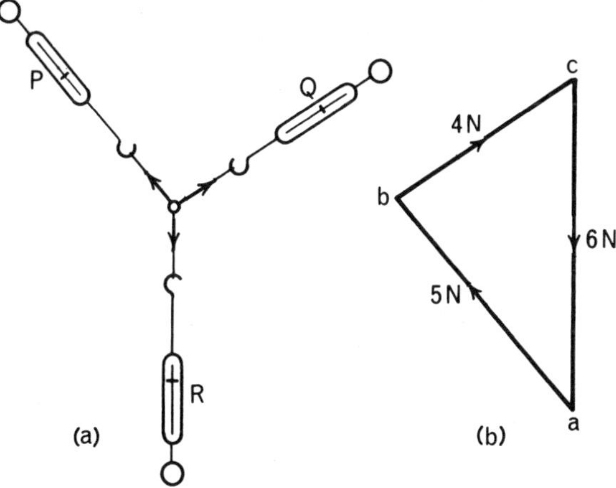


Fig. 2.5 Three inclined forces in equilibrium.

The relationship demonstrated* by this experiment is known as the Triangle of Forces Theorem which can be expressed thus:

*If three forces, acting at a point, can be represented in magnitude and direction by the sides of a triangle taken in a cyclic order around the triangle, the forces are in equilibrium.*

Conversely, *if three forces, acting at a point, are in equilibrium, they can be represented in magnitude and direction by the sides of a triangle taken in a cyclic order.*

* The Triangle of Forces Theorem can be *proved* by the aid of the Principle of Moments (section 3.4).

It is this converse theorem which is very useful for solving problems on the equilibrium of three non-parallel forces acting at a point.

## 2.5 Bow's notation

Before we proceed to the application of the triangular force diagram to problems, it is necessary to state the way in which forces are named by letters.

Suppose three inclined forces, $P$, $Q$ and $R$, acting at a point O, to be in equilibrium, and suppose the lines of action of these three forces to be as shown in fig. 2.6(a). Such a diagram is referred to as a *space* diagram. The forces in this space diagram can be designated by capital letters inserted in a clockwise direction in the *spaces* between the forces, as indicated in fig. 2.6 (a). Force $P$ can then be referred to as force AB, force $Q$ as force BC and force $R$ as force CA.

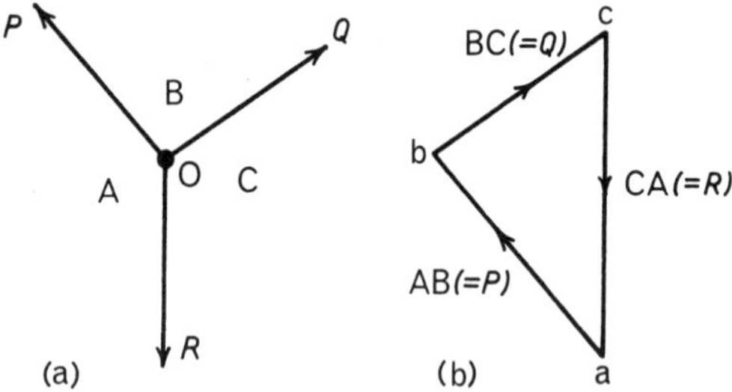


Fig. 2.6 Bow's notation.

The force or vector diagram of fig. 2.6(b) is constructed by drawing vector *ab* parallel to the line of action of force AB to represent to some convenient scale the magnitude of force AB, having its direction *a* to *b* (not *b* to *a*).

The remainder of the vector triangle of fig. 2.6(b) is constructed by drawing two lines, one from *b* parallel to force BC and the other from *a* parallel to force CA, meeting at point *c*.

From a comparison of figs. 2.6(a) and (b), it will be seen that forces AB, BC and CA in the space diagram are represented by vectors *ab*, *bc* and *ca* respectively in the force diagram. This system of lettering is known as Bow's notation.

## 2.6 Equilibrant and resultant

When three inclined forces meet at a point and are in equilibrium, we can say that any two of them, without the third, would not be in equilibrium. They require the third force, termed the *equilibrant*, to produce equilibrium. For example, in fig. 2.7, a body is suspended by two cords from D and E. Suppose the gravitational force on this

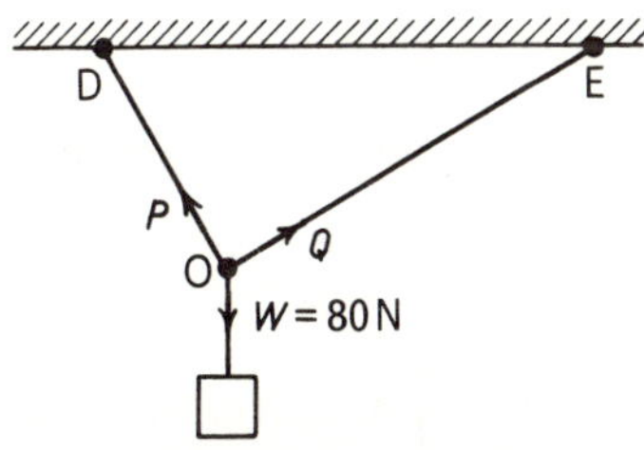


Fig. 2.7 Equilibrant of two inclined forces.

body (i.e. the weight $W$ of the body) to be 80 N. The tension in the cord supporting the body is therefore 80 N and the arrowhead on this cord indicates the direction of the force exerted *by* the cord on junction O. Suppose the corresponding tensions in cords OD and OE to be $P$ and $Q$ respectively. The arrowheads inserted on OD and OE in fig. 2.7 indicate the directions of the forces exerted *by* the cords on junction O.

If only forces 80 N and $P$ acted on O (e.g. if the cord from E were cut), the body and the other cord would not stay in their positions; they require force $Q$ to maintain equilibrium. In that case, force $Q$ is termed the *equilibrant* of the two forces 80 N and $P$. Similarly, force $P$ is the equilibrant of forces 80 N and $Q$, and force 80 N is the equilibrant of $P$ and $Q$.

Since the two forces $P$ and $Q$ balance the vertical downward pull of 80 N, they may be said to exert on junction O a vertical *upward* pull of 80 N which just balances the downward equilibrant of $P$ and $Q$, being equal and opposite to it and in the same straight line. Such a force is termed the *resultant* of forces $P$ and $Q$; for the purpose of calculating the effects of two forces, it is often convenient to replace the two forces by their resultant. Thus we can replace $P$ and $Q$ of fig. 2.7 by an upward resultant force $R$ of 80 N, as shown in fig. 2.8. This resultant exactly balances the downward force of 80 N, namely the weight of the body.

Fig. 2.9(a) represents the space diagram for the three forces of fig. 2.7, and fig. 2.9(b) represents the force or vector diagram for the *three* forces. The dotted line *bc* in fig. 2.9(b) represents the equilibrant of forces *P* and *Q*, namely the *third* force that must be added to the two forces *P* and *Q* in order to maintain equilibrium. The arrowhead on line *bc* is in the direction *b* to *c*, and is therefore in the same cyclic sequence around the triangle *abc* as those on *ca* and *ab*.

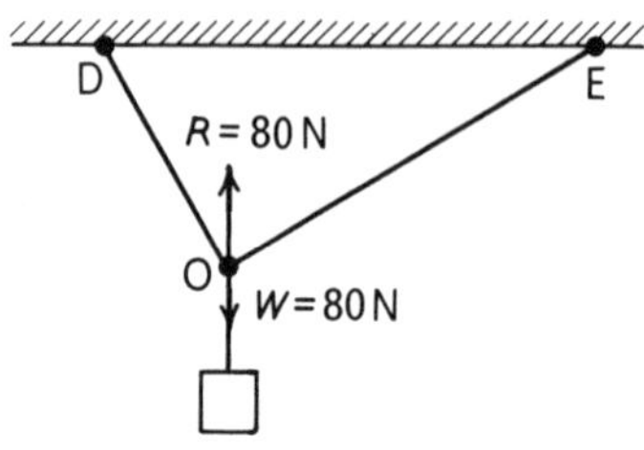


Fig. 2.8 Equilibrant and resultant.

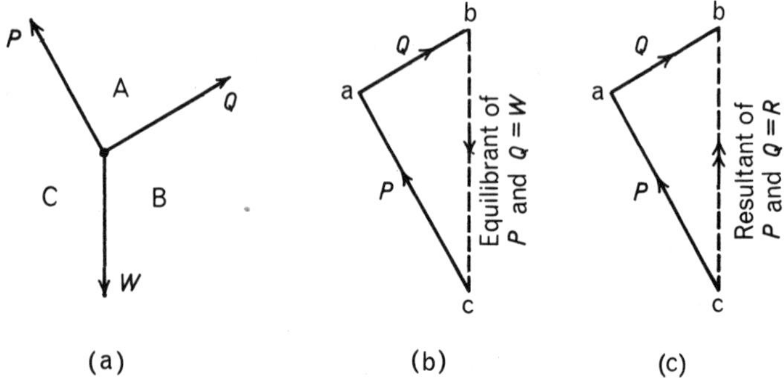


Fig. 2.9 Equilibrant and resultant.

The dotted line *bc* in fig. 2.9(c) has two arrowheads in the direction *c* to *b* to indicate that this vector represents the *resultant* *R* of the *two* forces *P* and *Q*. In other words, vector *cb* represents a *single* force *R* that will replace and produce the same effect as the combination of forces *P* and *Q*.

## 2.7 Parallelogram of forces

The magnitude and direction of the resultant of the two forces $P$ and $Q$ in fig. 2.7 can also be determined by the construction shown in fig. 2.10, where vectors OA and OB are drawn to scale to represent forces $P$ and $Q$ in magnitude and direction.

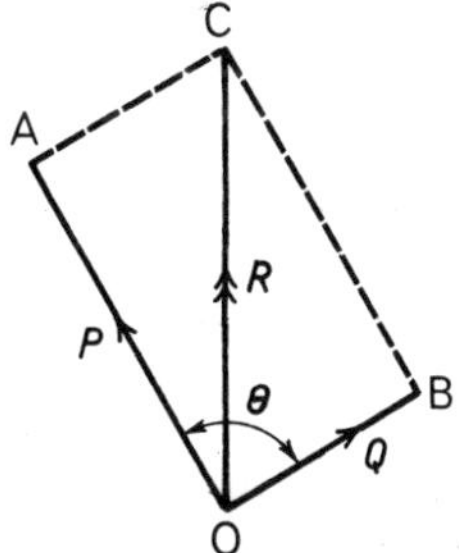


Fig. 2.10 Parallelogram of forces.

The parallelogram OACB is completed by drawing AC parallel to OB and BC parallel to OA. The diagonal OC of this *parallelogram of forces* represents in magnitude and direction the resultant of the two forces $P$ and $Q$.

Two arrowheads have been inserted on OC in fig. 2.10 to indicate that this vector is *not a third force* but is a *single force* that can replace the *two* forces $P$ and $Q$.

If $\theta$ be the angle between the directions of the two forces $P$ and $Q$, the resultant force is given by:

$$R^2 = P^2 + Q^2 + 2PQ \cos \theta$$

or

$$R = \sqrt{(P^2 + Q^2 + 2PQ \cos \theta)}$$

Comparison of figs. 2.9(c) and 2.10 shows that triangle *abc* of fig. 2.9(c) is exactly similar to triangle ACO of fig. 2.10. They differ only in one respect: the three lines OA, OB and OC in fig. 2.10 represent the relative lines of action, whereas line *ab* in fig. 2.9(c) shows force $Q$ displaced relative to its true line of action.

The choice of the triangle or the parallelogram for use in determining the resultant of two forces is largely a matter of personal preference.

**Example 2.1** *A mass of* 10 *kg is suspended by two cords from points D and E, as shown in fig.* 2.11(a). *Calculate the tension in each cord.*

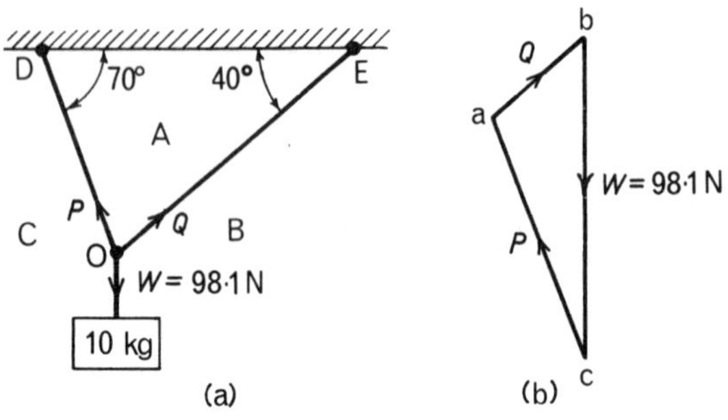


Fig. 2.11 Example 2.1.

$$\text{Weight of the 10-kg mass} = W \simeq 10 \times 9{\cdot}81 = 98{\cdot}1\ \text{N}.$$

Draw a vertical line $bc$ in fig. 2.11(b) to represent $W$ to a scale of, say, 1 mm to 1 N. Draw a line from $b$ parallel to force $Q$; and draw a line from $c$ parallel to force $P$ to meet the first line at $a$.

By measurement, $ca = 80$ mm and $ab = 35{\cdot}7$ mm.

Hence, tension in cord OD $= P = 80$ N

and tension in cord OE $= Q = 35{\cdot}7$ N.

**Example 2.2** *A jib crane has a jib* 5 *m long and a tie-rod* 3·5 *m long attached to a post* 2 *m vertically above the foot of the jib, as shown in fig.* 2.12. *Determine the forces transmitted by the jib and the tie-rod when a mass of* 3 *tonnes is suspended from the crane head.*

The problem is set out in fig. 2.12(a), and fig. 2.12(b) represents the force or vector triangle $abc$ for the equilibrium of the joint at the crane head for the three forces AB, BC and CA acting at that point.

Mass suspended at crane head $= 3\ \text{t} = 3000$ kg,

$\therefore$ force BC $=$ weight of suspended mass

$\simeq 3000 \times 9{\cdot}81 = 29\ 430$ N

$= 29{\cdot}43$ kN.

This force BC is set off to a scale of, say, 1 mm to 0·5 kN downward at *bc* in fig. 2.12(b), i.e. line *bc* is drawn 58·8 mm long. The directions of *ca* and *ab* are known and it is only necessary to draw lines parallel to the jib and the tie-rod from *c* and *b* respectively to meet at *a* to complete the force triangle *abc*. By measurement, *ca* is found to represent 73·6 kN and *ab* to represent 51·5 kN. Also, since the force of 29·43 kN (BC) acts downwards, i.e. from *b* to *c*, the sequence of letters in the force triangle is anticlockwise. Hence the force in the jib (at the crane head) acts in the direction *c* to *a* and *thrusts* at the head. Consequently the jib is in compression. The force in the tie-rod acts from *a* to *b* and therefore *pulls* at the head. Hence the tie-rod is in tension.

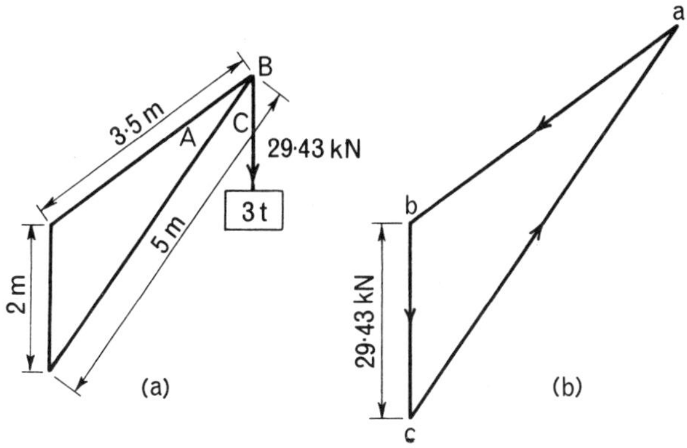


Fig. 2.12 Forces at head of crane.

It will be noticed that the triangular force diagram is geometrically similar to the triangular skeleton diagram of the crane in the space diagram. Hence, from a mere sketch of the force diagram, not drawn to scale, we could easily calculate the forces.

Thus, $ca = 29{\cdot}43\ [\text{kN}] \times 5/2 = 73{\cdot}6\ \text{kN}$

and $ab = 29{\cdot}43\ [\text{kN}] \times 3{\cdot}5/2 = 51{\cdot}5\ \text{kN}.$

**Example 2.3** *A piece of machinery having a mass of* 2 t *is lifted by a crane hook attached to the middle point of a rope sling as in fig.* 2.13. *The hook is vertically above the centre of gravity of the load and the sling is symmetrically placed as shown. Determine the tension in the two sides of the sling and examine the effect of shortening the sling.*

$$\text{Mass of machinery} = 2\ \text{t} = 2000\ \text{kg},$$

$$\therefore \quad \text{weight of machinery} \simeq 2000 \times 9{\cdot}81 = 19\,620\ \text{N} = 19{\cdot}62\ \text{kN}.$$

Insert the letters A, B and C in the spaces around the junction of the forces at the hook in fig. 2.13(a) and draw the force triangle, fig. 2.13(b), for the three forces which are in equilibrium at the hook. The tension in the rope above the hook must obviously be 19·62 kN.

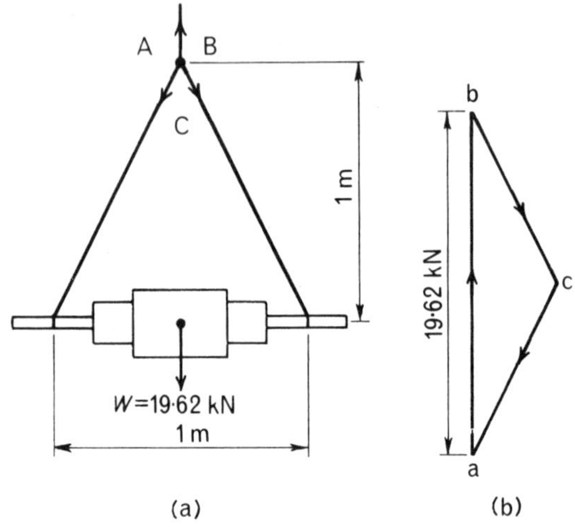


Fig. 2.13 Example 2.3

The force diagram is constructed by first drawing *ab* vertically to represent the force of 19·62 kN to a convenient scale and then drawing from *a* a line parallel to force AC and from *b* a line parallel to force BC. These two lines meet at *c*. The length *bc* represents to scale the pull BC on the hook. By measurement, it is found that *bc* represents 10·95 kN. By symmetry, the pull CA is also 10·95 kN.

If the rope sling were longer, the inclination of the sling to the horizontal would be steeper and *c* would lie nearer to *ab*. Consequently the tension in the sling would be reduced towards half the load. On the other hand, if the sling were shortened, the inclination would be smaller and *c* would lie further from *ab*, so that the tension in the rope would be greater. If the sling becomes nearly horizontal, the tension becomes very great (for *c* is then a long way to the right).

Thus a sling or a picture cord may easily be broken by being too nearly at right angles to the direction of the load.

**Example 2.4** *A triangular roof frame consists of a horizontal tie-rod and two rafters inclined at* 30° *and* 60° *respectively to the horizontal and is supported at the ends of its span as in fig.* 2.14(a). *If it carries a load of* 4 *kN at the apex, determine the resulting thrusts in the rafters, the tension in the tie-rod and the reactions at the supports.*

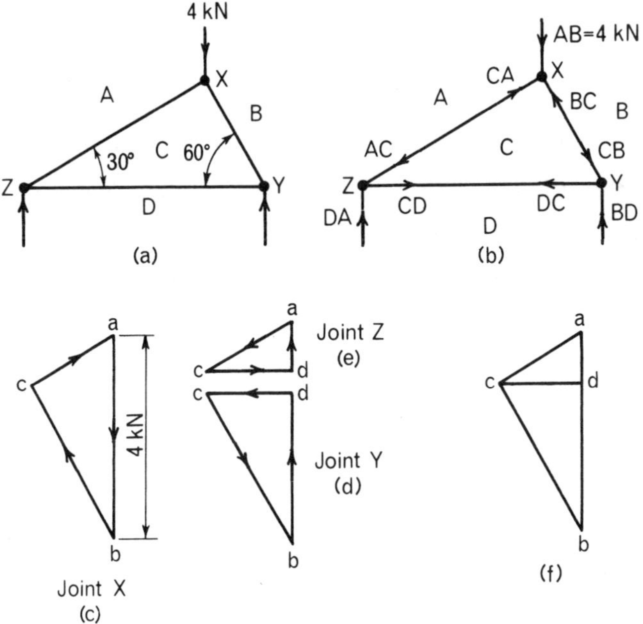


Fig. 2.14 Example 2.4.

Letter the spaces, A, B, C and D, as shown in fig. 2.14(a). The downward force AB on joint X is 4 kN and *this is the only force whose value is known.* In the space diagram of fig. 2.14(b), the arrow-heads indicate the directions of the various forces at each of the three joints; thus, BC represents the thrust in rafter XY, acting towards joint X. At the other end of this rafter, there is a thrust CB of equal magnitude, acting towards joint Y. Similarly DC represents the pull exerted by the tie-rod YZ on joint Y and CD represents the pull on joint Z. It will be noted that the sequence of each pair of letters is clockwise relative to the respective joints.

*Joint X.* The force triangle of fig. 2.14(c) is constructed by drawing vector *ab* to represent the downward load AB of 4 kN to a convenient scale. Vectors *bc* and *ac* are then drawn parallel to rafters XY and XZ respectively. By measurement,

thrust in rafter XY towards joint X = *bc* = 3·46 kN

and thrust in rafter XZ towards joint X = *ca* = 2 kN.

*Joints Y and Z.* By a procedure similar to that used for joint X, force triangles *cbd* and *acd* of figs. 2.14(d) and 2.14(e) are drawn to represent to scale the forces acting at joints Y and Z respectively. Vectors *cb* and *ac* in these diagrams have the same magnitude as vectors *bc* and *ca* respectively in fig. 2.14(c) and are drawn parallel to them. From triangle *cbd* it is found that for joint Y:

thrust in rafter XY towards joint Y = *cb* = 3·46 kN,
upward reaction of support at Y = *bd* = 3 kN
and pull of tie-rod YZ on joint Y = *dc* = 1·73 kN.

Similarly, from triangle *acd* it is found that for joint Z:

thrust in rafter XZ towards joint Z = *ac* = 2 kN,
upward reaction of support at Z = *da* = 1 kN
and pull of tie-rod YZ on joint Z = *cd* = 1·73 kN.

It is evident that the sum of the upward reactions at joints Y and Z, namely 3 kN + 1 kN = 4 kN, must be equal but opposite in direction to the vertical load of 4 kN at joint X.

The triangles of figs. 2.14(c), (d) and (e) can be combined in one diagram as shown in fig. 2.14(f); but it is not possible to insert arrowheads on these vectors since their directions are opposite for the three joints.

## 2.8 Equilibrium of three inclined forces: concurrency

If a body is in equilibrium under the action of three inclined forces in the same plane, the lines of action of the three forces must pass through one point, i.e. the forces are said to be concurrent.

The truth of the concurrency of three non-parallel forces acting in one plane can easily be demonstrated by attaching three strings, A, B and C, to a light sheet of metal or plywood D, as in fig. 2.15.

A piece of metal is suspended from string C, and strings A and B can be pulled in various directions. For each test, lines representing the directions of strings A, B and C are drawn on the sheet, as shown dotted in fig. 2.15. In every case it is found that these lines meet at a point. Similarly it may be demonstrated that concurrency is *not* necessary for the equilibrium of four or more forces in equilibrium.

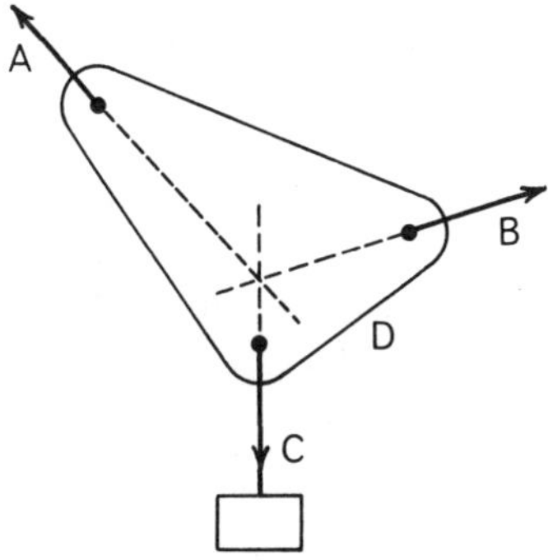


Fig. 2.15 Concurrency of three non-parallel forces in equilibrium.

As an example of the application of concurrency in the equilibrium of three inclined forces, let us consider a uniform ladder AB, fig. 2.16, resting against a *smooth* vertical wall with its lower end on rough ground.

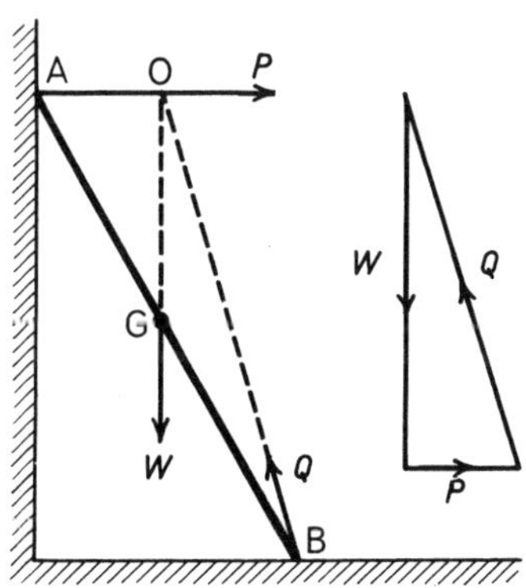


Fig. 2.16 Equilibrium of a ladder.

To determine the supporting forces exerted by the wall and the ground, we have to consider the following factors.

(*a*) Since the ladder is assumed uniform, its weight $W$ can be regarded as acting vertically downwards through a point G midway between its ends.

(*b*) The vertical wall is assumed perfectly smooth; consequently the reaction $P$ exerted by the wall must be perpendicular to the wall. Hence the direction of $P$ is horizontal. The line representing the direction of the reaction at A meets the vertical line through G at a point O.

(*c*) The third force acting on the ladder is the reaction $Q$ of the ground at B. For the ladder to be in equilibrium, it is necessary for the three forces $W$, $P$ and $Q$ to be concurrent. Consequently the line of action of $Q$ must pass through point O and the direction of $Q$ is therefore given by line BO.

## 2.9 Resolution of a force into two components

In sections 2.6 and 2.7 we considered how two forces could be replaced by a single force equal to the vector sum of the two forces. Conversely, we can replace a single force by two forces such that their vector sum is equal to the original force in magnitude and direction. Such a process is termed *resolution* of a force, i.e. splitting up the force into components.

The three diagrams of fig. 2.17 show how the two components $P$ and $Q$ of a force $R$ can be in any direction—the only condition being that the vector sum of $P$ and $Q$ must be the same as the vector representing $R$. In practice, the directions of the components are usually specified, and these directions are generally at right angles to each other, as shown in fig. 2.17(c). $P$ and $Q$ are then referred to as the *rectangular components* of $R$.

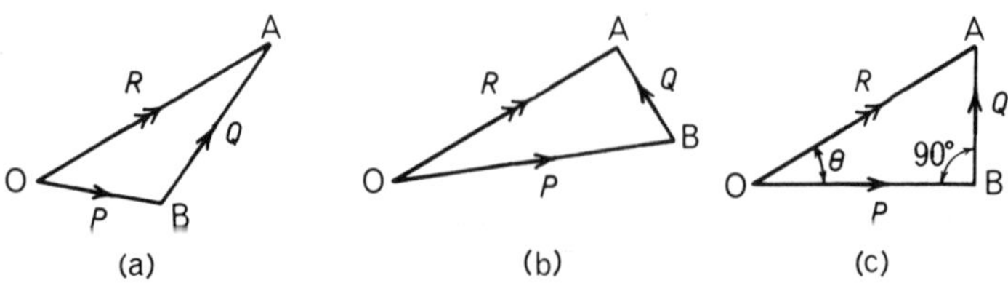


Fig. 2.17 Resolution of a force.

If $\theta$ be the angle between the original force $R$ and the horizontal component $P$ in fig. 2.17(c), then:

$$\text{horizontal component} = P = R \cos \theta$$

and $$\text{vertical component} = Q = R \sin \theta.$$

**Example 2.5** *A barge is being towed along a canal by a rope inclined at* 20° *to the direction of the canal. The tension in the rope is* 200 *N. Calculate the values of the rectangular components of this tension, one component being in the direction of the canal.*

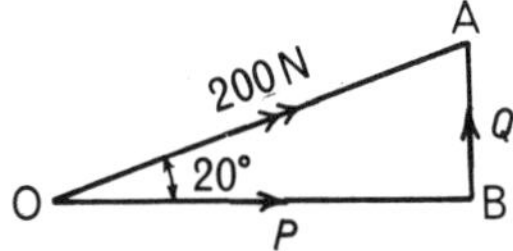


Fig. 2.18 Example 2.5.

In fig. 2.18, OA represents the tension of 200 N in the rope. Then component force $P$ in the direction of the canal

$$= \text{OB} = \text{OA} \cos 20°$$
$$= 200 \times 0{\cdot}94 = 188 \text{ N},$$

and component $Q$ perpendicular to the direction of the canal

$$= \text{BA} = \text{OA} \sin 20°$$
$$= 200 \times 0{\cdot}342 = 68{\cdot}4 \text{ N}.$$

**Example 2.6** *A force of* 60 *N acts horizontally. Resolve this force into two components, one of which acts at an angle of* 40° *above the horizontal and the other at* 20° *below the horizontal. Determine the value of each component.*

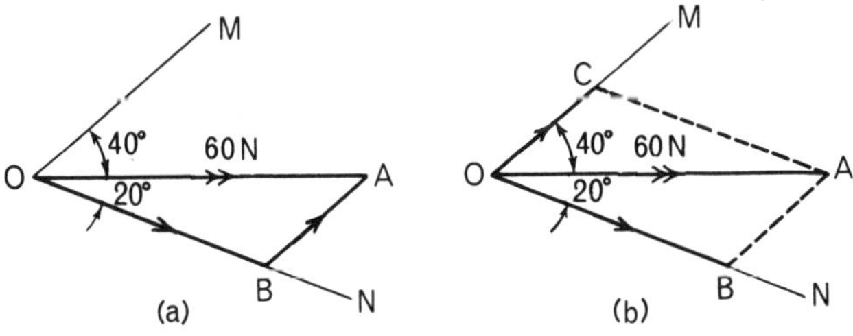


Fig. 2.19 Example 2.6.

Suppose vector OA in fig. 2.19(a) to represent the horizontal force of 60 N to a convenient scale. Lines OM and ON are drawn at 40° above and 20° below the horizontal respectively. From A, draw a line parallel to OM, cutting ON at B; then OB and BA represent the required components. From the scaled diagram it is found that component OB is 44·5 N and component BA is 23·7 N.

An alternative construction is to draw a line from A parallel to ON, cutting OM at C, and another line parallel to OM, cutting ON at B, as shown in fig. 2.19(b). The component forces are represented in magnitude and direction by OB and OC. Their values are, of course, the same as those already determined from fig. 2.19(a).

**Example 2.7** *Four co-planar forces act at a point O, the values and directions of the forces being as shown in fig.* 2.20. *Calculate the value and direction of the resultant force.*

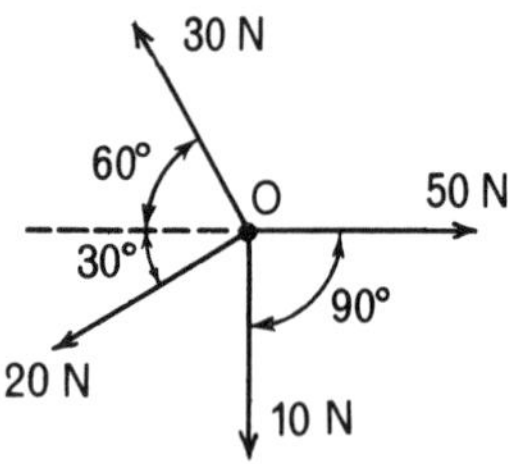


Fig. 2.20 Example 2.7.

It will be assumed that the horizontal components of the various forces are positive when they are acting towards the right of point O, and that their vertical components are positive when they are acting upwards from O.

For the 50-N force,

$$\text{horizontal component} = 50 \cos 0° = 50 \text{ N}$$

and $$\text{vertical component} = 50 \sin 0° = 0.$$

For the 10-N force,

$$\text{horizontal component} = 10 \cos 90° = 0$$

and $$\text{vertical component} = -10 \sin 90° = -10 \text{ N}.$$

For the 20-N force,

$$\text{horizontal component} = -20\cos 30^\circ = -17{\cdot}32\ \text{N}$$

and
$$\text{vertical component} = -20\sin 30^\circ = -10\ \text{N}.$$

For the 30-N force,

$$\text{horizontal component} = -30\cos 60^\circ = -15\ \text{N}$$

and
$$\text{vertical component} = 30\sin 60^\circ = 25{\cdot}98\ \text{N}.$$

Hence,

$$\begin{aligned}\text{resultant horizontal component} &= 50 - 17{\cdot}32 - 15\\ &= 17{\cdot}68\ \text{N}\end{aligned}$$

and
$$\begin{aligned}\text{resultant vertical component} &= -10 - 10 + 25{\cdot}98\\ &= 5{\cdot}98\ \text{N}.\end{aligned}$$

The above calculation can be tabulated thus:

| Force [N] | Horizontal component [N] | Vertical component [N] |
|---|---|---|
| 50 | 50 cos 0° = 50·0 | 50 sin 0° = 0 |
| 10 | 10 cos 90° = 0 | −10 sin 90° = −10·0 |
| 20 | −20 cos 30° = −17·32 | −20 sin 30° = −10·0 |
| 30 | −30 cos 60° = −15·0 | 30 sin 60° = 25·98 |
| Resultant | = 17·68 | = 5·98 |

The resultant horizontal and vertical components are represented by OA and AB respectively in fig. 2.21, and the resultant force $R$ is represented by OB.

Since
$$OB^2 = OA^2 + AB^2$$
$\therefore$
$$R^2 = (17{\cdot}68)^2 + (5{\cdot}98)^2 = 348{\cdot}4$$
so that
$$R = 18{\cdot}67\ \text{N}.$$

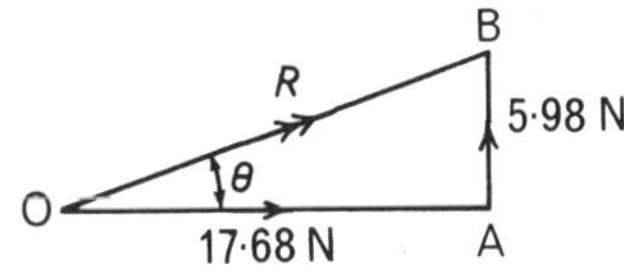


Fig. 2.21 Horizontal and vertical components.

If $\theta$ be the angle between the resultant and the horizontal component,

$$\tan\theta = \frac{AB}{OA} = \frac{5{\cdot}98}{17{\cdot}68} = 0{\cdot}338,$$

$$\therefore \theta = 18{\cdot}7^\circ$$

## 2.10 Polygon of forces

The Polygon of Forces Theorem is an extension of the Triangle of Forces Theorem and can be expressed thus:

*If four or more co-planar forces, acting at a point, can be represented in magnitude and direction by the sides of a polygon taken in a cyclic order around the polygon, the forces will be in equilibrium.*

Conversely, *if four or more co-planar forces acting at a point are in equilibrium, the magnitudes and directions of these forces can be represented by the sides of a polygon taken in a cyclic order around the polygon.*

Fig. 2.22(a) is a space diagram showing the directions of five forces, $P$, $Q$, $R$, $S$ and $T$, in one plane, acting at a point. Using Bow's notation, we can draw the force diagram shown in fig. 2.22(b), where vector $ab$ represents the value and direction of force $P$. From $b$, draw vector $bc$ to represent the value and direction of $Q$. Then the dotted line $ac$ represents the resultant of $P$ and $Q$, as already explained in section 2.6.

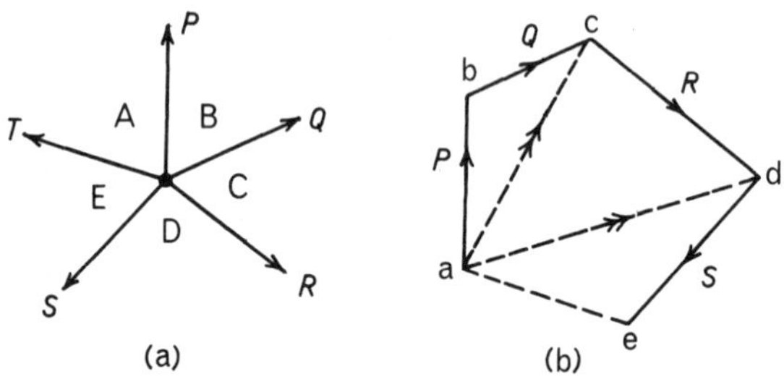


Fig. 2.22 Polygon of forces.

Similarly, from $c$, draw vector $cd$ to represent force $R$; then the dotted line $ad$ represents the resultant of $P$, $Q$ and $R$. Finally, from $d$, draw vector $de$ to represent force $S$. Vector $ae$ is the resultant of vectors $ad$ and $de$ and is therefore the resultant of forces $P$, $Q$, $R$ and $S$.

If this vector *ae* is *equal in magnitude* but *opposite in direction* to force *T* in fig. 2.22(a), then *T* is the equilibrant of forces *P*, *Q*, *R* and *S*, and the *five* forces are in equilibrium. In other words, if the five forces shown in the space diagram of fig. 2.22(a) are represented by vectors forming a *closed* polygon as in fig. 2.22(b), these forces are in equilibrium.

**Example 2.8** *Four forces, in one plane, act at a point O, the values and directions of the forces being as shown in fig.* 2.23(a). *Determine by the polygon of forces the value and direction of the resultant force.*

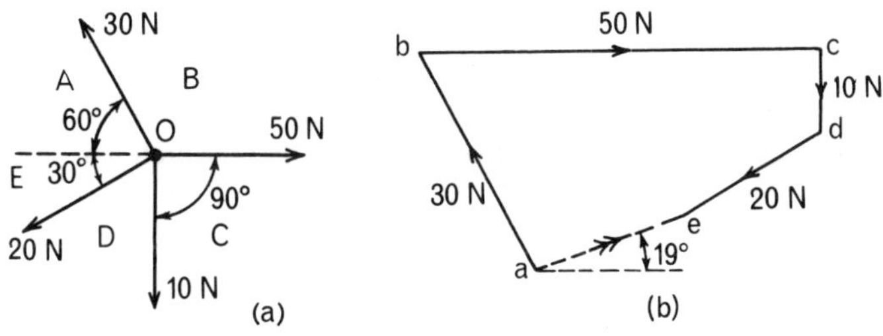


Fig. 2.23 Example 2.8.

The data are exactly the same as those of Example 2.7.

Fig. 2.23(a) is the space diagram of the four forces. Since these forces are not meant to be in equilibrium, it is necessary to add an additional capital letter E in the same space as that occupied by letter A, but A is placed near the 30-N force and E near the 20-N force.

Using a scale of, say 1 mm to 1 N, draw the force diagram of fig. 2.23(b), starting from *a*. The resultant of the four forces is represented by vector *ae*. By measurement, the value of this resultant is 18·6 N and its direction is 19° north-east of the 50-N force. These figures are in agreement with those calculated in Example 2.7.

The equilibrant of the four forces is represented by *ea* and its magnitude 19° south-west of the east-west line passing through point O. If an *additional* force of this magnitude were applied in this direction at O, the *five* forces would be in equilibrium.

## *Summary of Chapter 2*

A body cannot remain in equilibrium under the action of a *single* force.

A body is in equilibrium under the action of *two* forces provided that the forces are equal in magnitude and exactly opposite in direction.

*Triangle of Forces Theorem.* If *three* inclined forces, acting at a point, can be represented in magnitude and direction by the sides of a triangle taken in a cyclic order, the forces will be in equilibrium; or conversely, if the three inclined forces are in equilibrium, their magnitudes and directions can be represented by the sides of a triangle, taken in a cyclic order around the triangle.

If three inclined forces are in equilibrium, they must be concurrent, i.e. their lines of action must pass through the same point. Any one of the three forces is the equilibrant of the other two. The equilibrant, reversed in direction, is the resultant of the other two, and its magnitude and direction are given by their vector sum.

A force can be replaced by two components of which it would be the resultant. Usually the two components are at right angles to each other.

*Polygon of Forces Theorem.* If *four* or more co-planar forces, acting at a point, can be represented in magnitude and direction by the sides of a polygon taken in cyclic order, the forces will be in equilibrium; or conversely, if four or more co-planar forces are in equilibrium, their magnitudes and directions can be represented by the sides of a polygon, taken in a cyclic order around the polygon.

## EXAMPLES 2

1. Three forces, A, B and C, act in the same vertical plane from a point O. Force A is 30 N and acts horizontally to the left of O. Force B is 40 N and acts vertically upwards. Determine with the aid of a force diagram the value of force C and the direction in which it acts, if the forces are in equilibrium.
2. Three cords, A, B and C, are attached to one another at a point O. The angle between cords A and B is 80° and that between B and C is 150°. If the force exerted by cord A is 100 N, determine graphically the values of the forces exerted by cords B and C, assuming the three forces to be in equilibrium.
3. A ship is being towed at constant speed by two tugs. The angles made by the tow-ropes with the direction of motion of the ship are 60° and 30° respectively. The force opposing the motion of the ship is 60 kN, acting along the line of its motion. Determine graphically the pull in each of the tow-ropes.
4. A vertical load of 100 N is supported by two chains, A and B, in the same vertical plane. The force in A is 50 N and acts in a line at 30° to the

horizontal plane (i.e. at 120° to the 100-N load). Determine the force in chain B and the angle between A and B.

5. Two men carry a mass of 30 kg between them by means of two ropes fixed to the mass. One rope is inclined at 30° to the vertical and the other at 45°. Determine the tension in each rope.
6. A chain, 3 m long, has its ends fixed to a horizontal girder at points 2 m apart. A mass of 1 Mg is suspended from the chain at a point 1 m from one end. Determine the tensions in the two parts of the chain.
7. A mass of 30 kg is suspended by a cord attached to the ceiling, but is pulled to one side by a horizontal cord until the first cord makes an angle of 30° with the vertical. Draw a force diagram and determine the tensions in the two cords.
8. A pendant electric light fitting having a mass of 1 kg is to have its position adjusted by tying a cord to the flex and fastening the cord to a nail in a wall. If the cord makes an angle of 30° with the horizontal and the flex makes an angle of 60° with the horizontal, both measured above the horizontal, determine with the aid of a vector diagram the pull on the cord.
9. A chain AB is 2 m long and a hook is attached at a point C, 0·8 m from A. The ends A and B are attached to two points in the same horizontal line, 1·6 m apart. Determine the pull in each part of the chain when a mass of 2 t is suspended from the hook.
10. The jib and tie of a simple crane make angles of 60° and 30° respectively with the horizontal, both measured in an anticlockwise direction from the horizontal. Determine the force in each member when a mass of 5 t is suspended from the hook.
11. The vertical member of a jib crane is 3 m long. The jib has a length of 6 m and the length of the tie rod is 5 m. The compressive load in the jib is restricted to 50 kN. With the aid of a vector diagram drawn to scale, determine the maximum mass, in tonnes, that can be lifted by the crane and the corresponding tension in the tie rod.
12. A horizontal jib, 3 m long, is hinged to a vertical wall at its left-hand end. It is supported at its right-hand end by a stay wire making an angle of 60° with the horizontal, the upper end of the wire being secured to the same wall. If a mass of 5 t is hung on the jib, 2 m from the hinge, determine the magnitude and direction of the reaction at the hinge and the force in the stay wire.
13. A simple roof truss has a tie beam of 13 m and rafters of 10 m and 6 m. Taking the points as hinged, determine the forces in the three members of the truss when a vertical load of 6 kN is carried at the apex.
14. Determine the magnitude and nature of the forces in the three members of the pin-jointed frame shown on p. 38. The frame is simply supported at B and C, and carries a vertical load of 12 kN at A.
15. The figure on p. 38 shows a simple form of roof truss, freely supported at X and Y and carrying a vertical load of 2 kN at Z. Determine the reactions at X and Y and the force in each member, stating whether it is tensile or compressive.

16. Two forces of 8 N and 12 N act at a point, one force being inclined at an angle of 50° to the other. Determine the magnitude of the resultant force and its direction relative to the 8-N force.
17. The rope between a tug and the towed ship is inclined at an angle of 12° to the direction of travel of the ship. The tension in the rope is 500 N. Calculate the rectangular components of this tension, assuming one of the components to be in the direction of travel of the ship.

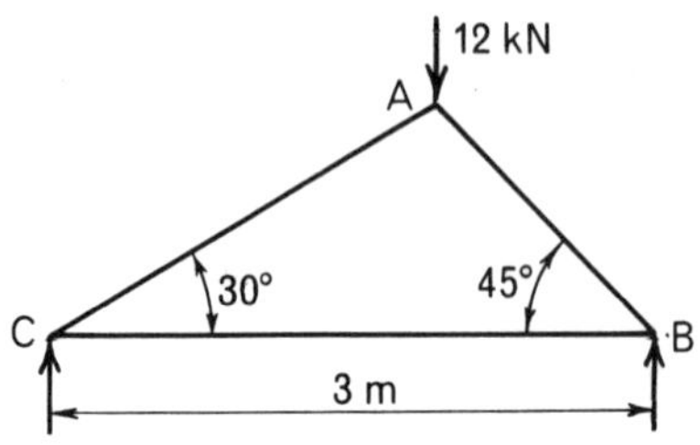


No. 14.– Examples 2.

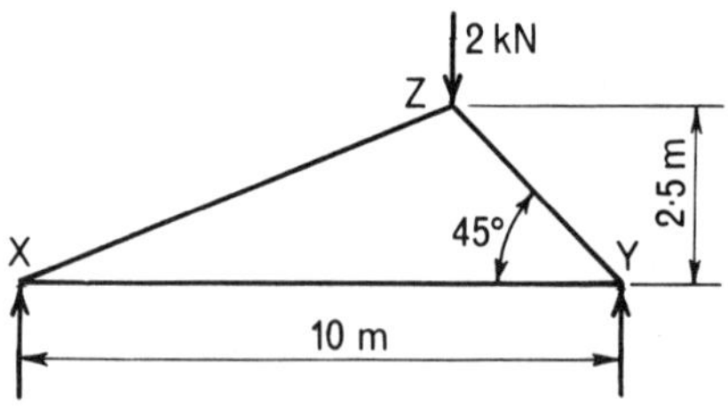


No. 15.– Examples 2.

18. A force of 30 N is inclined at an angle of 20° to the horizontal. Determine the component forces in the horizontal and vertical directions.
19. A truck is at rest on a level railway track. A horizontal pull of 250 N is applied at an angle of 30° with the direction of the rails. Calculate (a) the force tending to move the truck forward and (b) the sideways thrust on the rails.
20. A force of 20 N acts at an angle of 60° to a second force. If the resultant force is 28 N, determine graphically the magnitude of the second force.
21. The following horizontal forces act at a point:

    50 N in direction due east,
    80 N in direction due south,
    and 30 N in direction 20° north of west.

    Calculate the value and the direction of a fourth force necessary to maintain equilibrium.

22. The following horizontal forces act at a point:
 20 N in direction due north,
 30 N in direction 20° south of east,
 10 N in direction south-west,
 and 16 N in direction 5° south of west.

 Calculate the value and direction of the resultant force. Check your value by means of a force polygon drawn to scale.
23. A load $W$ is supported at a point O by three wires A, B and C, which act in the same vertical plane. The angles WOA, WOB and WOC, measured in a clockwise direction, are 120°, 210° and 240° respectively. With the aid of a force diagram drawn to scale, determine the value of $W$ and the force in C if the forces in A and B are 20 N and 15 N respectively.
24. A ladder, 5 m long, rests at an angle of 60° to the horizontal with its upper end against a smooth vertical wall and its lower end on rough ground. The ladder has a mass of 20 kg and the weight of the ladder may be assumed to act at a point 2 m from the lower end. Determine the reactions of the wall and of the ground on the ladder.
25. A uniform ladder, 4 m long and having a mass of 25 kg, rests against a smooth vertical wall at A and is supported on rough horizontal ground at B. The ladder is inclined at 60° to the horizontal. Determine the magnitude and direction of the reaction at the ground and at the wall respectively.
26. A ladder having a mass of 30 kg and a length of 6 m rests against a smooth vertical wall. The foot of the ladder stands on level ground 2·5 m away from the foot of the wall. If the whole weight of the ladder is assumed to act vertically downwards from a point one-third of its length from the end on the ground, determine the magnitude and direction of the reaction between (*a*) the ladder and the wall and (*b*) the ladder and the ground.
27. An overhanging horizontal platform AB is hinged at B and is supported by two chains attached to the outer edge and inclined 30° to the platform. The length AB is 4 m and the total weight of the platform and the load upon it is 10 kN and may be assumed to act at a distance of 3 m from the hinged side. Draw a force diagram to scale and determine the tension in each chain and the magnitude and direction of the reaction on the platform of the hinge at B.

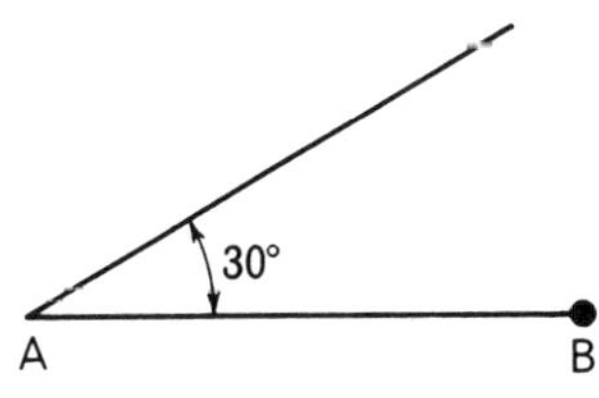


No. 27.– Examples 2.

28. A uniform steel bar is 4 m long and has a mass of 80 kg. Its lower end is hinged to a vertical wall and its upper end is supported by a horizontal chain, 2 m long, secured to a point in the wall directly above the hinge. Determine the tension in the chain and the value and direction of the reaction at the hinge.
29. A constant force of 40 N, acting at an angle of 20° with the horizontal, is applied to a body lying on a horizontal surface. Calculate the work done when the body is moved through a distance of 6 m along the surface. If the time taken is 3 s, what is the average value of the power?

## ANSWERS TO EXAMPLES 2

1. 50 N, 53° below horizontal.
2. 153·2 N, 197 N.
3. 30 kN, 52 kN.
4. 86·6 N, 90°.
5. 216 N, 153 N.
6. 8·87 kN, 2·54 kN.
7. 340 N, 170 N.
8. 4·9 N.
9. 17·75 kN in AC, 13·95 kN in BC.
10. Jib, 85 kN compression; tie, 49·05 kN tension.
11. 2·55 t, 41·7 kN.
12. 25 kN at 40·9° to jib, 37·8 kN.
13. 5·6 kN, 4·2 kN, 3·8 kN.
14. 10·75 kN compression in AB, 8·8 kN compression in AC, 7·6 kN tension in BC.
15. 0·5 kN at X, 1·5 kN at Y; 1·58 kN compression in XZ, 2·12 kN compression in YZ, 1·5 kN tension in XY.
16. 18·2 N, 30·3°.
17. 489 N, 104 N.
18. 28·2 N, 10·26 N.
19. 216·5 N, 125 N.
20. 12 N.
21. 73 N, 72·6° north of west.
22. 5·36 N, 13·8° north of east.
23. 28·7 N, 11·35 N.
24. 45·4 N; 202 N, 77° to ground.
25. 255 N, 73·9° to ground; 70·7 N.
26. 45 N, normal to wall; 298 N, 81·3° to ground.
27. 7·5 kN; 13·2 kN, 10·9° to horizontal.
28. 227 N; 818 N, 16·1° to vertical.
29. 225·6 J, 75·2 W.

CHAPTER 3

# Moments and centre of gravity

## 3.1 Moment of a force about an axis

Suppose that a uniform strip or rod of wood, such as a metre rule, is pivoted at its mid-point C, as in fig. 3.1. If no external force is applied to the rod, there is no tendency for it to rotate either clockwise or anticlockwise about C.

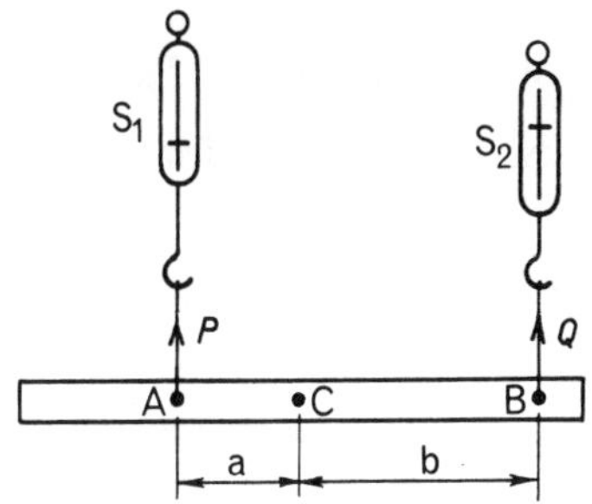


Fig. 3.1 Moments of parallel forces.

If we now attach the hook of a spring balance* $S_1$ to a cord passing through a hole at A, any upward force applied by this balance causes the rod to turn clockwise about pivot C. On the other hand, if we apply an upward force by a spring balance $S_2$ at point B ($S_1$ having been removed), the rod turns anticlockwise.

Let us next apply upward forces simultaneously by means of balances $S_1$ and $S_2$, of such values as to maintain the rod in a horizontal position. If $P$ and $Q$ be the forces exerted by $S_1$ and $S_2$ respectively, and if $a$ and $b$ be the distances of points A and B from pivot C, it is found that:

$$P \times a = Q \times b.$$

* The spring balances should be calibrated in newtons. The experiment can, however, be performed with known masses suspended from the rod; but in that case, it is the weight of each mass that must be used for calculating the moment. For example, if a mass of 3 kg is suspended from a horizontal rod at a distance of 0·6 m from the fulcrum, the weight of the mass $\simeq 3 \times 9{\cdot}81 = 29{\cdot}43$ N, and the moment about the fulcrum $= 29{\cdot}43$ [N] $\times$ 0·6 [m] $= 17{\cdot}66$ N m.

This relationship can be confirmed by noting the balance readings for various upward forces exerted by the balances and the experiment can be repeated with various values of $a$ and $b$.

The tendency of a force to produce rotation of a body about an axis is termed the *moment* of the force about that axis, and the magnitude of the moment is the product of the force and the perpendicular distance or *moment arm* of the line of action of the force from the axis. For instance, in fig. 3.1, if $P$ is the force, in newtons, exerted by spring balance $S_1$ tending to turn the rod clockwise about an axis through C perpendicular to the paper, and if $a$ is the distance, in metres, of the line of action of $P$ from the axis through C, then:

clockwise moment of force $P$ about this axis
$= P$ [newtons] $\times$ $a$ [metres] $= Pa$ newton metres.

Similarly, anticlockwise moment of force $Q$ about the same axis
$= Qb$ newton metres.

When the moments are such as to maintain the rod of fig. 3.1 horizontally, then:

clockwise moment due to $P$ = anticlockwise moment due to $Q$
i.e. $Pa = Qb$

This relationship is termed the *principle of moments* and is discussed more generally in section 3.4.

### 3.2 Levers

A lever is simply a rod or bar capable of turning about a fixed axis called the *fulcrum*, which may be a spindle as in fig. 3.1 or a knife edge as in fig. 3.2. The lever may be straight or curved or cranked (section 3.7), and the forces acting upon it may be parallel or otherwise.

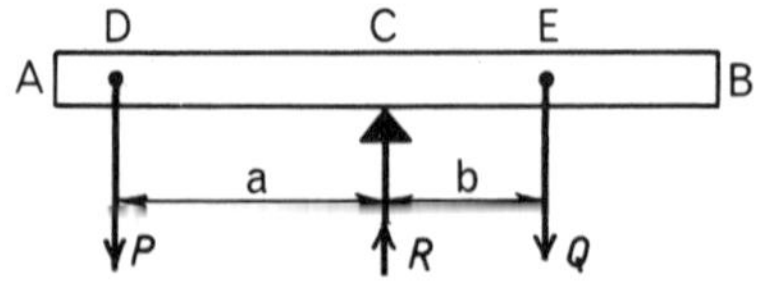


Fig. 3.2 A straight lever.

Let us first consider a straight uniform lever AB resting on a fulcrum at C, as in fig. 3.2. By arranging for the fulcrum to be midway along AB, we avoid any complication from the weight of the lever, since the gravitational pull on the right-hand half of the lever, tending to turn the lever clockwise about C, is balanced by that on the left-hand side tending to turn the lever anticlockwise. Consequently, when the lever is not loaded, there is no tendency for it to turn about the fulcrum C.

A downward force $P$, applied at a distance $a$ from the fulcrum, gives rise to an anticlockwise moment $Pa$ about C. Similarly, a downward force $Q$, applied at a distance $b$ from the fulcrum, gives rise to a clockwise moment $Qb$ about C. As already explained in section 3.1, the lever remains horizontal if the clockwise and anticlockwise moments are equal,

i.e. if $$Pa = Qb.$$

The supporting or upward force $R$ exerted by the fulcrum on the lever must be equal to the sum of the downward forces $P$ and $Q$,

i.e. $$R = P + Q.$$

Note that we have taken moments about the axis through which the supporting force $R$ acts on the lever. This force $R$ has no moment about axis C and therefore does not affect the balance of the lever. Had moments been taken about any other axis, the moment of $R$ about such an axis would have to be taken into consideration (see section 3.4).

**Example 3.1** *A uniform lever (fig. 3.3) is pivoted at its mid-point C. A body having a mass of* 5 *kg is suspended at a point E,* 180 *mm to the right of C. Calculate the mass to be suspended at a point D,* 400 *mm to the left of C to maintain the lever in balance.*

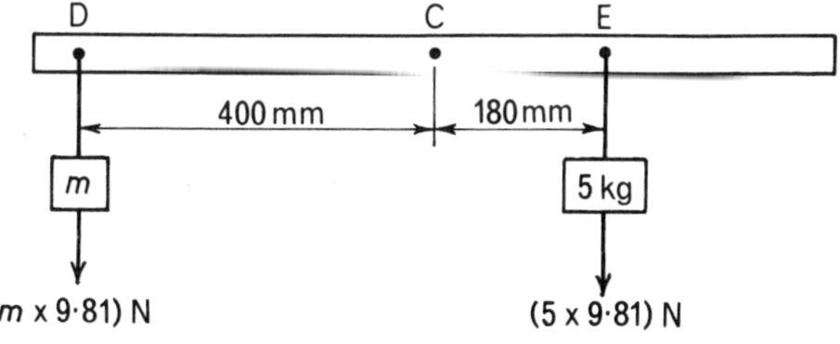


Fig. 3.3 Example 3.1.

Weight of the 5-kg mass $\simeq (5 \times 9{\cdot}81)$ N,

$\therefore$ clockwise moment of this force about C

$$= (5 \times 9{\cdot}81)\,[\mathrm{N}] \times 0{\cdot}18\,[\mathrm{m}].$$

If $m$ be the mass, in kilogrammes, to be suspended at D,

weight of this mass $\simeq (m \times 9{\cdot}81)$ N

and its anticlockwise moment about C

$$= (m \times 9{\cdot}81)\,[\mathrm{N}] \times 0{\cdot}4\,[\mathrm{m}].$$

For balance, the clockwise and anticlockwise moments must be equal,

i.e. $5\,[\mathrm{kg}] \times 9{\cdot}81\,[\mathrm{N/kg}] \times 0{\cdot}18\,[\mathrm{m}] = m \times 9{\cdot}81\,[\mathrm{N/kg}] \times 0{\cdot}4\,[\mathrm{m}]$

$$\therefore \qquad m = 2{\cdot}25\ \mathrm{kg}.$$

### 3.3 Equilibrant and resultant of parallel forces

In section 3.2 it was mentioned that the single upward force $R$ exerted by the fulcrum on the lever, in fig. 3.2, opposes or balances the two downward forces $P$ and $Q$. Force $R$ is therefore the *equilibrant* of the other two forces. (The equilibrant and resultant of inclined forces have already been discussed in section 2.6.)

Let us consider the slightly more involved case shown in fig. 3.4, where three parallel downward forces of 80 N, 40 N and 20 N act on a light rod. It is required to determine the magnitude and position of a single vertical upward force (applied, say, by a spring balance or by a knife-edge support underneath the rod) that will balance these three downward forces.

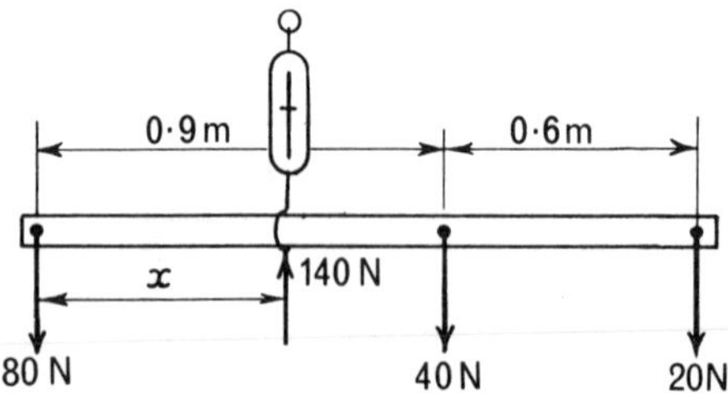


Fig. 3.4 Equilibrant of three parallel forces.

The upward force must obviously be equal to 80 + 40 + 20 = 140 N, as an experiment with a spring balance (fig. 3·4) would show; i.e. the *equilibrant* is 140 N. To determine the position of this equilibrant, let us take moments about an axis in the line of action of, say, the 80-N force. If $x$ is the distance between the line of action of the equilibrant and that of the 80-N force, as shown in fig. 3.4, then:

anticlockwise moment of equilibrant about the chosen axis

$$= 140\ [\mathrm{N}] \times x$$

and total clockwise moment of the other three forces about same axis

$$= 80\ [\mathrm{N}] \times 0\ [\mathrm{m}] + 40\ [\mathrm{N}] \times 0{\cdot}9\ [\mathrm{m}] + 20\ [\mathrm{N}] \times 1{\cdot}5\ [\mathrm{m}]$$

$$= 66\ \mathrm{N\ m}.$$

Since the clockwise and anticlockwise moments are equal,

$$140\ [\mathrm{N}] \times x = 66\ [\mathrm{N\ m}]$$

$$\therefore \qquad x = 0{\cdot}471\ \mathrm{m}.$$

Hence an upward force of 140 N in a line 0·471 m to the right of the 80-N force would balance the three downward forces. A *downward* force of 140 N in this line is the *resultant* of the three forces of 80 N, 40 N and 20 N. In other words, the three forces could be replaced by a single downward force of 140 N acting at a distance of 0·471 m from the line of action of the 80-N force. The resultant is exactly equal in magnitude and directly opposite to the equilibrant.

To take another example, suppose the 20-N force to be absent, i.e. consider only the downward forces 80 N and 40 N, 0·9 m apart, as shown in fig. 3.5. The magnitude of the equilibrant is 80 N + 40 N = 120 N. Suppose $x$ to be the distance between the equilibrant and the 80-N force.

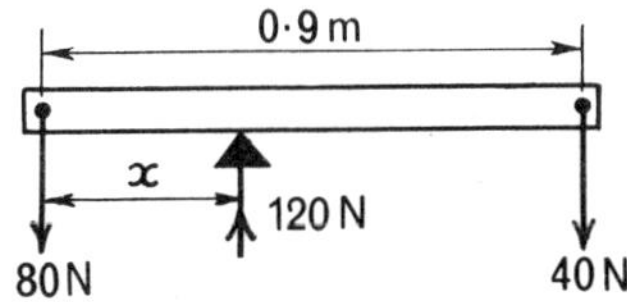


Fig. 3.5 Equilibrant of two parallel forces.

Taking moments about the line of action of the 80-N force, we have:

$$120\ [\text{N}] \times x = 40\ [\text{N}] \times 0{\cdot}9\ [\text{m}]$$

$$\therefore \quad x = 0{\cdot}3\ \text{m}.$$

It will be seen that the line of action of the resultant (and of the equilibrant) divides the distance (0·9 m) between the 80-N and the 40-N forces *inversely* as the values of the forces, i.e. the distance from the 80-N force is 40/(80 + 40) of 0·9 m, namely 0·3 m, and that from the 40-N force is 80/120 of 0·9 m, namely 0·6 m.

**3.4 General principle of moments**

In the preceding sections we have taken moments about an axis on the line of action of one of the forces, for instance, about the fulcrum C in fig. 3.2. We shall now show that the Principle of Moments applies to *any* axis.

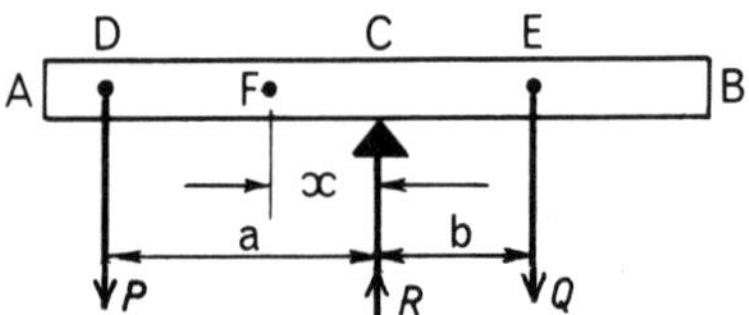


Fig. 3.6 General Principle of Moments.

Consider the arrangement shown in fig. 3.6 (similar to that of fig. 3.2), and let us take moments about *any* axis F situated at a distance $x$ from the fulcrum C.

Anticlockwise moment of $P$ about F $= P(a - x)$,

clockwise moment of $Q$ about F $= Q(b + x)$

and anticlockwise moment of $R$ about F $= Rx$.

Equating the clockwise and anticlockwise moments, we have:

$$Q(b + x) = P(a - x) + Rx$$

Since $$R = P + Q$$

$$\therefore \quad Qb + Qx = Pa - Px + Px + Qx$$

so that $$Qb = Pa$$

which is the condition of balance already derived in section 3.1. Hence we can express the General Principle of Moments thus:

*If a body is at rest under the action of several forces, the total clockwise moment of the forces about* **any** *axis is equal to the total anticlockwise moment of the forces about the same axis.*

If the forces are all in one plane, we can, for convenience in stating the principle, speak of moments about any *point* instead of about any *axis*.

### 3.5 Reactions on a horizontal beam supported at two points

A very common problem is to determine what upward forces must act on a beam at its supports when the beam is loaded and supported at two points. Often these two points are at its ends, but not necessarily so. Consider a uniform beam AB, 3 m long (fig. 3.7), resting on supports A and B at its ends. Suppose a body having a mass of 5 kg to be suspended at a point C, 2 m from A,

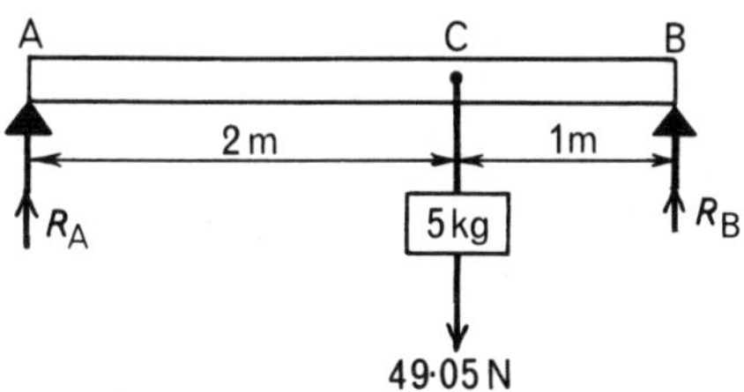


Fig. 3.7 Supporting forces at ends of a beam.

$$\text{then downward force at C} = \text{weight of the 5-kg mass}$$
$$\simeq 5 \times 9{\cdot}81 = 49{\cdot}05 \text{ N}.$$

The weight of the beam and of the suspended mass will exert downward forces on the supports at A and B; and the supports will exert exactly equal and opposite (upward) forces, or reactions, on the beam. These reactions are represented by $R_A$ at A and $R_B$ at B in fig. 3.7. For the present we shall ignore the weight of the beam, i.e. we shall regard the beam as of negligible weight or alternatively we shall regard the supporting forces as being the increases due to the load placed on the beam. In either case, it will be evident that the sum of the reactions $R_A$ and $R_B$ will be equal to the weight of the mass suspended at C,

i.e. $$R_A + R_B = 49{\cdot}05 \text{ N}.$$

The values of $R_A$ and $R_B$ can be determined by taking moments about any point along the lever; but the problem is simplified if we deal with only *one unknown force* at a time. This can be done by choosing an axis on one of the unknown reactions: for example, by choosing the axis through, say, A. The reaction $R_A$ has no moment about A.

We can now imagine the beam of fig. 3.7 about to turn anticlockwise about A due to the action of the upward force $R_B$ upon it. This effect, however, is exactly balanced by the action of the downward force of 49·05 N at C. In other words, the moment due to $R_B$ tending to turn the beam anticlockwise about A is balanced by the moment due to the force at C tending to turn the beam clockwise about A.

Hence, $R_B \times 3\ [\text{m}] = 49{\cdot}05\ [\text{N}] \times 2\ [\text{m}]$

$\therefore$ $R_B = 32{\cdot}7$ N.

Similarly, if we had chosen B as our axis, the clockwise moment due to reaction $R_A$ would be $R_A \times 3$ [m], and the anticlockwise moment due to the force at C would be 49·05 [N] $\times$ 1 [m].

Hence, $R_A \times 3\ [\text{m}] = 49{\cdot}05\ [\text{N m}]$

$\therefore$ $R_A = 16{\cdot}35$ N.

Alternatively, since $R_A + R_B = 49{\cdot}05$ N

$\therefore$ $R_A = 49{\cdot}05 - 32{\cdot}7 = 16{\cdot}35$ N.

Or, *after* having determined the value of $R_B$, we could have considered rotation about C, i.e. we could have taken moments about C thus:

$$R_A \times 2\ [\text{m}] = 32{\cdot}7\ [\text{N}] \times 1\ [\text{m}]$$

$\therefore$ $R_A = 16{\cdot}35$ N.

*Effect of weight of the beam.* The effect of the weight of the beam on the downward forces on the supports and on the equal upward forces of the supports on the beam is obviously to increase them by an amount equal to the weight of the beam. Thus, if the mass of the beam in fig. 3·7 is, say, 0·2 kg, its weight is approximately 0·2 $\times$ 9·81, namely 1·96 N; and since the beam is uniform, each reaction is increased by 0·98 N,

i.e. $R_A = 16{\cdot}35 + 0{\cdot}98 = 17{\cdot}33$ N

and $R_B = 32{\cdot}7 + 0{\cdot}98 = 33{\cdot}68$ N.

If the supports are not symmetrically placed with respect to the centre of a uniform beam, the two reactions due to the weight of the beam will not be equal. For the purpose of these simple problems in statics, we may take the weight of the beam (although actually distributed) as acting at a point in the beam called its *centre of gravity*. We shall defer more general consideration of centre of gravity until a little later (section 3.10), but for the present we may be satisfied by the fact that the centre of gravity of a straight uniform rod is midway between its ends and the weight of the rod may be taken as acting there.

**Example 3.2** *A uniform horizontal beam, 6 m long, rests on two supports A and B, 4 m apart, A being at one end of the beam. The mass of the beam is 20 kg. Calculate the reactions of the supports on the beam.*

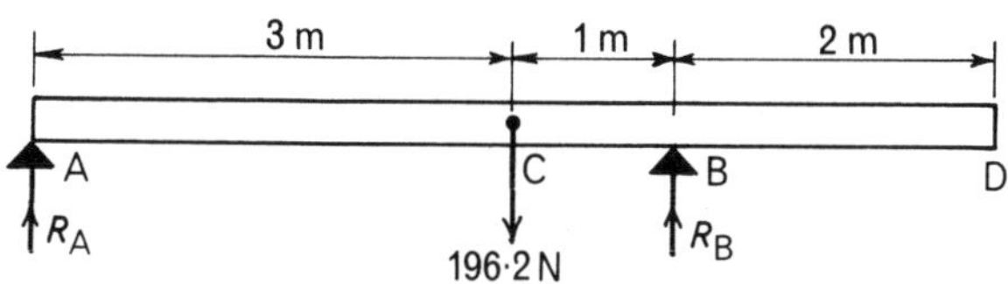


Fig. 3.8 Example 3.2.

Let $R_A$ and $R_B$ be the reactions at A and B respectively, as shown in fig. 3.8.

The gravitational force on the beam, i.e. the weight of the beam

$$\simeq 20 \times 9{\cdot}81 = 196{\cdot}2 \text{ N}.$$

Taking moments about A, we have:

$$\text{clockwise moment} = 196{\cdot}2 \text{ [N]} \times 3 \text{ [m]} = 588{\cdot}6 \text{ N m}$$

and

$$\text{anticlockwise moment} = R_B \times 4 \text{ [m]}.$$

Equating the clockwise and anticlockwise moments, we have:

$$R_B \times 4 \text{ [m]} = 588{\cdot}6 \text{ [N m]}$$

$$\therefore \quad R_B = 147{\cdot}15 \text{ N}.$$

$$\text{Total upward reactions} = R_A + R_B$$

$$= \text{weight of beam} = 196{\cdot}2 \text{ N},$$

$$\therefore \quad R_A = 196{\cdot}2 - 147{\cdot}15 = 49{\cdot}05 \text{ N}.$$

Alternatively, taking moments about B, we have:

$$R_A \times 4\ [\text{m}] = 196{\cdot}2\ [\text{N}] \times 1\ [\text{m}]$$

$\therefore$ $R_A = 49{\cdot}05$ N

and $R_B = 196{\cdot}2 - 49{\cdot}05 = 147{\cdot}15$ N.

**Example 3.3** *If a mass of* 8 *kg be hung from the beam of Example* 3.2 *at a distance of* 1 *m from A, calculate the reaction of the supports.*

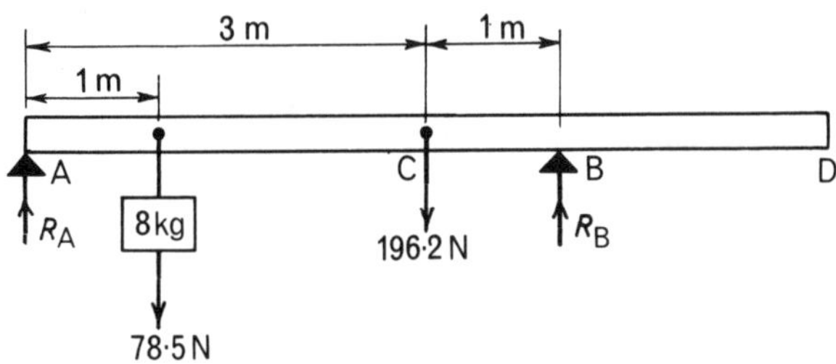


Fig. 3.9 Example 3.3.

Weight of the 8-kg mass $\simeq 8 \times 9{\cdot}81 = 78{\cdot}5$ N.

We proceed as in Example 3.2, but this time we have to take account of the two downward forces of 78·5 N and 196·2 N (fig. 3.9).

To find reaction $R_B$, we take moments about A thus:

$$\text{clockwise moment} = 78{\cdot}5\ [\text{N}] \times 1\ [\text{m}] + 196{\cdot}2\ [\text{N}] \times 3\ [\text{m}]$$
$$= 78{\cdot}5 + 588{\cdot}6 = 667{\cdot}1\ \text{N m}.$$

$$\text{Anticlockwise moment} = R_B \times 4\ [\text{m}].$$

Equating clockwise and anticlockwise moments, we have:

$$R_B \times 4\ [\text{m}] = 667{\cdot}1\ [\text{N m}]$$

$\therefore$ $R_B = 166{\cdot}8$ N

and $R_A = 196{\cdot}2 + 78{\cdot}5 - 166{\cdot}8 = 274{\cdot}7 - 166{\cdot}8$
$= 107{\cdot}9$ N.

Alternatively, taking moments about B, we have:

$$R_A \times 4\ [\text{m}] = 78{\cdot}5\ [\text{N}] \times 3\ [\text{m}] + 196{\cdot}2\ [\text{N}] \times 1\ [\text{m}]$$
$$= 431{\cdot}7\ \text{N m}$$

$\therefore$ $R_A = 107{\cdot}9$ N

and $R_B = 274{\cdot}7 - 107{\cdot}9 = 166{\cdot}8$ N.

**Example 3.4** *If a mass of* 6 *kg be suspended at the end of the projecting beam of Example* 3.2, *calculate the reactions at the supports.*

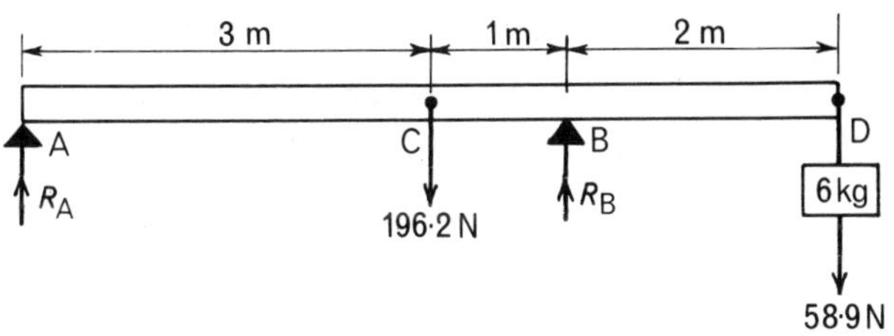


Fig. 3.10 Example 3.4.

$$\text{Weight of the 6-kg mass} \simeq 6 \times 9{\cdot}81 = 58{\cdot}9 \text{ N.}$$

Taking moments about A (fig. 3.10), we have:

$$\begin{aligned} \text{clockwise moment} &= 196{\cdot}2 \text{ [N]} \times 3 \text{ [m]} + 58{\cdot}9 \text{ [N]} \times 6 \text{ [m]} \\ &= 942 \text{ N m.} \\ \text{Anticlockwise moment} &= R_B \times 4 \text{ [m]} \\ \text{Hence} \qquad R_B \times 4 \text{ [m]} &= 942 \text{ [N m]} \\ \therefore \qquad R_B &= 235{\cdot}5 \text{ N} \\ \text{and} \qquad R_A &= 196{\cdot}2 + 58{\cdot}9 - 235{\cdot}5 = 255{\cdot}1 - 235{\cdot}5 \\ &= 19{\cdot}6 \text{ N.} \end{aligned}$$

Alternatively, taking moments about B, we have:

$$\begin{aligned} R_A \times 4 \text{ [m]} + 58{\cdot}9 \text{ [N]} \times 2 \text{ [m]} &= 196{\cdot}2 \text{ [N]} \times 1 \text{ [m]} \\ \therefore \qquad R_A &= 19{\cdot}6 \text{ N} \\ \text{and} \qquad R_B &= 255{\cdot}1 - 19{\cdot}6 = 235{\cdot}5 \text{ N.} \end{aligned}$$

**Example 3.5** *If the mass suspended at the projecting end in Example* 3.4 *is increased to* 12 *kg, calculate the reactions of the supports.*

$$\text{Weight of the 12-kg mass} \simeq 12 \times 9{\cdot}81 = 117{\cdot}7 \text{ N.}$$

Taking moments about A (fig. 3.10), we have:

$$\begin{aligned} 196{\cdot}2 \text{ [N]} \times 3 \text{ [m]} + 117{\cdot}7 \text{ [N]} \times 6 \text{ [m]} &= R_B \times 4 \text{ [m]} \\ \therefore \qquad R_B &= 323{\cdot}7 \text{ N} \\ \text{and} \qquad R_A &= 196{\cdot}2 + 117{\cdot}7 - 323{\cdot}7 \\ &= -\,9{\cdot}8 \text{ N.} \end{aligned}$$

This means 9·8 N in the opposite direction to the assumed upward direction, i.e. 9·8 N *downward.*

Alternatively, taking moments about B, we have:

$$R_A \times 4\,[\text{m}] + 117{\cdot}7\,[\text{N}] \times 2\,[\text{m}] = 196{\cdot}2\,[\text{N}] \times 1\,[\text{m}]$$

$\therefore$ $$R_A = -\,9{\cdot}8\text{ N; or } 9{\cdot}8\text{ N downward,}$$

and $$R_B = 196{\cdot}2 + 117{\cdot}7 + 9{\cdot}8$$

$$= 323{\cdot}7\text{ N, as before.}$$

## 3.6 Other forms of straight levers

At (*a*), (*b*) and (*c*) in fig. 3.11 are shown the three possible forms of straight lever in which an effort $F$ may overcome the force due to a load $W$ (both forces being perpendicular to the lever) by turning the lever about a fulcrum C.

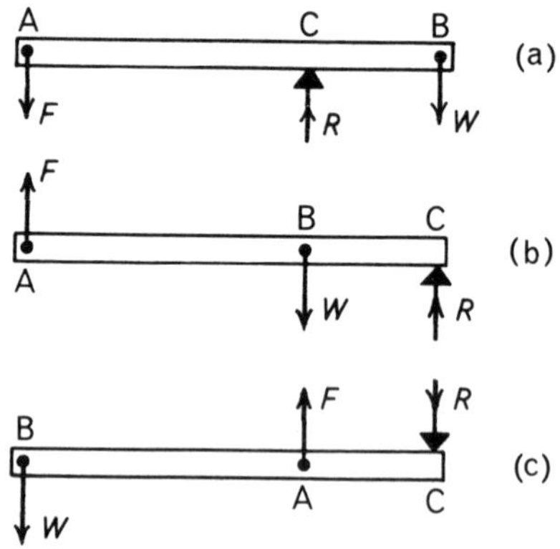


Fig. 3.11 Forms of straight lever.

Case (*a*) was dealt with in section 3.2, and examples of this class of lever are: beam balances in which the arms may be of equal lengths, as in a grocer's balance, or of unequal lengths, as in a steelyard—a device consisting of a mass which can be moved along a graduated beam to balance a relatively heavy body suspended near to the fulcrum; a seesaw, etc.

Case (*b*) is, in principle, identical with the beam carrying a load as dealt with in section 3.5; and examples of this class of lever are: a wheelbarrow, a nut cracker, etc.

Case (*c*) represents a large effort $F$ moving through a small distance and overcoming a small load $W$ which moves through a large distance. For example, the manipulation of a fishing rod; the large upward pull of the biceps muscle acting about the elbow as fulcrum and lifting a relatively small load held in the hand.

The principle involved is the same in each of the above cases; and by equating opposing moments of a known load $W$ and an unknown effort $F$ about the fulcrum C, we can determine the value of $F$. Also, if desired, we can find the reaction $R$ of the fulcrum C on the lever by taking moments about A or by finding what force is required to balance the sum or the difference of $F$ and $W$, according to whether they act in the same or in opposite directions.

### 3.7 Bell crank lever

A lever is not necessarily straight, but may be cranked as in the bell crank lever shown in fig. 3.12. The value of a force $P$ required to balance a force $Q$, neither of which need be perpendicular to the arm of the lever to which it is applied, can be found by considering

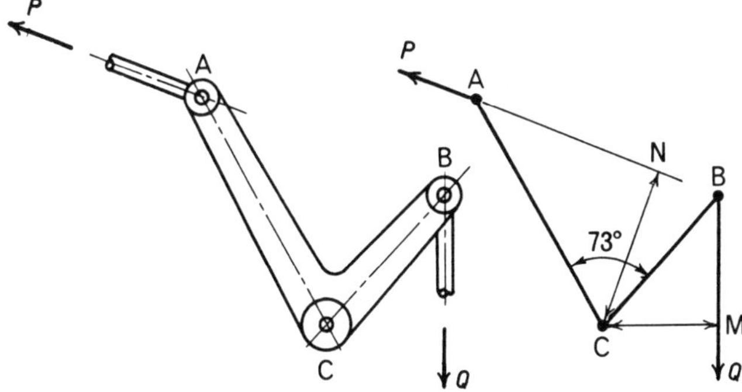


Fig. 3.12 Bell crank lever.

the opposing moments of $P$ and $Q$ about the axis C. The moment arms of $P$ and $Q$ about the axis C may readily be found by drawing the skeleton or centre-line diagram to scale as shown on the right-hand side of fig. 3.12. They are CN and CM respectively. Equating the anticlockwise moment of $P$ about C to the clockwise moment of $Q$, we have:

$$P \times \text{distance CN} = Q \times \text{distance CM}.$$

**Example 3.6** *In fig.* 3.12, *the length AC is* 180 *mm and that of BC is* 120 *mm, and angle ACB is* 73°. *Measuring the perpendicular distances of the lines of action of P and Q from axis C, we find CN* = 115 *mm and CM* = 82 *mm. If Q is* 60 *N, calculate the value of force P required to maintain a balance.*

Taking moments about C, we have:

$$\text{clockwise moment} = 60\ [\text{N}] \times 0{\cdot}082\ [\text{m}] = 4{\cdot}92\ \text{N m}$$

and

$$\text{anticlockwise moment} = P \times 0{\cdot}115\ [\text{m}].$$

Equating the clockwise and anticlockwise moments, we have:

$$P \times 0{\cdot}115\ [\text{m}] = 4{\cdot}92\ [\text{N m}]$$

$$\therefore \quad P = 42{\cdot}8\ \text{N}.$$

## 3.8 Application of the principle of moments

Many problems in statics, such as finding the value of an unknown force acting on a body at rest when its line of action is known, can easily be solved by this principle.

Two points are of great importance to the beginner. The first is to decide upon which body or part of a body or structure we are going to concentrate attention. In the case of simple levers or beams already considered, there was only one body and therefore no doubt about the body on which the various forces were exerted (though it was necessary to take the forces exerted *on* the lever and not those equal and opposite forces exerted *by* the lever or beam on the supports). But even in a simple structure, not only is the structure as a whole at rest under the action of forces but also its several parts are at rest, and we can apply the principle to the whole or to any part as may be most convenient.

The second point to be observed is that the point or axis about which we take the equal and opposite moments should, if possible, be chosen so that the moments include that of only *one* unknown force. This is generally done by choosing a point on the line of action of another unknown force if there are more than one. The method can best be understood by considering a few examples.

**Example 3.7** *A wall crane is represented diagrammatically by the centre lines of its members in fig.* 3.13 *and has a load of* 3 *kN at O. Determine the tension in the tie-rod QR. This tie-rod is freely jointed at Q and R.*

The unknown tension $T$ in the tie-rod RQ exerts a pull at the ends R and Q. Member PQ is at rest and the forces exerted upon it are: (*a*) the load of 3 kN, (*b*) the unknown tension $T$ (which must act along the tie-rod QR) and (*c*) whatever force may be exerted upon PQ through the pivot or hinge at the end P. This last force, like $T$, is an unknown one. In order that it shall not have any effect on any equation of moments, consider the forces tending to cause rotation and their moments about an axis perpendicular to the diagram and *through* P. Then any reaction or external force acting at P has no moment about this axis.

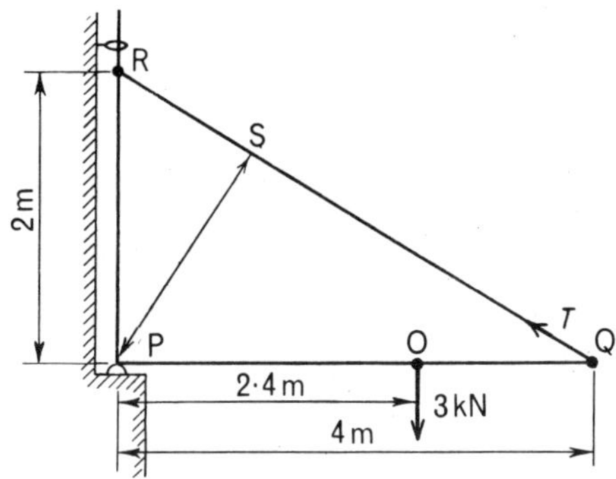


Fig. 3.13 Wall crane diagram

On drawing the diagram to scale or by calculation, we find that the distance PS, which is the moment arm of the tension $T$ about the axis selected, is 1·79 m. Hence,

anticlockwise moment about P $= T \times 1{\cdot}79$ [m]
and clockwise moment about P $= 3000$ [N] $\times$ 2·4 [m] $= 7200$ N m.

Equating the clockwise and anticlockwise moments, we have:

$$T \times 1{\cdot}79\ [\mathrm{m}] = 7200\ [\mathrm{N\ m}]$$

$$\therefore \qquad T = 4020\ \mathrm{N} = 4{\cdot}02\ \mathrm{kN}.$$

This problem could also have been solved graphically by means of a force diagram drawn to scale, as already explained for Example 2.2.

**Example 3.8** *The simple triangular roof frame or truss shown in fig.* 3.14*(a) carries a load of* 12 *kN at its top joint. Determine the forces in the members AB, BC and AC.*

Since the frame is symmetrical about a vertical axis through B, the vertical upward supporting forces at A and C are each 6 kN. To find the force which BC exerts on the joint at C, consider the fact that the tie-rod AC is at rest and the forces acting on it are in

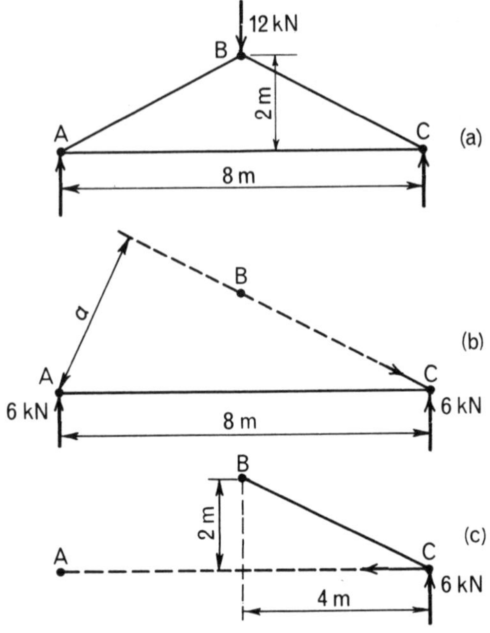


Fig. 3.14 Roof frame diagram

equilibrium, as shown in fig. 3.14(*b*). Then the moment tending to turn AC anticlockwise about an axis through A and perpendicular to the plane of the diagram

$$= \text{vertical reaction at C} \times \text{distance AC}$$
$$= 6000\ [\text{N}] \times 8\ [\text{m}] = 48\ 000\ \text{N m}.$$

Moment tending to turn AC clockwise about A

$$= \text{thrust of BC on joint C} \times \text{distance } a.$$

From a scaled diagram, distance *a* is found to be 3·58 m. Hence:

$$\text{thrust of BC} \times 3{\cdot}58\ [\text{m}] = 48\ 000\ [\text{N m}]$$
$$\therefore \quad \text{thrust of BC} = 13\ 400\ \text{N} = 13{\cdot}4\ \text{kN}.$$

This is the compressive force in the member BC; and by symmetry, the compressive force in the member BA must be the same.

In order to find the force in member AC, note that it exerts a force on C and therefore tends to rotate BC about an axis through B, perpendicular to the figure, as shown in fig. 3.14(c). Considering the forces acting on BC and having moment about B, we have:

moment tending to turn BC anticlockwise about B
= vertical reaction at C × *horizontal* distance of this reaction from B
= 6000 [N] × 4 [m] = 24 000 N m,

and moment tending to turn BC clockwise about B
= horizontal pull of AC × vertical distance from B to AC
= pull of AC × 2 [m].

Since these moments balance each other,

$$\text{pull of AC} \times 2\ [\text{m}] = 24\,000\ [\text{N m}]$$

$$\therefore \qquad \text{pull of AC} = 12\,000\ \text{N} = 12\ \text{kN}.$$

This is the tension in member AC, i.e. the pull which it exerts on the joints at C and A, and is equal to the pulls or reactions which the joints exert on the tie-rod AC. The arrowheads in fig. 3.14 (*b*) and (*c*) indicate the directions of the forces exerted *by* the members *on* the hinges or pin-joints, *not* the forces exerted by the joints *on* the members.

## 3.9 Couples

Fig. 3.15 shows two equal forces $P$ acting on a body such as a beam AB. The forces are in opposite directions but *their lines of action are not the same*. The only effect of these forces is to produce a turning effect.

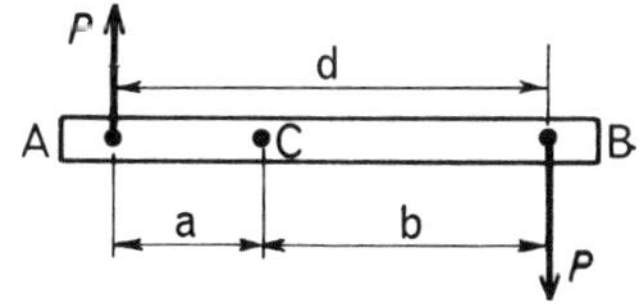


Fig. 3.15 A couple.

Let us consider any point C in the plane of the two forces, and suppose the distances between C and the lines of action of the upward and downward forces to be $a$ and $b$ respectively. Then:

clockwise moment about an axis through C due to the force acting upward $= Pa$.

Similarly the clockwise moment about the same axis due to the force acting downward $= Pb$,

$\therefore$ total clockwise moment about an axis through C

$$= Pa + Pb = Pd$$

where $d = a + b$

= perpendicular distance between the lines of action of the two forces.

Two equal parallel forces acting in opposite directions, as in fig. 3.15, are termed a *couple*; and the *moment of a couple* is the product of one of the forces and the perpendicular distance between the lines of action of the forces, i.e. $Pd$ in fig. 3.15.

A couple cannot be balanced by a *single* force; it can only be balanced by a couple of equal moment acting in the opposite direction.

One example of a couple is the tightening or loosening of a wing nut. Another example, shown in fig. 3.16($a$), consists of equal forces $P$ applied to two cords attached to the periphery of a drum (or wheel) of diameter $d$. As already explained, the moment of the

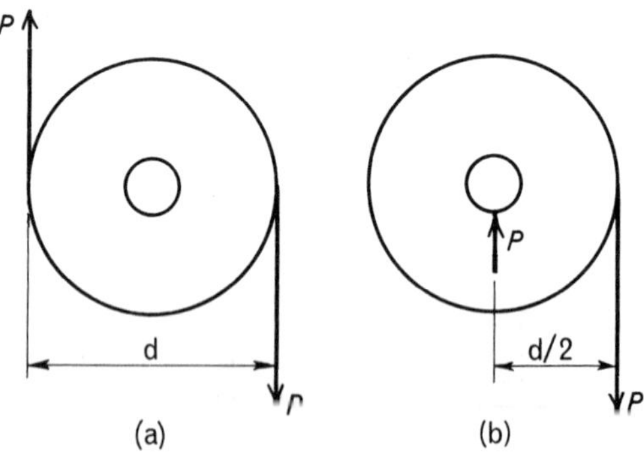


Fig. 3.16 A couple (a) with no reaction at support and (b) with reaction at support.

couple is $Pd$. In this case, the couple does not exert any force on the bearings supporting the shaft to which the drum is attached. (There will, of course, be an upward force exerted by the bearings to balance the weight of the drum and shaft, but this force is independent of the value of forces $P$.)

Let us next consider the case of a drum (or wheel) with only *one* cord attached to its periphery, as shown in fig. 3.16(*b*). When a downward force $P$ is applied to the cord, the bearings must exert an equal upward force $P$, and the two forces form a couple having a moment $= P \times d/2$, where $d$ is the diameter of the drum.

Well-known examples of this type of couple are the tightening or loosening of a nut by means of a spanner and the winding of a rope on a drum or a capstan.

## 3.10 Centre of gravity

Every particle of matter is attracted towards the centre of the earth and every object or body consists of a large number of particles. If the body is small compared with the earth, the gravitational forces on all the particles of the body can, for practical purposes, be regarded as being parallel with one another. These parallel forces can be replaced by a resultant force equal to the weight of the body and having its line of action passing through a point termed the *centre of gravity* (abbreviation, c.g.) of the body. In other words, the mass of a body may be regarded as being concentrated at its centre of gravity and its weight acts vertically through that point. The location of the centre of gravity is independent of the position of the body. Thus, if a body is suspended by a cord, the vertical

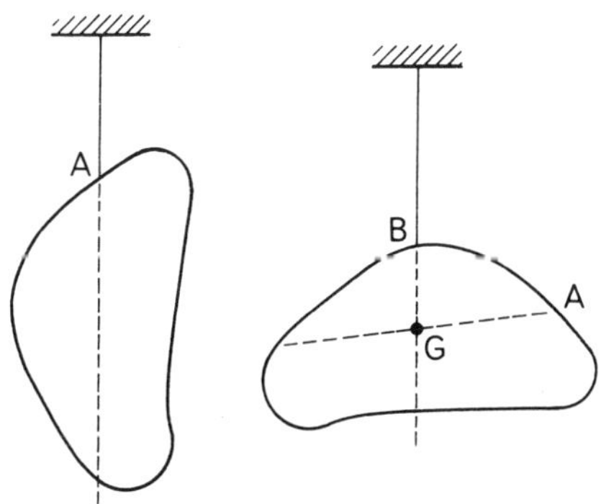


Fig. 3.17 Determination of the centre of gravity of a thin plate by suspension.

upward force exerted by the cord is equal to the weight of the body and its line of action passes through the centre of gravity of that body.

It is often possible to determine the position of the centre of gravity by a simple experiment. For instance, fig. 3.17 shows a thin plate of irregular shape. If the plate is suspended in turn from two points, A and B, and if vertical lines from A and B respectively are drawn as shown dotted in fig. 3.17, the intersection G of these lines gives the position of the centre of gravity of the plate.

In the case of each of the following symmetrically-shaped solids, the centre of gravity is at its geometrical centre. The plates are assumed to be of uniform thickness so that their centres of gravity are at mid-thickness.

*Rectangular plate:* c.g. at intersection of diagonals.

*Triangular plate:* c.g. at intersection of *medians* or lines joining an apex to the mid-point of the opposite side. It is one-third of the distance from the mid-point towards the apex.

*Circular plate:* c.g. at centre.

*Sphere:* c.g. at centre.

*Cylinder:* c.g. at mid-point of the axis.

*Ring:* c.g. at centre (which is not in the body of the ring).

We shall now consider some examples of bodies of less symmetrical shape which are divisible into parts. If we know the centre of gravity of each part and the relative *weights* of the parts, we can determine the centre of gravity by the principle of moments.

**Example 3.9** *Determine the centre of gravity of the T-shaped piece of uniform sheet metal shown in fig.* 3.18.

Let $w$ be the weight of the sheet in newtons per square millimetre.

$$\text{Area of cross-piece ABCD} = 60\,[\text{mm}] \times 20\,[\text{mm}] = 1200\,\text{mm}^2$$

$$\therefore \text{ weight of cross-piece ABCD} = 1200\,w \text{ newtons.}$$

$$\text{Area of stem KLFE} = 80\,[\text{mm}] \times 20\,[\text{mm}] = 1600\,\text{mm}^2$$

$$\therefore \qquad \text{weight of stem KLFE} = 1600\,w \text{ newtons,}$$

$$\text{and} \qquad \text{total weight of sheet} = 1200\,w + 1600\,w$$

$$= 2800\,w \text{ newtons.}$$

The c.g. of the cross-piece ABCD is at G in the centre line and is 10 mm from AB.

The c.g. of the stem KLFE is at H in the centre line and is $20 + 40 = 60$ mm from AB.

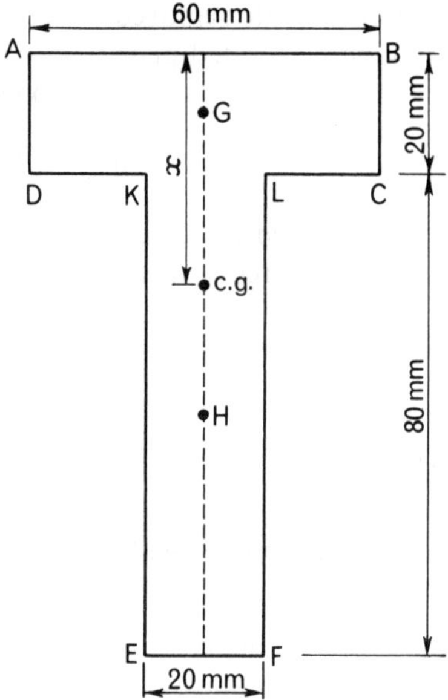


Fig. 3.18 Example 3.9.

If we take moments about AB as axis, the moment of the whole piece about AB must be equal to the sum of the moments of the two parts about AB. Hence, if $x$ be the distance of the c.g. from AB,

$$2800\ w\ [\text{N}] \times x = 1200\ w\ [\text{N}] \times 10\ [\text{mm}] + 1600\ w\ [\text{N}] \times 60\ [\text{mm}]$$

$$\therefore \qquad x = 38{\cdot}6\ \text{mm},$$

i.e. the c.g. is in line GH at a distance of 38·6 mm from AB.

**Example 3.10** *Determine the c.g. of the uniform rectangular plate,* 320 *mm* × 160 *mm, with a circular hole,* 120 *mm diameter, cut out in the position shown in fig.* 3.19.

Let $w$ be the weight of the plate in newtons per square millimetre.

$$\text{Total weight of plate with no hole} = w \times 320 \times 160$$
$$= 51\,200\ w \text{ newtons.}$$

$$\text{Area of circular plate removed} = (\pi/4) \times 120 \times 120$$
$$= 11\,300\ \text{mm}^2$$

$\therefore$ weight of circular plate removed $= w \times 11\,300$ newtons,
and weight of remaining perforated plate $= 51\,200\,w - 11\,300\,w$
$= 39\,900\,w$ newtons.

Let $x$ be the distance of the c.g. of the perforated plate from axis AD.

The c.g. of the unperforated plate is at O, 160 mm (= 0·16 m) from AD, and that of the circular piece is 240 mm (= 0·24 m) from AD.

Taking moments about axis AD, we have:

moment due to weight of perforated plate $= 39\,900\,w\,[\text{N}] \times x$
moment due to weight of circular plate $= 11\,300\,w\,[\text{N}] \times 0{\cdot}24\,[\text{m}]$
$= 2712\,w\,[\text{N m}]$
and moment due to weight of whole plate $= 51\,200\,w\,[\text{N}] \times 0{\cdot}16\,[\text{m}]$
$= 8192\,w\,[\text{N m}]$.

Equating the sum of the moments of the parts to the moment of the whole rectangular plate, we have:

$$39\,900\,w\,[\text{N}] \times x + 2712\,w\,[\text{N m}] = 8192\,w\,[\text{N m}]$$

$$\therefore \quad x = 0{\cdot}1373 \text{ m} = 137{\cdot}3 \text{ mm}.$$

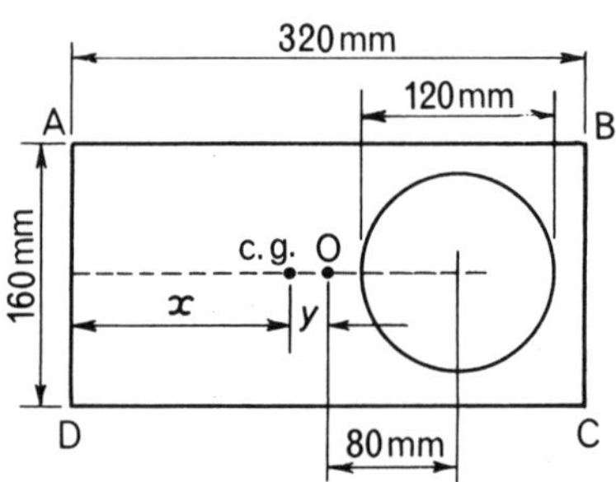


Fig. 3.19 Example 3.10.

Alternatively, and as a check, we can determine the distance of the c.g. of the perforated plate from the centre O by taking moments about an axis through O and parallel to AD.

Suppose $y$ to be the distance of the c.g. of the perforated plate from O. This c.g. must lie to the left of O.

The moment due to the weight of the unperforated plate about O is zero. Hence the clockwise moment due to the weight of the circular piece must balance the anticlockwise moment of the perforated plate,

i.e. $11\ 300\ w\ [\mathrm{N}] \times 0{\cdot}08\ [\mathrm{m}] = 39\ 900\ w\ [\mathrm{N}] \times y$

$\therefore \quad y = 0{\cdot}0227\ \mathrm{m}$

$= 22{\cdot}7\ \mathrm{mm},$

which corresponds to $160 - 22{\cdot}7 = 137{\cdot}3$ mm from AD, as previously calculated.

**Example 3.11** *The connecting rod of an engine has mass of* 200 *kg. It is placed on two knife-edge supports A and B,* 2 *m apart (fig.* 3.20). *The force on support A is found to be* 1·2 *kN. Calculate the distance of the centre of gravity of the rod from the other support B.*

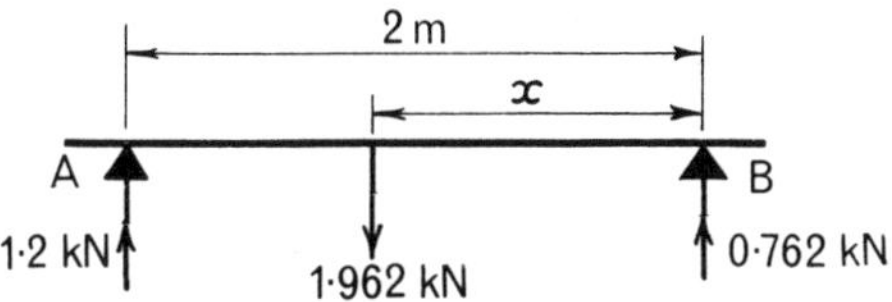


Fig. 3.20 Example 3.11.

Weight of connecting rod $\simeq 200 \times 9{\cdot}81 = 1962$ N.

Let $x$ be the distance of the centre of gravity from support B.

Taking moments about a horizontal axis perpendicular to the rod at support B, we have:

anticlockwise moment due to weight of rod $= 1962\ [\mathrm{N}] \times x$

and

clockwise moment due to supporting force at A $= 1200\ [\mathrm{N}] \times 2\ [\mathrm{m}]$

$= 2400\ \mathrm{N\ m}.$

Equating these moments, we have:

$1962\ [\mathrm{N}] \times x = 2400\ [\mathrm{N\ m}]$

$\therefore \quad x = 1{\cdot}223\ \mathrm{m}$

= distance of c.g. from support B.

## *Summary of Chapter* 3

The moment of a force about an axis is the product of the force and the perpendicular distance of the line of action of the force from the axis.

*General principle of moments.* If a body is at rest under the action of several forces, the total clockwise moment of the forces about any axis is equal to the anticlockwise moment of the forces about the same axis.

The principle of moments applies to parallel and non-parallel forces. It also applies to cranked levers and to framed structures, the joints of which can be assumed to act as frictionless hinges.

In the case of *parallel* forces, the principle of moments can be used to determine the magnitude and position of a single force, called the *equilibrant*, to balance several parallel forces. This force, reversed, is the *resultant* of the several parallel forces. An important application is the location of the resultant of the distributed weight of a body. It always acts through the centre of gravity of the body.

## EXAMPLES 3

1. A wooden rod, whose weight may be neglected, is pivoted at a point A. A spring balance is attached vertically to the rod at a point B, 200 mm from A, and another spring balance is similarly attached to the rod at a point C on the other side of A, the distance between C and A being 360 mm. What must be the reading on the balance at C when the balance at B reads 80 N, in order that the rod may remain in equilibrium? What is the downward force exerted by the pivot on the rod?
2. Where must the fulcrum be situated in a straight lever, 2 m long, if a downward effort of 200 N at one end is to lift a mass of 80 kg at the other end? State the distance from the effort.
3. A lever AB, 3 m long, rests on a fulcrum at C, 1 m from A. There is a downward force of 80 N acting on the lever at A. Calculate (*a*) the force required at B to maintain equilibrium and (*b*) the force exerted by the fulcrum on the lever. Neglect the weight of the lever.
4. A lever AB, 2 m long, is pivoted at end A. A mass of 3 kg is suspended at a point C, 0·8 m from A. Calculate the vertical force required at B to maintain the lever horizontally. Neglect the weight of the lever.
5. If the mass of 3 kg in Q. 4 were suspended at point B of the lever, what would be the vertical force required at point C to maintain the lever horizontally?
6. A rod AB of uniform section, 2 m long, has a mass of 5 kg. The rod rests on a fulcrum 0·6 m from A. Calculate (*a*) the vertical force required at B to maintain the rod in a horizontal position and (*b*) the value of the reaction at the fulcrum.

7. A uniform horizontal lever AE, 7 m long, is supported on a fulcrum at C, 3 m from A. There are downward forces of 20 N and 30 N at A and E respectively and another downward force of 10 N at a point B, 2 m from A. What vertical force must be applied at a point D, 1 m from E, in order that the lever may remain horizontal? Will this force act upward or downward? Neglect the weight of the lever.
8. A horizontal steel beam of uniform section, 4 m long, rests on two supports, one at the end of the beam and the second at a distance of 1 m from the other end of the beam. The mass of the beam is 80 kg. Calculate the upward reactions at the supports.
9. If a 50-kg mass is suspended midway between the supports of the beam in Q. 8, calculate the reactions of the supports.
10. If a 50-kg mass is suspended half-way along the overhanging length of the beam in Q. 8, calculate the reactions of the supports.
11. The mass of a uniform steel girder, 12 m long, is 3 Mg. It is supported at one end and at a point 5 m from the other end. Calculate the reactions of the two supports.
12. If a mass of 2 Mg is suspended at the end of the overhanging portion of the beam in Q. 11, calculate the reactions of the supports.
13. If the 2-Mg mass in Q. 12 were suspended from the beam midway between the supports, what would be the reactions of the supports?
14. A uniform horizontal beam, 6 m long, is supported at points A and B, 1 m and 4 m respectively, from one end. If the supporting force at B is 400 N, calculate the mass of the beam and the supporting force at A.
15. If a uniform beam, 5 m long, is supported 1 m from one end, how far from that end must a second support be placed in order that it shall support 60 per cent of the weight of the beam?
16. Two men carry a scaffolding pole, 10 m long, having a mass of 50 kg, each supporting one end. The centre of gravity of the pole is 4 m from one end. How much weight does each man support?

    If a bag of tools having a mass of 15 kg is to be carried by slinging it over the pole, at what distance from the lighter end of the pole should it be slung in order that the total load should be shared equally?
17. A beam AB is 15 m long and is freely supported at A and B. Loads of 60 kN, 100 kN and 80 kN are placed on the beam at distances of 3 m, 6 m and $x$ metres from A. Calculate the distance $x$ if the loads on the supports are equal. Neglect the weight of the beam.
18. A uniform beam AB has a mass of 1 Mg and is 4 m long. It is supported at A and B. A mass of 3 Mg is suspended at C, 1 m from A, and a rope exerting an upward pull of 5 kN is attached to the beam at D, 1·5 m from B. Calculate the reactions at A and B.
19. A square plate ABCD, of 100 mm side, is free to rotate in a vertical plane about a peg through its centre. Forces of 8 N, 20 N, 12 N and 16 N act along AB, CB, DC and DA respectively. Calculate the magnitude and sense of the resultant turning moment about the peg. What would be the length of the arm of a couple composed of 5-N forces which would maintain the plate in equilibrium?
20. In order to find the position of the centre of gravity of a small connecting

rod, it is suspended in a horizontal position from two spring balances, one at each end. The reading on the balance at the big end is 32 N and that on the other balance is 12 N. If the distance between the points of support is 180 mm, determine the distance of the centre of gravity from the big end.

21. A cranked lever ACB has a horizontal arm AC, 400 mm long, and an arm CB, 150 mm long. The lever can turn freely about a pin at C and the angle ACB is 120°. What vertical pull will be required at A to produce a horizontal pull of 200 N in a wire attached at B?

22. A uniform bar, 2 m long, has a mass of 25 kg. It is hinged at one end and is kept in a horizontal position by a rope attached to the free end, making an angle of 45° with the bar and in the same vertical plane as the bar. If a mass of 60 kg is hung on the bar 1·2 m from the hinge, what is the tension in the rope?

23. A wall crane has the dimensions shown in the diagram and a mass of 500 kg is suspended at C. Determine the forces in AC and BC, and state whether they are tensile or compressive.

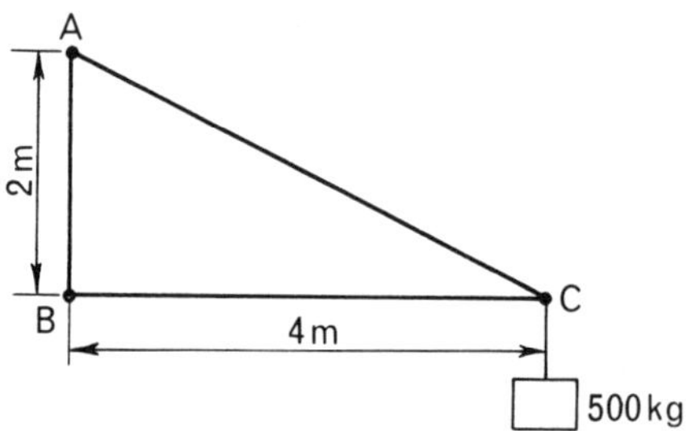


No. 23.– Examples 3.

24. A simple roof truss has the dimensions shown in the diagram and a load of 12 kN at A. Determine the forces in the members of the frame and state which are in tension and which are in compression.

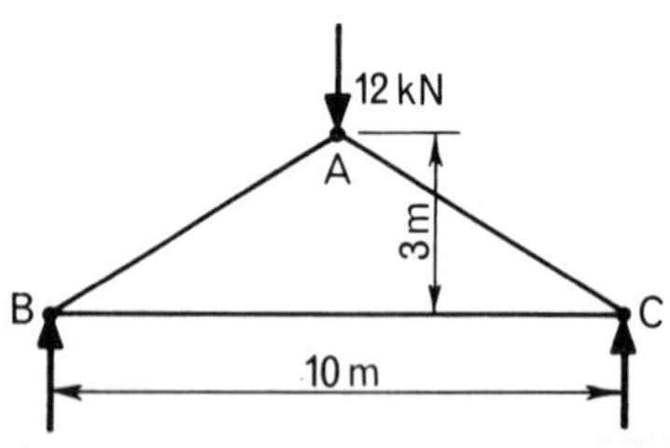


No. 24.– Examples 3.

25. A simple pin-jointed structure is in the form of an isosceles triangle ABC in which A is the apex and the supports are at B and C. If AB = AC = 5 m, and BC = 8 m, determine the force in each member when a vertical load of 4 kN is applied at A.
26. A metal plate of uniform thickness is in the form of an equilateral triangle of 2-m sides. If the plate has a mass of 200 $kg/m^2$ and is lying flat on the ground, calculate the value of the vertical force to be applied at one corner of the plate in order to lift it to an upright position by rotating it about the opposite side (as about a hinge).
27. A triangular plate ABC of wrought iron, 10 mm thick, has the following dimensions: AB, 2 m; BC, 3 m; CA, 2 m. The plate lies flat on a horizontal surface. What vertical force must be applied to the corner A in order to lift this corner? Density of wrought iron is 7 800 $kg/m^3$.
28. A T-shaped piece of metal of uniform thickness has a cross-piece 100 mm × 30 mm and a stem 140 mm × 25 mm. Determine the distance of the centre of gravity from the top edge of the T-piece.
29. An L-shaped piece of steel plate of uniform thickness can be divided into two rectangles, each 60 mm × 20 mm. Determine the distances of the centre of gravity from the left-hand and bottom edges respectively of the L-piece.
30. A rectangle, 90 mm × 60 mm, is cut from a thin rectangular plate, 100 mm × 70 mm, to form an L-section, the width of the legs being 10 mm. Determine the centre of gravity of the L-section.
31. A sheet of metal, 400 mm square, has centre lines drawn on it parallel to the sides. A circle, 250 mm diameter, is cut out of the square, the centre of the circle lying on one of the centre lines of the square. Determine the distance between the centres of the square and of the circle if the centre of gravity of the remaining plate is 30 mm from the centre of the square.
32. A hole having a diameter of 80 mm is cut out from a uniform plate of 250 mm diameter. The distance between the centre of the plate and the centre of the hole is 50 mm. Calculate the position of the centre of gravity of the remaining material.

## ANSWERS TO EXAMPLES 3

1. 44·4 N, 124·4 N.
2. 1·593 m.
3. 40 N, 120 N.
4. 11·77 N.
5. 73·6 N.
6. 14 N, 35·05 N.
7. 16·7 N, upward.
8. 261·6 N, 523·2 N.
9. 506·8 N, 768·5 N.
10. 179·8 N, 1095·5 N.
11. 4·2 kN, 25·23 kN.
12. —9·81 kN, 58·86 kN.
13. 14·02 kN, 35·03 kN.
14. 61·2 kg, 200 N.
15. 3·5 m.
16. 294·3 N, 196·2 N; 1·665 m.
17. 12·75 m.
18. 25·1 kN, 9·14 kN.
19. 0·4 N m, 40 mm radius.
20. 49·1 mm.
21. 64·95 N.
22. 673 N.
23. 11 kN tension in AC, 9·81 kN compression in BC.
24. 11·66 kN compression in AB and AC, 10 kN tension in BC.
25. 3·33 kN compression in AB and AC, 2·67 kN tension in BC.
26. 1132 N.
27. 506 N.
28. 60·8 mm.
29. $x = 20$ mm, $y = 30$ mm.
30. $x = 18·1$ mm, $y = 33·2$ mm.
31. 67·9 mm.
32. 5·7 mm from centre.

CHAPTER 4

# Inclined plane

## 4.1 Smooth inclined plane

The use of the inclined plane for lifting heavy bodies by means of small forces has been known since ancient times; for instance, the large stone blocks forming the Pyramids were lifted in this way.

In practice, it is not possible to make an inclined plane perfectly smooth, but much the same effect is obtained by the use of rollers or wheels under the body to be hauled up the plane.

It is instructive to employ more than one method to determine the force required to hold a body in equilibrium on an inclined plane or to haul it up the plane at a constant speed, and to see that all the methods lead to the same conclusions. The inclined plane thus forms an excellent example to illustrate matters dealt with in chapters 1, 2 and 3.

## 4.2 Effort parallel to inclined plane

Fig. 4.1(a) represents a metal roller attached by a cord to a spring

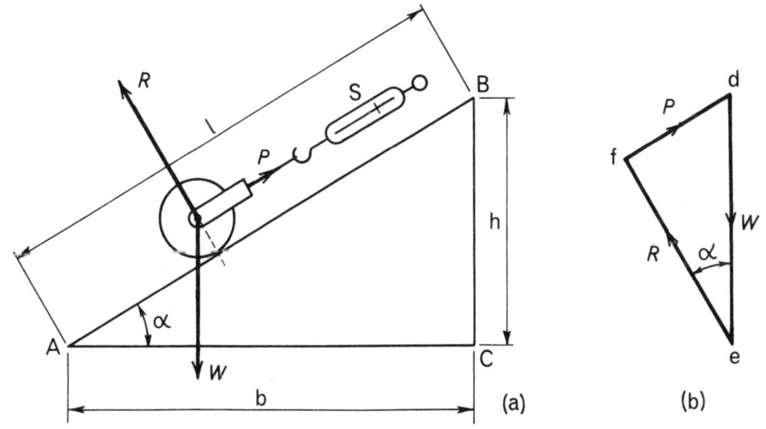


Fig. 4.1 Smooth inclined plane and force diagram.

balance S. The roller rests on a smooth surface AB inclined at an angle $\alpha$ to the horizontal. The weight $W$ of the roller can be determined by suspending it from a spring balance calibrated in newtons.

Suppose the reading on spring balance S to be $P$ newtons when the roller is stationary or is being hauled up the inclined surface at a steady speed.

The contact between a perfect cylindrical surface and a perfectly flat surface is a straight line parallel to the axis of the cylinder. Since the inclined surface AB in fig. 4.1 is being assumed smooth, the direction of the reaction $R$ between the cylinder and that surface is represented by a straight line drawn at right angles to the surface and passing through the line of contact and also through the axis of the cylinder, as shown in fig. 4.1(a).

The relationship between $P$ and $W$ can be derived by several methods, utilizing principles discussed in earlier chapters.

*(a) By force diagram*

The three forces $P$, $W$ and $R$, acting on the body, are in equilibrium and can therefore be represented by the triangular force diagram shown in fig. 4.1(b), where vectors $de$, $ef$ and $fd$ are parallel to $W$, $R$ and $P$ respectively, and the lengths of the vectors represent to scale the magnitudes of the respective forces.

Hence,
$$\frac{fd}{de} = \frac{P}{W}$$

or
$$P = \frac{fd}{de} \times W.$$

Since the line of action of $W$ is at right angles to AC and that of $R$ is at right angles to AB, the angle between the directions of $W$ and $R$ must be equal to angle $\alpha$ between AC and AB. Also, the angle between the lines of action of forces $P$ and $R$ is 90°. Hence the triangle formed by $fd$, $de$ and $ef$ in fig. 4.1(b) is similar to triangle ABC in fig. 4.1(a), so that:

$$\frac{fd}{de} = \frac{\text{BC}}{\text{AB}}$$

$$\therefore \quad P = \frac{fd}{de} \times W = \frac{\text{BC}}{\text{AB}} \times W$$

$$= \frac{h}{l} \times W = W \sin \alpha \qquad (4.1)$$

where $l$ = length of plane
$h$ = height of plane
and $\alpha$ = angle between the inclined plane and the horizontal.

(*b*) *By moments*

If we choose any point O on the line of action of $R$, fig. 4.2, distant ON from the centre of gravity of the roller; then the anticlockwise moment of $P$ about an axis through O, perpendicular to the diagram, must balance the clockwise moment of $W$ about the same

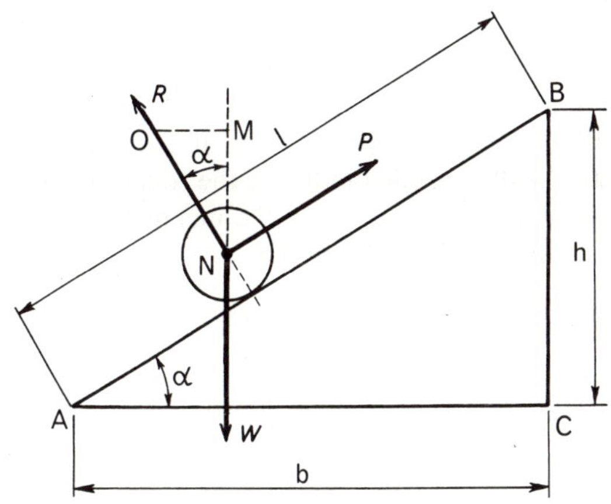


Fig. 4.2 Moment solution for smooth inclined plane.

axis (the moment of $R$ about axis O is zero),

hence $$P \times \text{ON} = W \times \text{OM}$$

$$\therefore \quad P = \frac{\text{OM}}{\text{ON}} \times W.$$

Since triangle NOM is similar to triangle ABC,

$$\therefore \quad \frac{\text{OM}}{\text{ON}} = \frac{\text{BC}}{\text{AB}}$$

so that $P = \dfrac{\text{OM}}{\text{ON}} \times W = \dfrac{\text{BC}}{\text{AB}} \times W = \dfrac{h}{l} \times W = W \sin \alpha$, as before.

(*c*) *By resolution into components*

Let us resolve force $W$ (fig. 4.1) into two components, one downward along the plane and the other perpendicular to the plane. The component downward along the plane is:

$$W \cos \text{ABC} = W \times h/l.$$

This component, being the only component acting downward along the plane, must balance force $P$ acting upward along the plane, (all other forces being perpendicular to the plane);

hence, $P = W \times h/l = W \sin \alpha$, as before.

The other component of $W$, namely $W \cos \alpha$, must be equal in magnitude and opposite in direction to reaction $R$.

*(d) By equating the work done by effort to the work done on load*

If the friction at the bearings of the roller in fig. 4.1(a) is negligible, then the work done by the effort $P$ in hauling the roller up the inclined plane is all expended in lifting the roller. Thus, if the roller is hauled a distance $l$ metres along the inclined plane by an effort $P$ newtons acting parallel to the plane,

work done *by* effort $= Pl$ joules.

If $h$ metres be the height through which the roller has been raised, and if the weight of the roller is $W$ newtons,

work done in lifting the roller $= Wh$ joules.

Since the effect of friction is being assumed negligible,

work done by effort = work done in lifting the roller

i.e. $Pl = Wh$

$\therefore$ $P = W \times h/l = W \sin \alpha$, as before.

The work done by or against reaction $R$ is zero since the direction of $R$ is at right angles to the direction of motion of the roller.

### 4.3 Effort horizontal

Suppose the roller to be hauled up the inclined plane at a constant speed by a horizontal effort $P$, as shown in fig. 4.3(a). The three forces, $P$, $W$ and $R$, can be represented by the triangular force diagram of fig. 4.3(b). From this diagram it is seen that:

$$\frac{P}{W} = \frac{fd}{de} = \tan \alpha$$

$$\therefore \quad P = W \tan \alpha = W \times h/b \qquad (4.2)$$

where $b$ is the distance through which $P$ moves *in its own direction* while the roller is raised through a distance $h$.

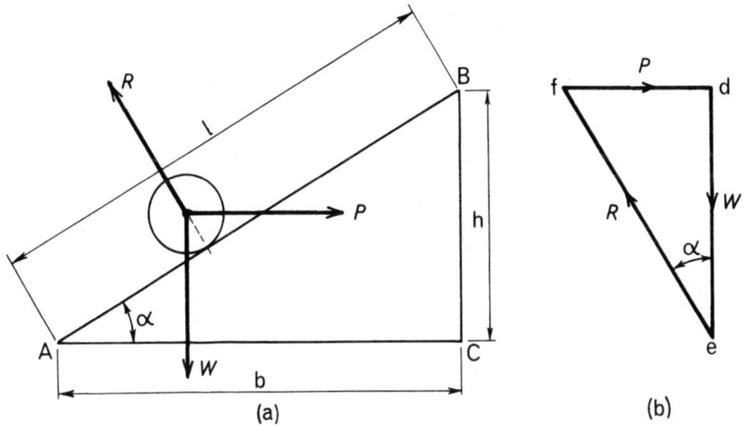


Fig. 4.3 Horizontal effort on smooth inclined plane.

Alternatively,

work done by effort $P = Pb$

and work done in lifting roller $= Wh$.

Equating the work done by the effort to the work done in lifting the roller, we have:

$$Pb = Wh$$

$$\therefore \quad P = W \times h/b = W \tan \alpha, \text{ as before.}$$

Reaction $R$ in fig. 4.3(a) does no work since its direction is perpendicular to the direction of motion of the roller.

## 4.4 Effort at any angle

If the effort $P$ is applied at any angle $\theta$ to the inclined plane, as in fig. 4.4(a), to haul the roller up the plane at a constant speed, the value of $P$ can be found by drawing the vector diagram shown in fig. 4.4(b). Vector *de* is drawn vertically to represent $W$ to scale, and from $d$ and $e$, lines are drawn parallel to $P$ and $R$ respectively, to meet at $f$. The value of $P$ is represented to scale by the length of vector *fd*.

An expression for the value of effort $P$ can be derived thus:

$$\text{component of effort } P \text{ acting along plane} = P \cos \theta.$$

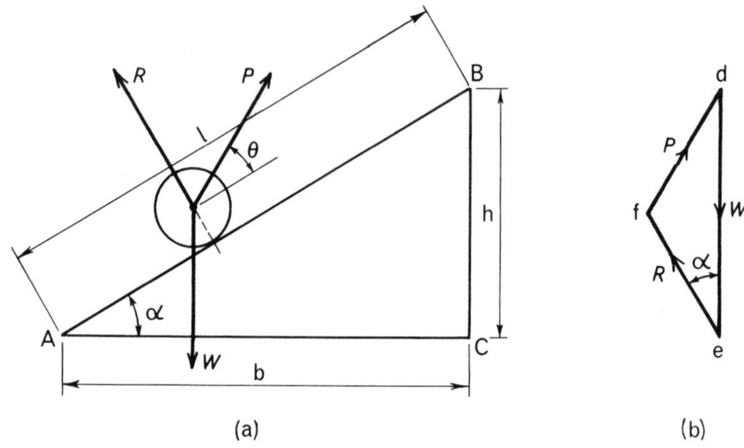


Fig. 4.4 Effort at any angle on smooth inclined plane.

Work done by this component in hauling the roller through a distance $l$ along the inclined plane $= (P \cos \theta) \times l$.

If the roller is raised through a height $h$,

$$\text{work done in lifting roller} = W \times h.$$

Since the reaction $R$ is at right angles to the direction of motion of the roller,

$$\text{work done on or by reaction } R = 0.$$

Since friction is assumed to be negligible,

$$\text{work done by effort } P = \text{work done in lifting roller}$$

i.e. $$(P \cos \theta) \times l = W \times h$$

$\therefore$ $$P = \frac{W}{\cos \theta} \times \frac{h}{l}$$

$$= W \times \frac{\sin \alpha}{\cos \theta} \qquad (4.3)$$

The same result can be obtained by resolving $P$ and $W$ along the inclined surface, thus:

$$\text{component of } P \text{ acting up the incline} = P\cos\theta$$

and $$\text{component of } W \text{ acting down the incline} = W\sin\alpha.$$

Since the component of $R$ acting parallel to the inclined plane is zero, the components of $P$ and $W$ must be equal and opposite,

hence $$P\cos\theta = W\sin\alpha$$

$\therefore$ $$P = W\frac{\sin\alpha}{\cos\theta}, \text{ as before.}$$

**Example 4.1** *A smooth inclined plane has a gradient of* 1 *in* 4. *Calculate the force required to haul a mass of* 10 *kg up the gradient at a constant speed when the direction of the force is* (a) *parallel to the plane,* (b) *horizontal and* (c) *at an angle of* 20° *with the plane.*

A gradient of 1 in 4 means that there is a rise of 1 m for every 4 m travelled up the inclined plane.

$$\text{Weight of the 10-kg mass} = W \simeq 10 \times 9{\cdot}81$$
$$= 98{\cdot}1 \text{ N}.$$

(*a*) If the mass is hauled a distance of 4 m along the inclined plane by an effort $P$ parallel to the plane,

$$\text{work done by effort} = P \times 4 \text{ [m]}.$$

During this time the mass has been raised to a height of 1 m above its original position against a gravitational force (or weight) of 98·1 N,

$\therefore$ $$\text{work done on the mass} = 98{\cdot}1 \text{ [N]} \times 1 \text{ [m]}.$$

Since the plane is assumed to be smooth,

$$\text{work done by the effort} = \text{work done on the mass}$$

i.e. $$P \times 4 \text{ [m]} = 98{\cdot}1 \text{ [N]} \times 1 \text{ [m]}$$

$\therefore$ $$P = 24{\cdot}5 \text{ N}.$$

(*b*) Let $b$ be the horizontal base corresponding to a 4-m length of plane, then:

$$4^2 = b^2 + 1^2$$

$\therefore$ $$b^2 = 15$$

so that $$b = 3{\cdot}87.$$

Hence, while the horizontal effort $P$ travels 3·87 m, the mass is lifted 1 m. Equating the work done by the effort to the work done in lifting the mass, we have;

$$P \times 3{\cdot}87\ [\text{m}] = 98{\cdot}1\ [\text{N}] \times 1\ [\text{m}]$$

$$\therefore \quad P = 25{\cdot}4\ \text{N}.$$

(*c*) Component of $P$ along the plane $= P \cos 20^\circ$
$= P \times 0{\cdot}94,$

$\therefore$ work done by effort in hauling the mass through a distance of 4 m along the inclined plane

$$= (P \times 0{\cdot}94) \times 4\ [\text{m}].$$

During this time the mass has been lifted 1 m,

$\therefore$ work done on the mass $= 98{\cdot}1\ [\text{N}] \times 1\ [\text{m}]$.

Equating the work done by the effort to the work done on the mass, we have:

$$(P \times 0{\cdot}94) \times 4\ [\text{m}] = 98{\cdot}1\ [\text{N}] \times 1\ [\text{m}]$$

$$\therefore \quad P = 26{\cdot}1\ \text{N}.$$

**Example 4.2** *A train having a mass of* 300 *t is hauled up a gradient of* 1 *in* 250. *The tractive resistance on the level is* 4·5 *mN/N (or* 4·5 *N/kN), i.e. for every newton of train weight, the tractive force required to overcome wind resistance and rail and axle friction is* 4·5 *mN. Calculate the tractive effort required to haul the train at a constant speed. Also, calculate the tractive effort required to haul the train down the gradient at a constant speed.*

Mass of train $= 300$ t
$= 300\,000$ kg,

$\therefore$ weight of train $\simeq 300\,000 \times 9{\cdot}81$
$= 2{\cdot}943 \times 10^6$ N
$=$ vertical force on track.

Since the gradient is only 1 in 250, the force perpendicular to the track is practically the same as the vertical force on the track, namely $2{\cdot}943 \times 10^6$ N.

$$\left.\begin{array}{r}\text{Hence, effort to overcome}\\ \text{tractive resistance}\end{array}\right\} = (2{\cdot}943 \times 10^6)\ [\text{N}] \times 4{\cdot}5\ [\text{mN/N}]$$

$$= 13{\cdot}25 \times 10^6\ \text{mN}$$

$$= 13\ 250\ \text{N}.$$

From expression (4.1),

effort to haul train up gradient if there were no friction

$$= (2{\cdot}943 \times 10^6)\ [\text{N}] \times 1/250 = 11\ 770\ \text{N}.$$

$$\therefore \qquad \text{total tractive effort} = 13\ 250 + 11\ 770 = 25\ 020\ \text{N}$$

$$= 25{\cdot}02\ \text{kN}.$$

When the train is being hauled down the gradient, gravitation, in the form of the component of the weight of the train along the gradient, is providing a force of 11 770 N towards overcoming the tractive resistance of 13 250 N opposing motion.

$$\left.\begin{array}{r}\text{Hence the force required to}\\ \text{haul the train down the gradient}\end{array}\right\} = 13\ 250 - 11\ 770 = 1480\ \text{N}$$

$$= 1{\cdot}48\ \text{kN}.$$

## *Summary of Chapter* 4

By the use of an inclined plane, a heavy body can be lifted by a relatively small force. The ratio of the effort to the weight of the body lifted can be found by: (a) the force diagram, (b) moments, (c) the resolution of forces acting on the body and (d) equating the work done by the effort to the work done in lifting the body.

When the effort is along the inclined plane,

$$P = W \sin \alpha \tag{4.1}$$

When the effort is horizontal,

$$P = W \tan \alpha \tag{4.2}$$

When the effort makes an angle $\theta$ with the plane,

$$P = W \frac{\sin \alpha}{\cos \theta} \tag{4.3}$$

## EXAMPLES 4

1. A body having a weight of 500 N is placed on a smooth inclined plane, 5 m long. The top of the plane is 2 m above the lower end. Calculate the

force required to hold the body in equilibrium if it is applied (*a*) parallel to the plane, (*b*) horizontally.

2. A body having a mass of 20 kg is pulled at a uniform speed up a smooth inclined plane having a gradient of 1 in 3. Calculate the force required if it acts (*a*) parallel to the plane, (*b*) horizontally.
3. A smooth inclined plane has a gradient of 1 in 6. Calculate the mass of a body that can be held in equilibrium on the plane by a force of 80 N when the force acts (*a*) parallel to the plane, (*b*) horizontally.
4. A body having a mass of 5 t is hauled at a constant speed up a smooth plane which is inclined at 10° to the horizontal. Calculate the value of the force required parallel to the plane.
5. An inclined plane rises 1 m for every $x$ metres of its length. A force of 60 N, acting parallel to the plane, is required to haul a body having a weight of 450 N at a uniform speed up the plane. Calculate the value of $x$, assuming friction resistance to be negligible.
6. A body having a mass of 30 kg rests on a smooth plane inclined at 20° to the horizontal and held in equilibrium by a horizontal force $P$. Sketch the force diagram and determine graphically the magnitudes of force $P$ and of the reaction between the load and the inclined plane.
7. A body having a weight of 60 N rests on a smooth plane inclined at 30° to the horizontal. Determine (*a*) the horizontal force required to prevent the body from sliding down the plane, (*b*) the value and direction of the least force required to pull the body up the plane.
8. A smooth inclined plane has a gradient of 1 in 5. A body having a mass of 60 kg rests on the plane. Draw the force diagram to scale and, from the diagram, determine the effort required to pull the body at a uniform speed up the plane if the effort makes an angle of 30° with the plane.
9. The slope of a smooth inclined plane is 20° to the horizontal. A force of 400 N, inclined at an angle of 15° to the plane, holds a body in equilibrium. Determine, graphically or otherwise, the mass of the body.
10. A train having a mass of 400 t travels at a constant speed along a straight track. At that speed, the tractive force required to overcome friction and wind resistance is 5 mN/N (or 5 N/kN). Calculate the tractive effort required to maintain the same speed: (*a*) on a level track, (*b*) up an incline of 1 in 150 and (*c*) down an incline of 1 in 300.
11. A train is hauled up an incline of 1 in 40 by a tractive effort of 20 kN acting parallel to the slope. Calculate the distance travelled and the work done, in megajoules, when the train has been raised through a vertical distance of 30 m.

## ANSWERS TO EXAMPLES 4

1. 200 N, 218 N.
2. 65·4 N, 69·3 N.
3. 48·9 kg, 48·2 kg.
4. 8·51 kN.
5. 7·5 m.
6. 107 N, 313 N.
7. 34·6 N, 30 N parallel to plane.
8. 136 N.
9. 115 kg.
10. 19·62 kN, 45·78 kN, 6·54 kN.
11. 1·2 km, 24 MJ.

CHAPTER 5

# Friction

## 5.1 Friction

The term 'friction' has been referred to in earlier chapters and its general meaning is widely understood, but here we consider its effects as a definite force affecting problems of equilibrium.

Suppose a body to be resting on a horizontal table. It is acted upon by a downward force $W$ due to the gravitational pull of the earth upon it (i.e. its weight). There must also be a reaction $N$ normal to the surface of the table. This reaction, exerted *by* the table *on* the body, is equal and opposite to force $W$, as shown in fig. 5.1, where—for clarity—the line of action of $N$ is shown displaced slightly from that of $W$. Actually, *they are in direct opposition.*

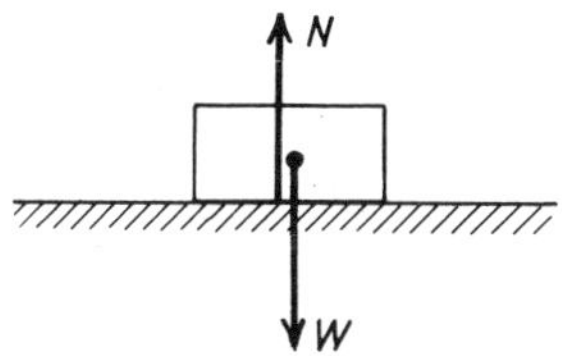


Fig. 5.1 Normal forces only.

If the body were resting on a surface inclined to the horizontal, the value of $N$ would be less than that of $W$ and the line of action of $N$ would not be directly opposite that of $W$.

If a horizontal force $P$ is applied to the body, pulling towards the right as in fig. 5.2, the body will not move unless $P$ is large enough. This is due to the presence of a force of friction $F$ exerted on the body *by the table* towards the left and just sufficient to balance force $P$. This force of friction $F$ has the feature of *adjustment* so as to be exactly equal and opposite to $P$, i.e. if $P$ is increased, $F$ increases by exactly the same amount.

While the body is at rest, we may regard it as being in equilibrium under the action of the *four* forces shown in fig. 5.2. The table top

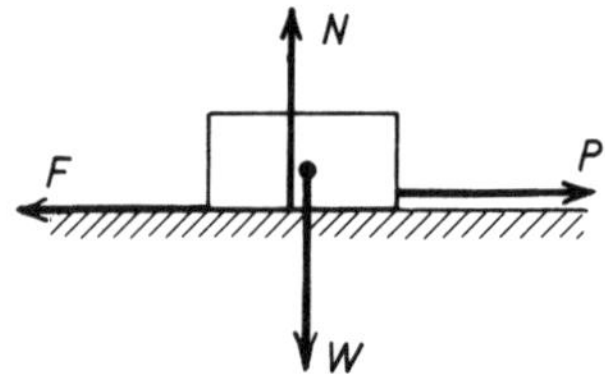


Fig. 5.2 Frictional drag.

exerts *two* forces, $N$ vertically (equal and opposite to $W$) and $F$ horizontally (equal and opposite to $P$), as shown in fig. 5.3(a). From the force diagram of fig. 5.3(b), it is seen that these two forces can

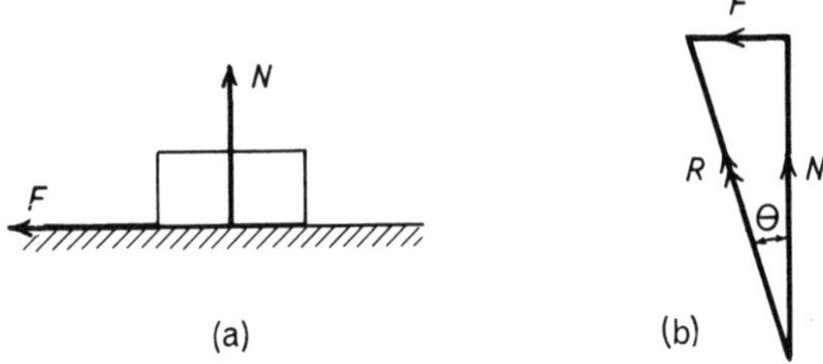


Fig. 5.3 Resultant reaction.

be replaced by a resultant force $R$* acting at an angle $\theta$ to the left of the line of action of force $N$, the value of $\theta$ being given by:

$$\tan \theta = F/N \tag{5.1}$$

We may therefore regard the body as being in equilibrium under the action of *three* forces, namely $W$, $P$ and $R$, shown in fig. 5.4(a). These forces can be represented vectorially by the triangular force diagram of fig. 5.4(b).

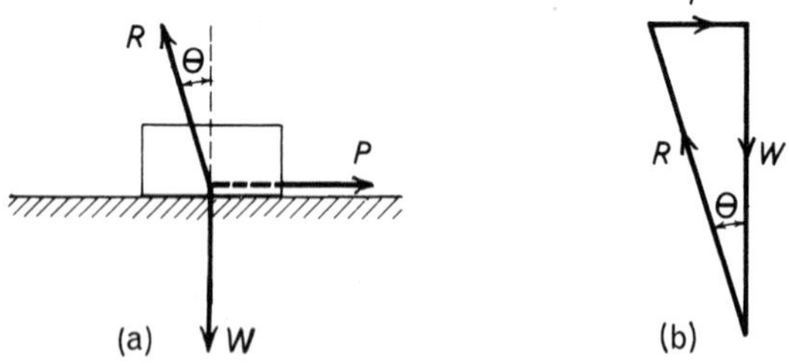


Fig. 5.4 Three forces acting on body.

* The double arrowhead on $R$ indicates that $R$ represents the resultant of two forces and *not* the equilibrant of those forces.

As the pulling force $P$ is increased, the friction force $F$ (equal and opposite to $P$) also increases and the resultant $R$ inclines more to the left, i.e. angle $\theta$ increases. There is, however, a *limit* to this adjustment of $F$ to resist a growing value of $P$. When this limit is reached, the body is on the point of motion towards the right; and if started in the direction of $P$, it will continue to move. Thus, when motion is taking place, the friction operating against motion is the maximum or *limiting* friction.

Strictly, this may be a little less than the limiting friction exerted while the body remains at rest. In other words, the friction of motion is often less than the limiting or maximum static friction or friction of rest. It is a matter of common knowledge that bodies may 'stick' or adhere somewhat when at rest and glide one over the other more easily after a start has been made. We shall neglect this difference and concern ourselves with the limiting friction operative during *steady* motion.

This condition (i.e. when component $F$ has reached its limiting value, or what is the same thing, when $R$ has reached its greatest inclination to the vertical) is shown in fig. 5.5(a). The corresponding angle between $R$ and the perpendicular or normal to the sliding surface is termed the *angle of friction* and is represented by $\varphi$.

From expression (5.1) it follows that:

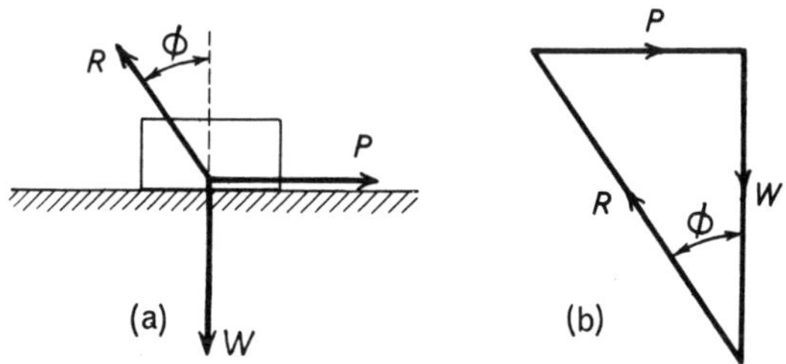


Fig. 5.5 Angle of friction.

$$\tan \varphi = \frac{\text{sliding friction force } F}{\text{normal reaction } N}$$

The ratio of the sliding friction force to the normal reaction of the supporting surface is termed the *coefficient of friction* and is represented by $\mu$,

i.e. $$\mu = \frac{\text{sliding friction force}}{\text{normal reaction}} = \tan \varphi \qquad (5.2)$$

For a body, of weight $W$, pulled along a *horizontal* surface by a *horizontal* force $P$, $P$ and $W$ are equal and opposite to the sliding friction force and the normal reaction respectively; hence for this condition:

$$\mu = P/W \tag{5.3}$$

Expression (5.3) is not applicable if the pulling force $P$ is inclined to the surface or if the surface is not horizontal (see Example 5.2).

## 5.2 Determination of the coefficient of friction

Fig. 5.6 shows a simple apparatus by which experiments on the amount of friction between two surfaces can be performed. A wooden board A is supported horizontally on a table or bench; and on A rests a rectangular block B of known weight.

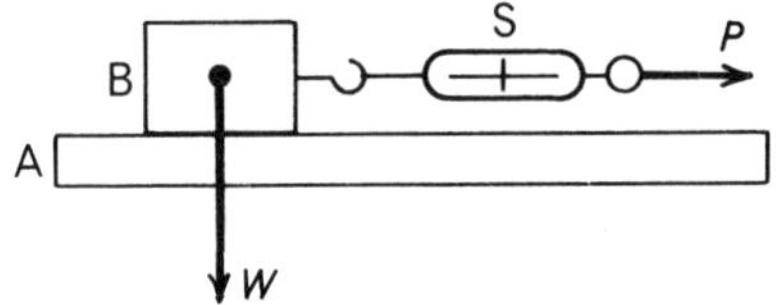


Fig. 5.6 Experiment on friction.

A spring balance S, calibrated in newtons, is attached to B. The horizontal pull $P$ applied to S is increased until B begins to move, and the reading on S is noted: (*a*) just before the movement begins and (*b*) while B is being pulled along at a steady speed. As mentioned in section 5.1, the force to overcome *sliding friction* is slightly less than that required to overcome *static friction.* The test is repeated with various values of load $W$ by placing metal blocks of known weight on top of block B.

Experiments between pairs of solids with *dry* surfaces show that if the surfaces are of uniform character throughout in regard to finish and condition, the force $P$ required to maintain steady slow motion is approximately proportional to the perpendicular force $W$ between the two sliding surfaces, i.e. the ratio $P/W$, namely the coefficient of friction, is approximately constant.

## 5.3 Other experimental results

It is also found from experiment that for *dry* solid surfaces the amount of friction

(*a*) depends upon the nature of the surfaces in contact;
(*b*) is independent of the area of the surfaces in contact;
(*c*) is independent of the speed of sliding.

The above must be regarded as only rough empirical rules applicable in regard to relatively slow motion and moderate pressures and not necessarily correct at high speeds or high pressures or in regard to any but dry surfaces. Actually it is very difficult to exclude some form of lubrication, however slight its amount may be. Consequently experiments on friction are difficult and the results are not simple.

**Example 5.1** *A block of metal having a mass of* 60 *kg requires a horizontal force of* 140 *N to drag it at a constant speed along a horizontal floor. Calculate (a) the coefficient of friction and (b) the angle of friction.*

$$\text{Since mass of block} = 60 \text{ kg.}$$

$$\therefore \quad \text{weight of block} \simeq 60 \times 9{\cdot}81 = 588{\cdot}6 \text{ N.}$$

$$\text{Friction force} = 140 \text{ N,}$$

$$\therefore \quad \text{coefficient of friction} = \frac{140 \text{ [N]}}{588{\cdot}6 \text{ [N]}} = 0{\cdot}238$$

$$= \tan \varphi$$

From trigonometrical tables, $\varphi = 13{\cdot}4°$.

**Example 5.2** *Determine the force, inclined at an angle of* 20° *to the horizontal, required to drag the* 60-*kg metal block of Example* 5.1 *along the floor at a uniform speed.*

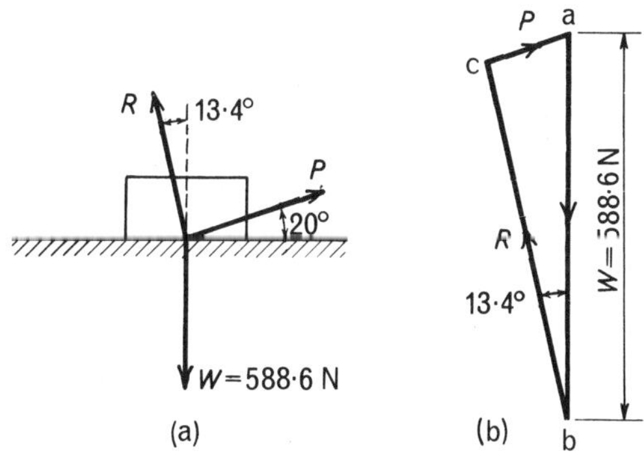


Fig. 5.7 Diagrams for Example 5.2.

Fig. 5.7(a) represents the three forces $P$, $W$ and $R$ acting on the metal block. The triangular force diagram of Fig. 5.7(b) may be constructed by drawing the vertical line *ab* 117·7 mm long to represent the weight $W$ to scale of 1 mm to 5 N. Line *bc* is then drawn to represent $R$, making an angle of 13·4° with *ba*, where 13·4° is the angle of friction $\varphi$ already determined in Ex. 5.1. Line *ac* is drawn parallel to force $P$ to form the force triangle *abc*.

From the diagram it is found that the length of line *ac* representing $P$ is 27·3 mm. Hence $P = 27{\cdot}3\ [\text{mm}] \times 5\ [\text{N/mm}] = 136{\cdot}5\ \text{N}$.

It will be seen that the value of $P$ is slightly less than when it is acting horizontally. The value is a minimum when $P$ is acting at right angles to $R$, i.e. when $P$ is inclined at the angle of friction (13·4°) to the horizontal.

**Example 5.3.** *A block of stone having a mass of* 75 *kg is hauled along a horizontal floor for a distance of* 100 *m in* 2 *min. The coefficient of friction is* 0·3. *Calculate (a) the horizontal force required, (b) the work done and (c) the power.*

$$\text{Weight of block} \simeq 75 \times 9{\cdot}81 = 736\ \text{N},$$

$$\therefore \quad \text{horizontal force required} = 736 \times 0{\cdot}3 = 220{\cdot}8\ \text{N}.$$

$$\text{Work done} = 220{\cdot}8\ [\text{N}] \times 100\ [\text{m}]$$

$$= 22\,080\ \text{J} = 22{\cdot}08\ \text{kJ},$$

$$\text{and power} = \frac{22\,080\ [\text{J}]}{(2 \times 60)\ [\text{s}]} = 184\ \text{W}.$$

## 5.4 Lubrication

Sometimes friction force is desirable as, for instance, in the grip of a tyre on a road, or a belt on a pulley or a brake block on a drum. But in machinery, where one piece has sliding or turning motion on or in another, any friction force acting against the motion results in loss of energy which is converted into heat. It is therefore desirable to reduce friction, and this is done by interposing a lubricant between the solid surfaces, thereby keeping them out of direct contact with each other.

Oils of various kinds are the best known lubricants; but for heavy loads, greases of various degrees of viscosity, which can resist pressure and avoid being squeezed out, are used. The lubricant employed depends upon many circumstances, such as the pressure in a bearing, and the speed as well as the cost. For high speeds, forced lubrication is used in which oil is supplied to bearings under pressure from a pump, the ideal condition being that the metal surfaces shall be completely separated by a film of oil in which a rotating shaft floats.

The resistance offered to lateral motion by a liquid is quite unlike that between solid surfaces. It increases with speed and with area of contact, and pressure has little or no effect on it. In practice friction resistance is often a complicated matter of forces exerted by solids and intervening fluids, on which not very much light is thrown by simple experiments on the sliding resistance between solid bodies, and even this is affected by variable amounts of films of grease, moisture, or even air partially separating the surfaces.

## *Summary of Chapter* 5

Friction is a force opposing the sliding of one body over another. It acts at the surface of contact in a direction opposite to that of motion.

For a horizontal surface,

$$\text{coefficient of friction} = \mu = P/W \qquad (5.3)$$

$$= \tan \varphi$$

where $P$ = horizontal force to overcome friction force,

$W$ = weight of body

and $\varphi$ = angle of friction.

The force required to overcome sliding friction is slightly less than that required to overcome static friction.

The friction between one body and another body sliding over it can be reduced by separating the two surfaces by means of a lubricant.

## EXAMPLES 5

1. A sledge having a mass of 400 kg requires a horizontal force of 620 N to keep it moving at a uniform speed along a horizontal surface. Calculate (*a*) the coefficient of friction and (*b*) the direction and magnitude of the resultant reaction of the surface.
2. A 2-t mass rests on a horizontal surface. Calculate the horizontal force, in kilonewtons, required to move this mass at a uniform speed along the surface if the coefficient of friction is 0·35.

3. A body having a mass of 40 kg rests on a horizontal table and it is found that the least horizontal force required to move the body is 90 N. What is the coefficient of friction?

   How much work is done in moving the body along the surface of the table through a distance of 3 m?
4. A wooden tray rests on a horizontal surface and it is found that with a mass of 3 kg in the tray, a horizontal force of 10 N is just sufficient to cause the tray to slide. When the 3-kg mass is replaced by a mass of 7 kg, it is found that the horizontal force required to cause the tray to slide is doubled. Calculate the coefficient of friction between the tray and the surface, and the mass of the tray.

   What would be the work done in moving the tray through a distance of 5 m with an 11-kg mass in the tray?
5. A block having a weight of 300 N rests on a flat horizontal surface, the coefficient of friction between the block and the surface being 0·25. Calculate the magnitude of the horizontal force which will just cause the block to slide over the surface.

   If the block slides in the direction of the applied force with a uniform velocity of 1·5 m/s, calculate the heat generated by friction, in joules per minute.
6. The sliding face of a slide valve of a steam engine is 150 mm by 300 mm, and the steam pressure on the back of the valve is 1200 kN/m$^2$. If the coefficient of friction is 0·02, what is the force required to move the valve?
7. A certain man has a mass of 70 kg. What is the magnitude of the largest mass he can pull by a horizontal rope along a horizontal floor if the coefficient of friction between the mass and the floor is 0·23 and that between his boot soles and the floor is 0·5?
8. A block of stone is hauled along a horizontal floor by a force inclined 20° to the horizontal. If the block has a mass of 40 kg and the coefficient of friction between the stone and the floor is 0·3, determine graphically or otherwise the effort required.
9. A mass of 30 kg resting on a rough horizontal surface is moved at a constant speed by a force of 40 N acting at 30° to the horizontal. Determine the coefficient of friction.
10. A body having a weight of 150 N is resting on a horizontal table and can just be moved by a horizontal force of 25 N. Calculate the coefficient of friction and the direction and magnitude of the resultant reaction of the table.

    If the force were acting at an angle of 20° with the horizontal, what would be the value of the force that could just move the body and what would be the resultant reaction of the table?
11. A mass of 12 t is dragged over a rough horizontal surface at a uniform speed. The load is moved a distance of 50 m in 12 min. If the coefficient of friction is 0·4, determine (*a*) the horizontal force required, (*b*) the work done, in kilojoules and (*c*) the power, in kilowatts.

## ANSWERS TO EXAMPLES 5

1. 0·158, 3970 N, 9° with vertical.
2. 6·86 kN.
3. 0·229, 270 J
4. 0·255, 1 kg, 150 J.
5. 75 N, 6750 J/min.
6. 1080 N.
7. 152 kg.
8. 113 N.
9. 0·126.
10. 0·167, 152 N, 9·5° with vertical; 25·2 N, 143·5 N.
11. 47·1 kN, 2355 kJ, 3·27 kW.

CHAPTER 6

# Velocity and acceleration

## 6.1 Displacement

When we say that a person has walked 7 kilometres, we give no indication either of the actual distance between his initial and final positions or of the direction of the final position relative to the initial position. He could be at any point within a radius of 7 km: he could even be back at his initial position. *Distance* is therefore a scalar quantity, i.e. it has magnitude only. If, however, the person has walked 3 km eastwards, as represented by AB in fig. 6.1, and then walked 4 km northwards, as represented by BC, his final position C is 5 km away from his initial position A. This *change of position* is termed the *displacement* and is independent of the path

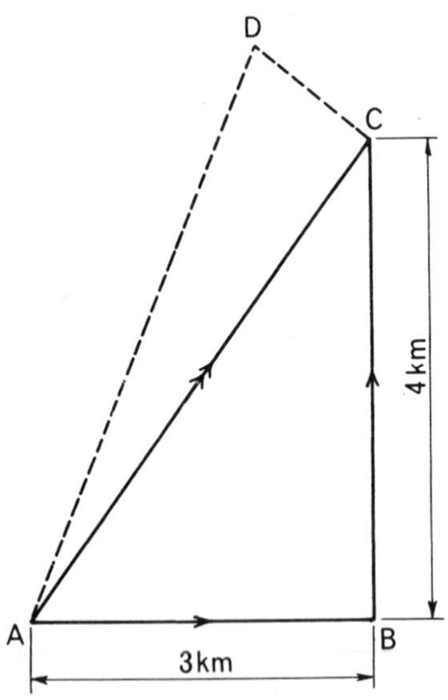


Fig. 6.1 Displacement.

followed and of the time taken; thus, he could have reached C by going north-north-east from A to D and then south-east from D to C, as shown by the dotted lines in fig. 6.1.

Since displacement has both magnitude and direction, it is a *vector* quantity and can be represented by a straight line drawn to scale in the direction of the displacement. From fig. 6.1, it will be be seen that during the first part of the 7-km journey from A to C, the displacement is 3 km in an easterly direction and this can be represented by a horizontal vector AB, drawn to a scale of, say, 1 cm to 1 km. During the second part of the journey, the displacement is 4 km in a northerly direction and is represented in fig. 6.1 by vector BC, 4 cm long, drawn vertically at B.

The resultant displacement is represented to scale by the straight line AC. By measurement, the length of AC is found to be 5 cm and the angle between AC and AB is 53°; i.e. the resultant displacement is 5 km in a direction 53° north of east and can be determined by adding vectorially the component displacements.

## 6.2 Speed

If a motor car travels 1·6 km in 2 minutes, its speed is 0·8 kilometres per minute, or $0{\cdot}8 \times 60 = 48$ kilometres per hour. In general, we can say that the speed of a body is the distance traversed in unit time, or the *rate* at which distance is traversed. The distance can be expressed in any convenient unit such as a metre, a kilometre, etc., and the unit of time can also be any convenient value such as an hour, a minute or a second.

If a motor car travels a distance of 48 km in one hour, its *average* speed is 48 km/h, but it is extremely unlikely that the car will travel at exactly this speed during the whole hour—its speed will be at times higher and at other times lower than this value. A body has *constant* speed only if it moves over equal distances in equal intervals of time—however short the interval.

The average speed of a body is the total distance divided by the time; thus, if a body travels a distance $s$ metres in $t$ seconds, the average speed, $v$ metres per second, is given by:

$$v = \frac{s\ [\text{metres}]}{t\ [\text{seconds}]} = \frac{s}{t}\ \text{metres/second} \qquad (6.1)$$

**Example 6.1** *If a motor car is travelling at a speed of* 100 *km/h, what is the speed in metres/second?*

$$\text{Since } 1 \text{ km} = 1000 \text{ m}$$
$$\text{and } 1 \text{ h} = 3600 \text{ s},$$
$$\therefore \quad 1 \text{ km/h} = \frac{1000 \text{ [m]}}{3600 \text{ [s]}} = 0{\cdot}278 \text{ m/s}.$$

$$\text{Hence} \quad 100 \text{ km/h} = 0{\cdot}278 \times 100 = 27{\cdot}8 \text{ m/s}.$$

**Example 6.2** *If an aeroplane travels a distance of* 8000 *km at a constant speed in* 12 *h, calculate (a) its speed in metres/second, (b) the number of kilometres travelled in* 20 *min and (c) the time taken to travel* 100 *km.*

(a)
$$\text{Distance travelled} = 8000 \times 1000 = 8 \times 10^6 \text{ m}$$
$$\text{and time taken} = 12 \times 3600 = 43\,200 \text{ s},$$
$$\therefore \quad \text{speed} = \frac{8 \times 10^6 \text{ [m]}}{43\,200 \text{ [s]}} = 185 \text{ m/s}.$$

(b)
$$\text{Since speed} = \frac{8000 \text{ [km]}}{12 \text{ [h]}} = 667 \text{ km/h}$$
$$\therefore \text{ distance travelled in 20 min} = 667 \text{ [km/h]} \times (20/60) \text{ [h]}$$
$$= 222 \text{ km}.$$

$$\text{Alternatively, since speed} = 185 \text{ m/s}$$
$$\therefore \text{ distance travelled in 20 min} = 185 \text{ [m/s]} \times (20 \times 60) \text{ [s]}$$
$$= 222\,000 \text{ m}$$
$$= 222 \text{ km}.$$

(c) From expression (6.1),

$$\text{time to travel 100 km} = \frac{100 \text{ [km]}}{667 \text{ [km/h]}} = 0{\cdot}15 \text{ h}$$
$$= 0{\cdot}15 \times 60 = 9 \text{ min}.$$

It will be noted that we have not taken any account of direction of motion of the aeroplane—in other words, the speed of a body is independent of the direction of motion.

### 6.3 Graphs relating distance, time and speed

The relationships between distance and time and between speed and time can usefully be represented by simple graphs.

*Distance/time and speed/time graphs for constant speed.* Fig. 6.2(a) is a graph showing the distance travelled during a period of 20 s by a body moving at a constant speed. The *slope* of this straight-line graph, calculated in terms of the units used for the two axes, is 100 m divided by 20 s, i.e. 5 m/s, and therefore represents the *speed* at which the body is moving.

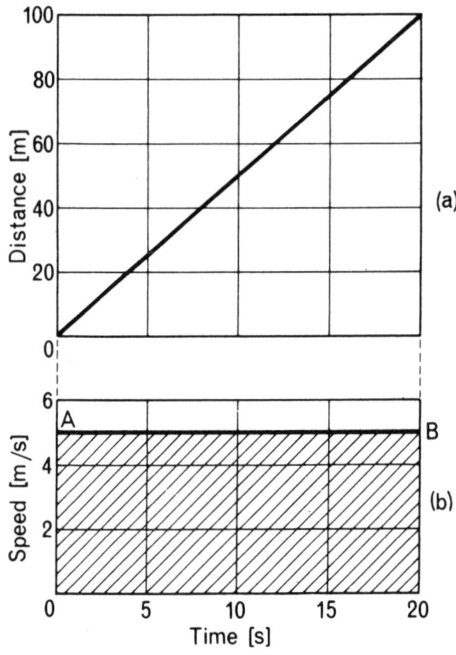


Fig. 6.2 Distance/time and speed/time graphs for constant speed.

The horizontal line **AB** in fig. 6.2(b) is a graph representing the constant speed of 5 m/s derived from fig. 6.2(a)

The area shown shaded in fig. 6.2(b)

$$= \text{(speed in metres/second)} \times \text{(time in seconds)}$$

$$= 5\ [\text{m/s}] \times 20\ [\text{s}] = 100\ \text{m}$$

$$= \text{distance travelled.}$$

*Distance/time and speed/time graphs for varying speed.* Fig. 6.3(a) is a distance/time graph for a body which travels a distance of 100 m in 20 s at varying speeds. The *average* speed is 100 [m]/20 [s] = 5 m/s. The initial and the final speeds are zero; consequently the slopes of the graph at the beginning and the end of the period must

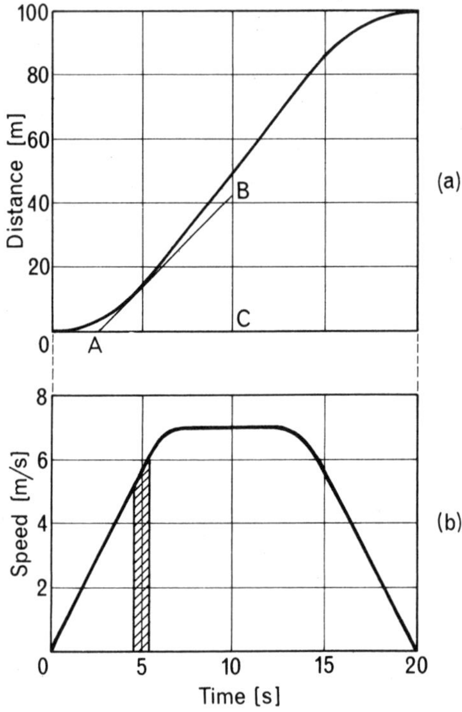


Fig. 6.3 Distance/time and speed/time graphs for varying speed.

be zero. The slope at any intermediate instant is obtained by drawing a tangent to the graph at that instant; for example, at 5 seconds, the tangent is shown by line AB.

$$\text{Slope of AB} = \frac{\text{BC}}{\text{AC}} = \frac{42\ [\text{m}]}{(10 - 2{\cdot}5)\ [\text{s}]} = 5{\cdot}6\ \text{m/s}.$$

Similarly, it is found that the slope of the graph between 7 s and 13 s is constant at about 7·1 m/s. Thus, by drawing tangents at various points of the distance/time graph of fig. 6.3(a), the speed/

time graph of fig. 6.3(b) can be derived.

During the 1 second from 4·5 s to 5·5 s, the average speed is practically 5·6 m/s, so that the distance travelled during that 1 second = 5·6 [m/s] × 1 [s] = 5·6 m, and is represented by the area of the shaded strip in fig. 6.3(b). Similarly for all the thin strips into which we might divide the area under the graph. Hence the total area enclosed by the speed/time graph of fig. 6.3(b) represents the total distance travelled.

If the only information available about the movement of a body was the speed/time graph such as that shown in fig. 6.4, the simplest method of determining the average speed is by means of mid-ordinates. Thus, if the base line is divided into, say, 6 equal lengths, as in fig. 6.4, and the mid-ordinates $v_1$, $v_2$, etc are drawn and measured, then:

$$\text{average speed} = v = \frac{v_1 + v_2 + v_3 + v_4 + v_5 + v_6}{6}$$

If $t$ be the length of the base of the speed/time graph,

$$\begin{aligned}\text{distance travelled} &= \text{area enclosed by graph} \\ &= \text{average speed} \times \text{time} \\ &= vt.\end{aligned}$$

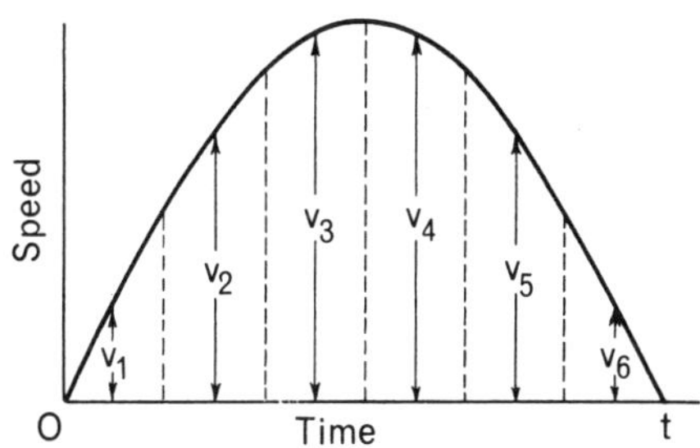


Fig. 6.4 Graphical determination of the average speed.

The larger the number of ordinates used, the more accurate the result.

## 6.4 Linear velocity

It was pointed out in section 6.2 that the speed of a body can be stated without any reference to the direction of movement of that body. Consequently, *speed* is a *scalar* quantity. If, however, we specify the *direction* of motion as well as the *speed* of the body, the quantity is then termed the *velocity* of the body; thus, if a car is travelling in a northerly direction at a speed of 40 kilometres/hour, the velocity is said to be 40 km/h northwards. Since velocity has both magnitude and direction, it is a *vector* quantity and can be represented by a straight line drawn to scale in the direction of the velocity. If a body travels a distance $s$ in a constant direction in time $t$, and if $v$ is the average velocity, then:

$$v = s/t \tag{6.1}$$

It follows that speed and velocity are numerically the same.

## 6.5 Acceleration

When the velocity of a body is increasing, the body is said to be accelerating, whereas if the velocity is decreasing, the body is said to be retarding. Retardation may be regarded as negative acceleration. Suppose that the velocity of a train on a straight horizontal track increases by 1·5 m/s every second from standstill until the train attains a speed of 30 m/s, then at the end of 1 second the speed is 1·5 m/s, at the end of the next second it is 3 m/s, at the end of the third second it is 4·5 m/s, etc., until at the end of 20 s the velocity is 30 m/s, and the variation of velocity with time can be represented by the straight line OA in fig. 6.5. It follows that *acceleration* can be defined as *the rate of change of velocity*; and

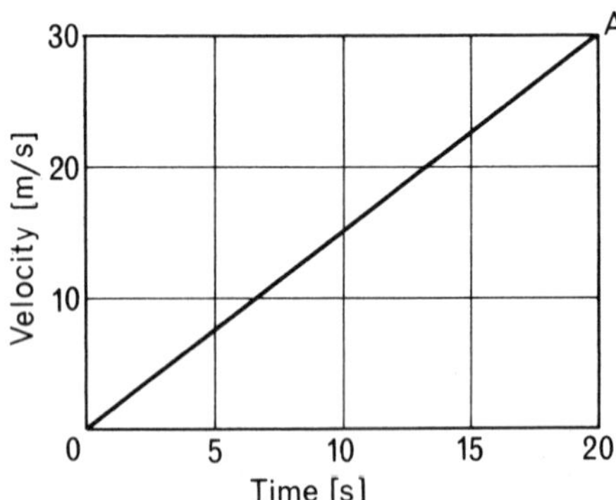


Fig. 6.5 Uniform acceleration.

when the rate of change remains constant, as in fig. 6.5, the acceleration is said to be *uniform*.

Suppose the initial velocity of a body moving in a straight line to be $u$, as shown in fig. 6.6, and suppose the velocity to increase at a uniform rate to $v$ in time $t$; then, if the acceleration is represented by the symbol $a$,

$$\text{change of velocity} = v - u$$

and

$$\text{acceleration} = a = \text{rate of change of velocity}$$

$$= \frac{\text{change of velocity}}{\text{time}}$$

$$= \frac{v - u}{t}$$

$$\therefore \qquad v = u + at \qquad (6.2)$$

If $u$ and $v$ are expressed in metres/second and $t$ is in seconds, then the acceleration is in metres per second every second, i.e. metres per second squared, the symbol being m/s$^2$ (not m/s/s).

Since the velocity is assumed to vary at a uniform rate between $u$ and $v$, it is represented by the straight line AB in fig. 6.6. The average velocity is the mean of $u$ and $v$, namely $\frac{1}{2}(u + v)$; hence, if $s$ represents the distance travelled, we have:

$$s = \text{average velocity} \times \text{time}$$

$$= \tfrac{1}{2}(u + v)t$$

Substituting the value for $v$ from equation (6.2), we have:

$$s = \tfrac{1}{2}(u + u + at)t$$

$$= ut + \tfrac{1}{2}at^2 \qquad (6.3)$$

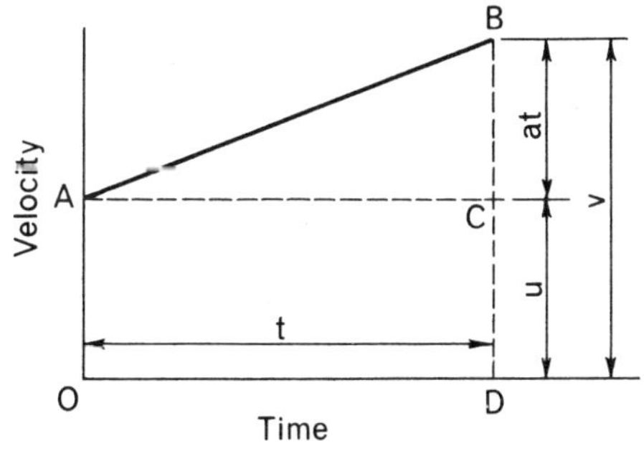

Fig. 6.6 Uniform acceleration.

We can derive an expression for $v$ in terms of $u$, $a$ and $s$ by squaring the two sides of equation (6.2) and then substituting the value of $s$ from equation (6.3), thus:

$$\begin{aligned} v^2 &= (u + at)^2 \\ &= u^2 + 2uat + (at)^2 \\ &= u^2 + 2a(ut + \tfrac{1}{2}at^2) \\ &= u^2 + 2as \qquad (6.4) \end{aligned}$$

It should be noted that in the above expressions, $a$ is positive when the body is accelerating and negative when it is retarding.

## 6.6. Acceleration of a falling body

When a body falls, the force of attraction between the body and the earth causes its velocity to increase. The pull with which a body is attracted towards the earth's centre is referred to as *gravitational force*. It is well known that a feather falls less rapidly than a piece of metal, but this difference is due to the relatively large surface of the feather so that the air resistance is comparatively large. It was Sir Isaac Newton who first proved experimentally that all bodies fall with the same acceleration so long as their movement is not impeded by any resistance. He dropped a feather and a coin in a long vertical glass tube from which practically all the air had been extracted and showed that they fell at the same rate.

The acceleration of a freely falling body is referred to as the acceleration due to gravity and is represented by the symbol $g$. It has already been stated in section 1.8 that owing to the radius of the earth being slightly smaller at the north and south poles than it is at the equator, the gravitational pull is slightly higher at the poles than it is at the equator. Consequently, the value of $g$ is about 9·832 m/s$^2$ at the poles and about 9·780 m/s$^2$ at the equator. In London, at sea level, the value of $g$ is almost exactly 9·81 m/s$^2$.

It follows from equation (6.2) that if a body, *initially at rest*, falls freely for time $t$, the final velocity $v$ is given by

$$v = gt \qquad (6.5)$$

If $s$ is the distance travelled, then from equation (6.3):

$$\begin{aligned} s &= \text{average velocity} \times \text{time} \\ &= \tfrac{1}{2}vt = \tfrac{1}{2}gt^2 \qquad (6.6) \end{aligned}$$

and from equation (6.4):

$$v = \sqrt{(2gs)} \tag{6.7}$$

**Example 6.3** *A train has a uniform acceleration of* 0·2 *m/s*$^2$ *along a straight track. Calculate* (a) *the velocity after an interval of* 16 *s from standstill,* (b) *the time required to attain a velocity of* 50 *km/h,* (c) *the distance travelled from standstill until the train attains a velocity of* 50 *km/h and* (d) *the time taken for the velocity to increase from* 30 *km/h to* 50 *km/h and the distance travelled during that time.*

(*a*) Since $v = u + at$ (6.2)

$\therefore$ $v = 0 + 0{\cdot}2\ [\text{m/s}^2] \times 16\ [\text{s}] = 3{\cdot}2\ \text{m/s}.$

(*b*) $50\ \text{km/h} = 50\,000\ [\text{m}]/3600\ [\text{s}] = 13{\cdot}9\ \text{m/s}.$

Substituting in expression (6.2) given above, we have:

$$13{\cdot}9\ [\text{m/s}] = 0 + 0{\cdot}2\ [\text{m/s}^2] \times t$$

$\therefore$ $t = 69{\cdot}5\ \text{s}.$

(*c*) Since the acceleration is *uniform,*

$$\begin{aligned} s &= \tfrac{1}{2}\,(u + v)t \\ &= \tfrac{1}{2}\,(0 + 13{\cdot}9)\ [\text{m/s}] \times 69{\cdot}5\ [\text{s}] \\ &= 483\ \text{m} = 0{\cdot}483\ \text{km}. \end{aligned}$$

(*d*) $30\ \text{km/h} = 30\,000\ [\text{m}]/3600\ [\text{s}] = 8{\cdot}33\ \text{m/s}.$

Substituting in expression (6.2) given above, we have:

$$13{\cdot}9\ [\text{m/s}] = 8{\cdot}33\ [\text{m/s}] + 0{\cdot}2\ [\text{m/s}^2] \times t$$

$\therefore$ $t = 27{\cdot}85\ \text{s}.$

Since the acceleration is uniform,

$$\begin{aligned} s &= \tfrac{1}{2}\,(u + v)t \\ &= \tfrac{1}{2}\,(8{\cdot}33 + 13{\cdot}9)\ [\text{m/s}] \times 27{\cdot}85\ [\text{s}] \\ &= 310\ \text{m} = 0{\cdot}31\ \text{km}. \end{aligned}$$

**Example 6.4** *A motor car travelling at* 50 *km/h on dry level surface should be able to stop in* 14 *metres. Assuming the retardation to be uniform, calculate* (a) *its value and the corresponding braking time and* (b) *the distance travelled during the first second after the application of the brakes.*

(*a*) 50 km/h $=$ 50 000 [m]/3600 [s] $=$ 13·9 m/s.

$$\text{Since } v^2 = u^2 + 2\,as \tag{6.4}$$

$$\therefore \quad 0 = (13{\cdot}9)^2\ [\text{m/s}]^2 + 2a \times 14\ [\text{m}]$$

$$\text{so that} \quad a = -\ 6{\cdot}9\ \text{m/s}^2,$$

i.e. the retardation is 6·9 m/s².

$$\text{Since} \quad v = u + at \tag{6.2}$$

$$\therefore \quad 0 = 13{\cdot}9\ [\text{m/s}] + (-\ 6{\cdot}9)\ [\text{m/s}^2] \times t$$

$$\text{so that} \quad t = 2{\cdot}02\ \text{s}.$$

Alternatively, since the retardation is uniform,

$$s = \tfrac{1}{2}\,(u + v)t$$

$$\therefore \quad 14\ [\text{m}] = \tfrac{1}{2}\,(13{\cdot}9 + 0)\ [\text{m/s}] \times t$$

$$\text{so that} \quad t = 2{\cdot}02\ \text{s}.$$

$$(b)\ \text{Since} \quad s = ut + \tfrac{1}{2}at^2 \tag{6.3}$$

$$\therefore \quad s = 13{\cdot}9\ [\text{m/s}] \times 1\ [\text{s}] + \tfrac{1}{2} \times (-\ 6{\cdot}9)\ [\text{m/s}^2] \times 1^2\ [\text{s}]^2$$

$$= 10{\cdot}45\ \text{m}.$$

**Example 6.5** *A stone is dropped from a tower,* 100 *m high. Assuming* $g = 9{\cdot}81$ *m/s² and the air resistance to be negligible, calculate* (a) *the time taken to reach the ground,* (b) *the velocity of the stone when it hits the ground and* (c) *the distance through which the stone falls during the first* 2 *seconds.*

$$(a)\ \text{Since} \quad s = \tfrac{1}{2}gt^2 \tag{6.6}$$

$$\therefore \quad 100\ [\text{m}] = \tfrac{1}{2} \times 9{\cdot}81\ [\text{m/s}^2] \times t^2$$

$$\therefore \quad t^2 = 20{\cdot}4\ \text{s}^2$$

$$\text{and} \quad t = 4{\cdot}52\ \text{s}.$$

$$(b)\ \text{Since} \quad v = gt \tag{6.5}$$

$$\therefore \quad v = 9{\cdot}81\ [\text{m/s}^2] \times 4{\cdot}52\ [\text{s}]$$

$$= 44{\cdot}3\ \text{m/s}.$$

(*c*) Substituting in expression (6.6) given above, we have:

$$s = \tfrac{1}{2} \times 9{\cdot}81\ [\text{m/s}^2] \times (2)^2\ [\text{s}]^2$$

$$= 19{\cdot}62\ \text{m}.$$

**Example 6.6** *A cricket ball is thrown vertically upwards at a velocity of* 20 *m/s. Calculate* (a) *the time taken to reach the maximum height and* (b) *the maximum height attained. Assume* $g = 9{\cdot}81$ *m/s² and the air resistance to be negligible.*

(*a*) While the ball is moving upwards, the retardation is 9·81 m/s², i.e. the acceleration is — 9·81 m/s².

Since $$v = u + at \tag{6.2}$$

∴ $$0 = 20\ [\text{m/s}] + (-\ 9{\cdot}81)\ [\text{m/s}^2] \times t$$

so that $$t = 2{\cdot}04\ \text{s}.$$

(*b*) Since the retardation is uniform,

$$\begin{aligned} s &= \tfrac{1}{2}\,(u + v)t \\ &= \tfrac{1}{2}\,(20 + 0)\ [\text{m/s}] \times 2{\cdot}04\ [\text{s}] = 20{\cdot}4\ \text{m}. \end{aligned}$$

## 6.7 Relative velocity

When we speak of the velocity of a body, we generally mean its velocity relative to the earth, which is itself moving at a high speed; i.e. we think of the rate of displacement as if the earth were at rest. Similarly we sometimes speak of the velocity of one body relative to another body which may be moving at a known velocity relative to the earth. For simplicity, we shall limit our discussion of relative velocity to velocities in, or parallel to, a single straight line.

If a train A is travelling, say, east at 80 km/h, then relative to a second train B travelling east at 50 km/h on a parallel track, the first train is moving at:

$$80 - 50 = 30\ \text{km/h}.$$

To an observer in the second train, A would *appear* to be travelling at 30 km/h eastward; while to an observer on A, train B would appear to be travelling westward at 30 km/h.

If the second train B had been travelling *westward* at 50 km/h, the first train A would be travelling eastward *relative* to the second train at:

$$80 - (-\ 50) = 130\ \text{km/h},$$

and would *appear* to be travelling eastward at 130 km/h to an observer in the second train B. To an observer in A, train B would appear to be travelling westward at 130 km/h.

## 6.8 Resultant of two velocities

Suppose an aeroplane to be flying at a constant velocity such that in *perfectly still air* it would travel northward at 500 km/h relative to the ground, and suppose the air to have a constant velocity of

80 km/h in an easterly direction relative to the ground. Under such circumstances, the plane will still be travelling at 500 km/h northward relative to the air.

We can represent these two velocities vectorially as in fig. 6.7, where vector OA represents the air velocity of 80 km/h easterly relative to the *ground*, and vector **AB** represents the plane velocity of 500 km/h northerly relative to the *air*. The vector sum of OA and AB is OB, i.e. OB represents the velocity of the aeroplane relative to the *ground*.

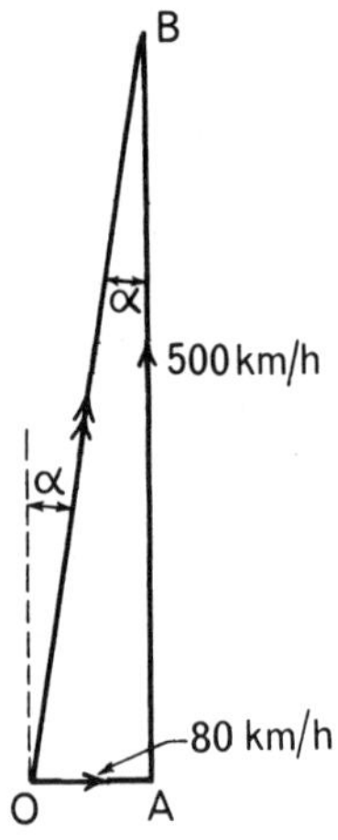


Fig. 6.7 Resultant of two velocities.

It will be noted that we started from point O with the vector OA representing a velocity component relative to the *ground*. Consequently the vector OB gives the resultant velocity also relative to the *ground*, i.e. relative to the object we regard as stationary.

From fig. 6.7 it will be seen that:

$$\begin{aligned} OB &= \sqrt{(OA^2 + AB^2)} \\ &= \sqrt{(80^2 + 500^2)} = 506 \text{ km/h}. \end{aligned}$$

If $\alpha$ is the angle between OB and AB,

$$\tan\alpha = OA/AB = 80/500 = 0{\cdot}16,$$

$$\therefore \quad \alpha = 9{\cdot}1^\circ.$$

Hence the resultant velocity of the aeroplane is 506 km/h, in a direction 9·1° east of north.

Let us now consider the general case of a body possessing two simultaneous velocities represented by vectors OA and AB in fig. 6.8, and suppose $\theta$ to be the angle between the directions of the two velocities. The resultant velocity is the vector sum of OA and AB, namely OB in fig. 6.8. Double arrowheads have been inserted on OB to indicate that it represents the resultant of *two* velocities.

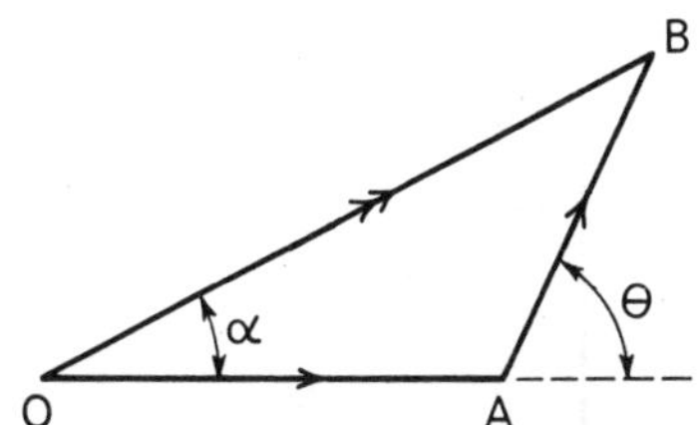


Fig. 6.8 Resultant of two velocities.

If both the magnitude and the direction of the resultant velocity is required, the simplest method at this stage is to draw the diagram to a large scale and measure the length of OB and the value of angle $\alpha$. If, however, only the magnitude of the resultant is required, this can be calculated from the relationship:

$$OB = \sqrt{(OA^2 + AB^2 + 2 \times OA \times AB \cos \theta)}$$

**Example 6.7** *A river is flowing at* 1 *m/s and is* 100 *m wide. If a man rows a boat at* 2 *m/s in still water, determine the direction in which he must row at the same pace in order to reach a point D on the other bank exactly opposite the starting point C (fig.* 6.9). *Also, calculate the time taken to cross the river.*

If vector OA in fig. 6.9 represents the velocity of the river, then in order that the boat may travel from point C on one bank to point D on the opposite bank, the vector representing the resultant velocity must lie along line CD.

A graphical solution can be obtained by using a scale of, say, 1 mm to 0·05 m/s. Thus OA is drawn 20 mm long to represent the velocity of the river relative to its banks. With point A as centre, draw an arc having a radius of 40 mm, cutting CD at B, to represent the speed of the boat in still water. Then line AB represents the velocity of the boat relative to the *river* and OB represents the resultant velocity of the boat relative to the *river banks*. From the diagram it is found that angle $\theta$ is 30°, i.e. the man must row the boat in a direction making an angle of 30° with CD.

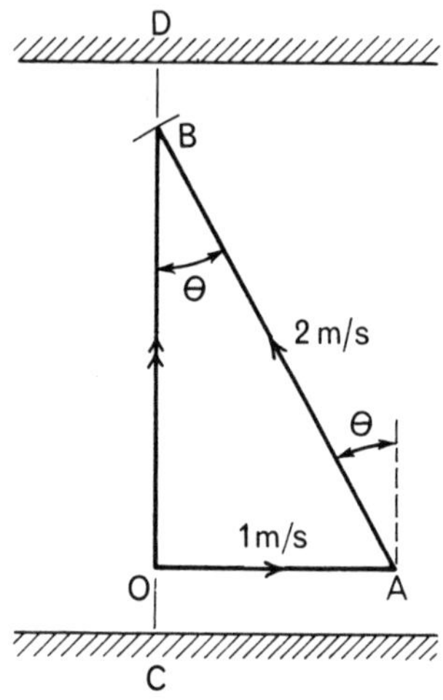


Fig. 6.9 Example 6.7.

Alternatively, angle $\theta$ can be calculated from the fact that angle AOB is a right angle; thus:

$$\sin\theta = OA/AB = 1\,[m/s]/2\,[m/s] = 0{\cdot}5$$

$$\therefore \qquad \theta = 30°.$$

The resultant velocity is represented by vector OB and its value can be determined graphically or calculated thus:

$$\begin{aligned} OB &= AB\cos 30° \\ &= 2\,[m/s] \times 0{\cdot}866 = 1{\cdot}732 \text{ m/s}. \end{aligned}$$

Hence,

$$\left.\begin{matrix}\text{time taken to}\\ \text{cross river}\end{matrix}\right\} = \frac{\text{width of river}}{\text{velocity in direction CD}} = \frac{100\,[m]}{1{\cdot}732\,[m/s]} = 57{\cdot}7 \text{ s}.$$

## 6.9 Resolution of a velocity into two component velocities

In the preceding section it was shown how the resultant of two simultaneous velocities in different directions can be determined. Conversely, a velocity can be resolved into two components, the only condition being that the resultant of the two vectors representing the component velocities must be the same in magnitude and direction as the vector representing the original velocity. The procedure is similar to the resolution of a force discussed in section 2.9.

In practice, the *directions* of the components are generally specified; and these directions are usually at right angles to each other, in which case they are referred to as *rectangular* components. Thus, if vector OB in fig. 6.10 represents the magnitude and direction of

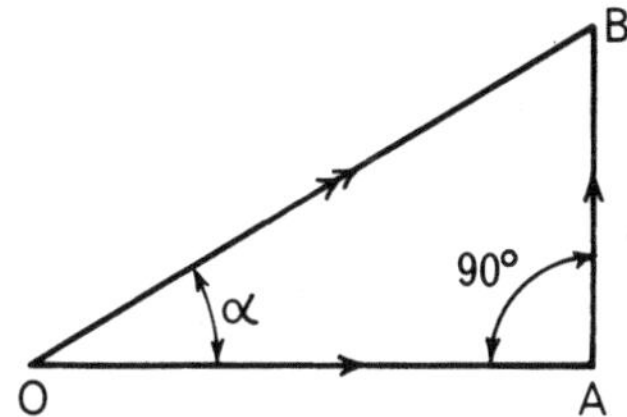


Fig. 6.10 Resolution of velocity into rectangular components.

the velocity with which, say, a cricket ball is thrown into the air, this velocity can be resolved into a horizontal component OA and a vertical component AB. If $\alpha$ be the angle between OB and the horizontal component,

$$\left.\begin{array}{r}\text{horizontal component}\\ \text{of the velocity}\end{array}\right\} = OA = OB\cos\alpha \qquad (6.8)$$

and

$$\left.\begin{array}{r}\text{vertical component}\\ \text{of the velocity}\end{array}\right\} = AB = OB\sin\alpha \qquad (6.9)$$

**Example 6.8** *If a ball is thrown with a velocity of* 20 *m/s in a direction inclined at* 60° *to the horizontal, determine* (a) *the horizontal and vertical components of the velocity,* (b) *the time taken for the ball to return to the ground and* (c) *the horizontal distance travelled before the ball touches the ground. Assume the ground to be level, g to be* 9·81 $m/s^2$ *and the air resistance to be negligible.*

(*a*) From expressions (6.8) and (6.9),

horizontal component of velocity $= 20 \cos 60° = 10$ m/s
and vertical component of velocity $= 20 \sin 60° = 17{\cdot}32$ m/s.

(*b*) Let us consider the vertical component of the velocity.

Since $v = u + at$ (6.2)

and $a = -9{\cdot}81 \text{ m/s}^2$,

$$0 = 17{\cdot}32\,[\text{m/s}] + (-9{\cdot}81)\,[\text{m/s}^2] \times t$$

so that
$$\left.\begin{array}{r}\text{time taken to reach}\\ \text{maximum height}\end{array}\right\} = t = \frac{17{\cdot}32\,[\text{m/s}]}{9{\cdot}81\,[\text{m/s}^2]} = 1{\cdot}77 \text{ s},$$

∴
$$\left.\begin{array}{r}\text{total time for ball to}\\ \text{return to ground}\end{array}\right\} = 2 \times 1{\cdot}77 = 3{\cdot}54 \text{ s}.$$

(*c*) Since the ball has a horizontal velocity of 10 m/s and remains in the air for 3·54 s,

$$\text{horizontal distance travelled} = 10 \times 3{\cdot}54 = 35{\cdot}4 \text{ m}.$$

This example illustrates the usefulness of resolving a velocity into rectangular components.

### 6.10 Angular velocity

The speed of rotation of, say, an engine shaft is usually expressed in revolutions per minute; but for many purposes in engineering science, it is more useful to express it in *radians per second*, a *radian* being the angle at the centre of a circle subtended by an arc equal in length to the radius of the circle, as shown in fig. 6.11. Since the circumference of a circle of radius $r$ is $2\pi r$,

$$\left.\begin{array}{r}\text{number of radians subtended by}\\ \text{the whole circle}\end{array}\right\} = 2\pi r/r$$
$$= 2\pi \text{ radians}.$$

$$\left.\begin{array}{r}\text{But the number of degrees subtended}\\ \text{by the whole circle}\end{array}\right\} = 360°,$$

∴ $2\pi$ radians $= 360°$

or 1 radian $= 360/2\pi = 57{\cdot}3°$.

Let $x$ be the length of an arc subtending an angle $\theta$ radians at the centre of a circle of radius $r$, as in fig. 6.11. Since an arc of length $r$ subtends an angle of 1 radian,

∴ an arc of length $x$ subtends an angle $x/r$ radians,

i.e. $$\theta = x/r \text{ radians}$$

or $$x = r\theta \tag{6.10}$$

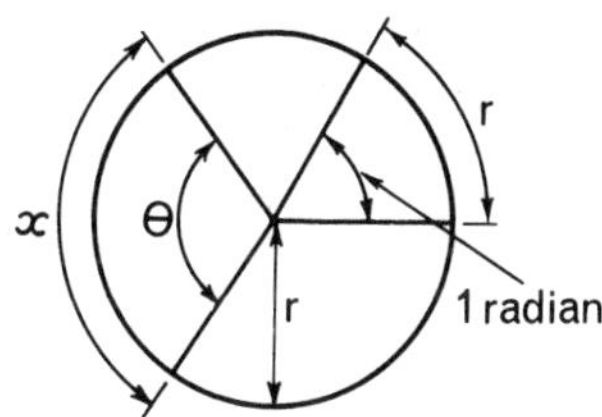


Fig. 6.11 Angular displacement.

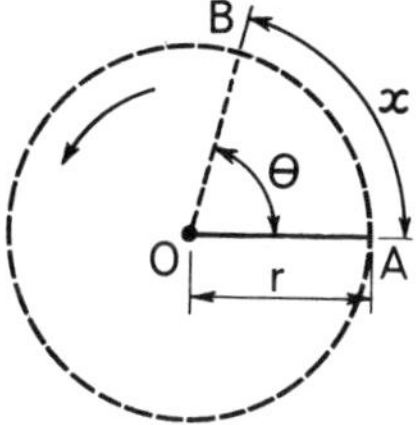


Fig. 6.12 Angular velocity.

Obviously, the lengths $x$ and $r$ must be expressed in the same unit.

Suppose an arm OA, fig. 6.12, to be pivoted at O and to be rotating anticlockwise at a constant speed of $N$ revolutions/minute. Then,

$$\text{number of revolutions/second} = N/60$$

and since the arm rotates through $2\pi$ radians in one revolution,

$$\text{angular velocity} = 2\pi \times N/60 \text{ radians/second.}$$

The symbol for angular velocity is the Greek letter $\omega$ (omega), and the abbreviation for *radian* is 'rad'. Hence,

$$\omega = 2\pi\, N/60 \text{ rad/s} \tag{6.11}$$

$$= (2\pi \times \text{revolutions/second}) \text{ rad/s.}$$

If the arm takes $t$ seconds to rotate through an angle $\theta$ in fig. 6.12,

$$\text{angular velocity} = \frac{\text{angular displacement } \theta \text{ [radians]}}{\text{time } t \text{ [seconds]}}$$

i.e. $$\omega = \theta/t \text{ radians/second} \tag{6.12}$$

If the length of arm OA is $r$, and the length of the arc corresponding to the angular displacement $\theta$ is $x$, then the *speed* of the outer end of arm OA is given by:

$$v = x/t = r\theta/t = r\omega$$

or $$\omega = v/r \tag{6.13}$$

When the arm is in position OA, fig. 6.13, its *velocity at that instant* is $v$ in direction AC, tangential to the circle at A. When the arm has moved to position OB, the speed is unchanged, but the velocity is now $v$ in direction BD, tangential to the circle at B. Hence the *velocity* of the outer end of the arm is continually

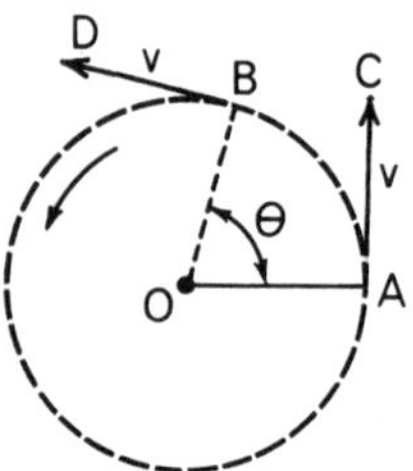


Fig. 6.13 Change of velocity on circular path.

changing though the *speed* remains constant. Further consideration of this matter is, however, beyond the scope of this volume.

## *Summary of Chapter* 6

The slope of a distance/time graph at any instant represents the speed at that instant, and the area enclosed by a speed/time graph represents the distance travelled.

Acceleration is the rate of change of velocity and the value of the acceleration due to gravity at sea level in the vicinity of London is almost exactly 9·81 m/s². If the velocity of a body is accelerated uniformly from $u$ to $v$ in time $t$, and if $s$ is the distance travelled and $a$ is the acceleration,

$$\text{average velocity} = \tfrac{1}{2}(u + v)$$

$$v = u + at \qquad (6.2)$$

$$s = \tfrac{1}{2}(u + v)t$$

$$= ut + \tfrac{1}{2}at^2 \qquad (6.3)$$

and

$$v^2 = u^2 + 2as \qquad (6.4)$$

If the two simultaneous velocities of a body are represented in magnitude and direction by two vectors, the resultant velocity is represented in magnitude and direction by the vector sum of the two vectors.

A velocity can be resolved into two component velocities. If the components are at right angles to each other, they are termed rectangular components.

Angular velocity is the rate of angular displacement about an axis and is expressed in radians/second. If the speed of rotation is $N$ revolutions/minute,

$$\text{angular velocity} = \omega = 2\pi N/60 \text{ rad/s.} \qquad (6.11)$$

## EXAMPLES 6

1. A car travels 20 km southwards and then travels another 30 km westwards. What is the displacement of the car from its initial position?
2. A person walks a distance of 8 km in a direction 20° east of north, and then walks another 6 km in a direction 50° south of east. Determine graphically the value and direction of the final displacement relative to the starting point.
3. An aeroplane flies 160 km in a southerly direction. Its direction is then changed to SE. With the aid of a vector diagram drawn to scale, determine the distance travelled in the new direction when its resultant displacement is 260 km.
4. A boat is rowed a distance of 2 km in a direction 40° north of east. Its direction is then changed to 20° west of south. Determine graphically the distance it has to travel in that direction for the final displacement to be 30° south of east. What is the value of that final displacement?
5. Convert the following speeds into metres/second: (*a*) 40 km/h, (*b*) 5000 mm/min, (*c*) 15 km/min.
6. Convert the following speeds into kilometres/hour: (*a*) 20 m/s, (*b*) 90 mm/s, (*c*) 3 km/s, (*d*) 15 m/min.
7. If a train is travelling at 90 km/h, what is its speed in (*a*) kilometres/minute, (*b*) metres/second?
8. If a man is walking at 1·8 m/s, calculate the distance covered in 2 hours.
9. If a person runs a distance of 0·7 km in 3 min, what is his average speed in (*a*) kilometres/hour, (*b*) metres/second?
10. If a car is travelling at 25 m/s, calculate (*a*) the speed in kilometres/hour and (*b*) the distance, in kilometres, covered in 12 min.
11. A cage is lowered down a pit shaft, 600 m deep, in 46 s. What is its average speed in (*a*) metres/second, (*b*) kilometres/hour?
12. The minute hand of Big Ben is 4·27 m long. Calculate the speed of the tip of this hand in (*a*) metres/minute, (*b*) millimetres/second.
13. A car travels at 50 km/h for the first half hour and at 80 km/h for the following hour. Calculate (*a*) the total distance travelled and (*b*) the average speed.
14. A train travels the first 80 km of its journey at an average speed of 60 km/h. What must be its average speed over the remaining 50 km in order that the average speed over the whole journey may be 70 km/h?
15. A cyclist covers the first 24 km of his journey in 80 min. After a rest of 10 min, he covers the remaining 20 km in 70 min. What is his average speed, in kilometres/hour, for the whole journey?

16. A train, starting from rest, covers the following distances $x$ metres in times $t$ seconds:

| $t$ | 0 | 5 | 11 | 18 | 22 | 27 | 31 | 38 | 46 | 50 |
|---|---|---|---|---|---|---|---|---|---|---|
| $x$ | 0 | 3 | 16 | 52 | 79 | 115 | 137 | 158 | 168 | 170 |

Plot the distance/time graph and determine the approximate speeds in metres/second, after 5, 15, 25, 35 and 45 seconds from the start. Using this data, plot the speed/time graph for the whole period.

What is the average speed, in metres/second, over the 50 s?

17. A car, starting from rest, accelerates uniformly at 2 m/s² for 10 s. Calculate (*a*) the distance travelled, in metres, and (*b*) the final speed, in metres/second.
18. The speed of a motor vehicle falls uniformly from 100 km/h to zero at 3 m/s². Calculate (*a*) the time taken for the vehicle to stop, (*b*) the distance travelled.
19. A train, starting from rest, reaches a speed of 18 m/s in 2 min. Assuming the acceleration to be uniform, calculate (*a*) the value of the acceleration in metres per second squared and (*b*) the distance travelled, in kilometres.
20. The velocity of a body increases uniformly from 15 km/h to 40 km/h while it travels 100 m. Calculate (*a*) the acceleration, in metres per second squared and (*b*) the time taken, in seconds.
21. A stone is dropped down a shaft, 160 m deep. Calculate (*a*) the time taken for the stone to reach the bottom and (*b*) the velocity of the stone as it reaches the bottom. Assume $g = 9{\cdot}81$ m/s².
22. A cricket ball, thrown vertically upwards, returns to the ground in 4 s. Calculate (*a*) the height, in metres, reached by the ball and (*b*) the velocity, in metres per second, with which it is thrown.
23. A ball is thrown vertically upwards at 40 m/s. Calculate (*a*) the greatest height attained, (*b*) the time taken to return to the ground, (*c*) its velocity after 3 s and after 6 s and (*d*) the distance travelled during the first 3 s.
24. The following table shows how the speed of a car varied with time:

| Speed (km/h) | 0 | 19 | 35 | 48 | 48 | 27 | 0 |
|---|---|---|---|---|---|---|---|
| Time (s) | 0 | 3 | 6 | 9 | 12 | 15 | 18 |

Assuming these speed values to be joined by straight lines, plot a graph of speed against time and estimate the distance travelled. Also determine the values of the acceleration during the first 3 s and of the retardation during the last 3 s, in metres per second squared. What is the average speed in kilometres per hour?

25. A train makes a journey between two stations in 8 min. The speed of the train at 1-minute intervals is:

| Time (min) | 0 | 1 | 2 | 3 | 4 | 5 | 6 | 7 | 8 |
|---|---|---|---|---|---|---|---|---|---|
| Speed (km/h) | 0 | 18 | 34 | 45 | 48 | 48 | 42 | 24 | 0 |

Draw the speed/time diagram and determine from the diagram (*a*) the approximate value of the acceleration 2 min after starting and (*b*) the distance between the stations.

26. From a point A, a body travels for 2 s at a constant velocity of 12 m/s due north. It then travels for 5 s due east whilst its velocity decreases uniformly from 12 m/s to 6 m/s. What is the resultant displacement from A? Also, determine the average velocity, in magnitude and direction, of the body for it to return along a straight path to point A in 3 s.
27. A railway carriage is travelling along a straight track at 12 m/s and a ball is rolled at 4 m/s across the floor of the carriage at right angles to the direction of motion of the train. Calculate the magnitude of the resultant velocity of the ball and its direction relative to that of the carriage.
28. A boat is rowed with a velocity of 7 km/h at right angles to a river flowing at 3 km/h. The river is 60 m wide. How far down the river will the boat reach the opposite bank below the point at which it was originally directed and what is the actual velocity of the boat?
29. A boat is rowed on a river so that its speed in still water would be 3 m/s. If the river flows at the rate of 2 m/s, determine at what inclination to the direction of flow of the river must the boat be headed so that the motion of the boat may be at right angles to the current. If the width of the river is 40 m, how long will it take for the boat to cross from one bank to the other?
30. A body is travelling in a straight line at 15 m/s. Determine the rectangular components of the velocity such that one component is inclined at an angle of 30° to the direction of motion.
31. A shell is fired with a velocity of 600 m/s in a direction inclined at 70° to the horizontal. Calculate (*a*) the horizontal and vertical components of the velocity, (*b*) the greatest height above ground attained by the shell and (*c*) the horizontal distance travelled by the shell before it touches ground. Assume the ground to be level and the air resistance to be negligible.
32. A body is travelling in a straight line at 40 m/s. Determine graphically or otherwise the component velocities such that one component is making an angle of 20° with the direction of motion and the angle between the two components is 70°.
33. Two trains are travelling on parallel tracks at speeds of 30 km/s and 80 km/s respectively. Calculate the relative speed of the trains if they are travelling (*a*) in the same direction and (*b*) in opposite directions.
34. Convert the following into radians/second: (*a*) 500 rev/min; (*b*) 8 rev/s.
35. If a wheel has an angular velocity of 70 rad/s, what is the speed in revolutions/minute?
36. A pulley, 800 mm diameter, rotates at 200 rev/min. Calculate (*a*) the angular velocity and (*b*) the linear speed of a point on the rim in metres/second.
37. A flywheel, 2 m diameter, has an angular velocity of 40 rad/s. Calculate (*a*) the speed in revolutions/minute and (*b*) the distance travelled in 5 s by a point on the rim of the flywheel.
38. A body moves round a circular path of 4 m diameter with a constant angular velocity of 100 rad/s. What is the magnitude and direction of the velocity of the body at any instant? (U.E.I., G2)

## ANSWERS TO EXAMPLES 6

1. 36·1 km, 33·7° south of west.
2. 7·21 km, 23·9° north of east.
3. 121 km.
4. 1·9 km, 1·02 km.
5. 11·1 m/s, 0·0833 m/s, 250 m/s.
6. 72 km/h, 0·324 km/h, 10 800 km/h, 0·9 km/h.
7. 1·5 km/min, 25 m/s.
8. 12·96 km.
9. 14 km/h, 3·89 m/s.
10. 90 km/h, 18 km.
11. 13·04 m/s, 47 km/h.
12. 0·447 m/min, 7·45 mm/s.
13. 105 km, 70 km/h.
14. 95·5 km/h.
15. 16·5 km/h.
16. 1·1 m/s, 5·3 m/s, 7·0 m/s, 2·8 m/s, 0·8 m/s; 3·4 m/s.
17. 100 m, 20 m/s.
18. 9·3 s, 129 m.
19. 0·15 m/s$^2$, 1·08 km.
20. 0·53 m/s$^2$, 13·1 s.
21. 5·71 s, 56 m/s.
22. 19·62 m, 19·62 m/s.
23. 81·5 m; 8·15 s; 10·57 m/s upward, 18·9 m/s downward; 75·8 m.
24. 147·5 m, 1·76 m/s$^2$, 2·5 m/s$^2$, 29·5 km/h.
25. 0·068 m/s$^2$ (approx.), 4·4 km (approx.).
26. 51 m, 17 m/s, 61·9° west of south.
27. 12·65 m/s, 18·4°.
28. 25·7 m, 7·62 km/h.
29. 131·8°, 17·9 s.
30. 13 m/s, 7·5 m/s.
31. 205 m/s, 564 m/s; 16·2 km; 23·6 km.
32. 32·6 m/s, 14·55 m/s.
33. 50 km/h, 110 km/h.
34. 52·4 rad/s, 50·3 rad/s.
35. 668 rev/min.
36. 20·95 rad/s, 8·38 m/s.
37. 382 rev/min, 200 m.
38. 200 m/s tangentially.

CHAPTER 7

# Laws of motion

## 7.1 Newton's laws of motion

These laws were known before the time of Sir Isaac Newton but he was the first to express them in their present form. These laws cannot be proved experimentally or otherwise, but calculations based upon these laws give results that agree with actual experience. For instance, predictions relating to the positions of the moon and the planets, calculated with the aid of these laws, are found to agree with observations.

## 7.2 First law of motion

*A body continues in its state of rest or of uniform motion in a straight line unless it is compelled by an external force to change that state.*

This law may be called the 'law of inertia'. A body will not change its state of rest or of motion in a straight line unless compelled to do so; i.e. it resists any change of velocity in magnitude or direction. From this law, we may define a *force* as *any push or pull which changes or tends to change the state of rest of a body or its uniform motion in a straight line.*

## 7.3 Momentum

This is the name given to the product of the mass $m$ of a body and its velocity $v$,

i.e. $$\text{momentum} = mv.$$

Mass has no direction and is therefore a scalar quantity; but velocity has magnitude and direction and is therefore a vector quantity. Hence momentum must also be a vector quantity; and a vector which represents the velocity of a body can also, to a different scale, represent the momentum of that body. Values of momentum can therefore be added and resolved vectorially in the same way as for velocity (sections 6.8 and 6.9).

## 7.4 Second law of motion

*When a body is acted upon by an external force, the rate of change of momentum is proportional to the force and takes place in the direction of the force.*

This law is in effect the definition of a force in respect to its effect on a body *which it can move.* It does not apply to a body that is immovable.

If a force $F$ acts upon a body of mass $m$ for a time $t$ and causes its velocity in the direction of the force to increase from $v_1$ to $v_2$,

$$\text{initial momentum} = mv_1$$

and

$$\text{final momentum} = mv_2,$$

$$\therefore \quad \left.\begin{array}{r}\text{average rate of change}\\ \text{of momentum}\end{array}\right\} = (mv_2 - mv_1)/t$$

$$= m\,(v_2 - v_1)/t = ma,$$

where $a$ = average acceleration during time $t$.

Hence, from Newton's Second Law,

$$F \propto ma$$

$$= ma \times \text{a constant} \qquad (7.1)$$

It has already been stated in section 1.8 that the SI unit of force is the *newton*, namely *the force required to give a mass of* 1 *kg an acceleration of* 1 *m/s²*. Substituting $m = 1$ kg, $a = 1$ m/s² and $F = 1$ N in expression (7.1), we have:

$$1\ [\text{N}] = 1\ [\text{kg}] \times 1\ [\text{m/s}^2] \times \text{a constant.}$$

Hence the constant is unity.

It follows that if $F$ be the force, in newtons, required to give a mass $m$, in kilograms, an acceleration $a$, in metres per second squared, then:

$$F = ma \qquad (7.2)$$

**Example 7.1** *A force of* 50 *N is applied to a mass of* 200 *kg. Calculate the acceleration.*

Substituting for $F$ and $m$ in expression (7.2), we have:

$$50\ [\text{N}] = 200\ [\text{kg}] \times a$$

$$\therefore \quad a = 0{\cdot}25\ \text{m/s}^2.$$

**Example 7.2** *A mass of 3 Mg is to be given an acceleration of 2 m/s². Calculate the force required.*

$$\text{Mass} = 3\ \text{Mg} = 3000\ \text{kg}$$

$$\therefore \quad F = 3000\ [\text{kg}] \times 2\ [\text{m/s}^2]$$

$$= 6000\ \text{N} = 6\ \text{kN}.$$

**Example 7.3** *A Diesel engine pulling a train along a level track has its oil supply cut off when the train is travelling at 60 km/h. It is observed that the speed falls to 40 km/h after the train has travelled a distance of 1200 m. The mass of the engine and carriages is 80 Mg. Assuming the retardation to be uniform, calculate the total force resisting motion.*

$$60\ \text{km/h} = 60\,000\ [\text{m}]/3600\ [\text{s}] = 16{\cdot}67\ \text{m/s}$$

and $$40\ \text{km/h} = 40\,000\ [\text{m}]/3600\ [\text{s}] = 11{\cdot}11\ \text{m/s}.$$

Since $$v^2 = u^2 + 2\,as \qquad (6.4)$$

$$\therefore \quad (11{\cdot}11)^2\ [\text{m/s}]^2 = (16{\cdot}67)^2\ [\text{m/s}]^2 + 2a \times 1200\ [\text{m}]$$

so that $$a = -\,0{\cdot}0642\ \text{m/s}^2.$$

$$\text{Mass of train} = 80\ \text{Mg} = 80\,000\ \text{kg}.$$

Since $$F = ma$$

$$\therefore \quad \textit{retarding}\ \text{force} = 80\,000\ [\text{kg}] \times 0{\cdot}0642\ [\text{m/s}^2]$$

$$= 5136\ \text{N} = 5{\cdot}136\ \text{kN}.$$

## 7.5 Third law of motion

*To every force there is an equal and opposite force.* This law may be stated in another way. If any body A exerts a certain force on another body B, then B exerts on A a force of equal magnitude but in the opposite direction. This is equally true if A and B are two parts of the same body, as already discussed in section 2.1.

This law applies to bodies whether they are at rest or in motion. Thus, if a beam or other body exerts a certain downward force upon a support, the support exerts an equal upward force on the beam—the latter force being often referred to as the reaction of the support (section 3.5).

The fact that this relationship is equally true in regard to two bodies in motion is not so widely realized, e.g. that the backward pull of a trailer on a motor vehicle is equal to the forward pull of the vehicle on the trailer. Let us look a little more closely at the state of affairs when a motor tractor is pulling a trailer along a level road. If the tractor is moving at a constant velocity, the forward pull exerted by the tractor exactly balances the backward pull of the trailer due to friction and wind resistance.

If, however, the forward pull exerted by the tractor exceeds the backward pull due to friction and wind resistance, acceleration occurs. In this case, the forward pull exerted by the tractor may be regarded as being the sum of the following two components: (*a*) the pull $P_1$ required to haul the trailer at a constant velocity, (*b*) the pull $P_2$ $(= ma)$ required to give the mass $m$ of the trailer an acceleration $a$. Hence the total *forward* pull exerted *by the tractor on the trailer* is $(P_1 + P_2)$ and is exactly equal to the total *backward* pull exerted *by the trailer on the tractor*.

Another example of the application of Newton's Third Law is the recoil of a gun firing a bullet or a shell. The force responsible for the recoil of the gun is exactly equal to that acting on the bullet or shell.

**Example 7.4** *A rope supports a mass of* 50 *kg. Calculate the pull on the rope when the mass is being* (a) *raised,* (b) *lowered, with an acceleration of* 0·5 *m/s²*.

When the mass is *stationary* or moving at a *uniform* velocity,

$$\text{pull on rope} = \text{weight of the mass}$$
$$\simeq 50 \times 9{\cdot}81 = 490{\cdot}5 \text{ N.}$$

(*a*) If $F$ be the upward pull, in newtons, when the mass is being raised with an acceleration of 0·5 m/s², the force available for upward acceleration is $(F - 490{\cdot}5)$ newtons. Hence,

$$(F - 490{\cdot}5)\ [\text{N}] = 50\ [\text{kg}] \times 0{\cdot}5\ [\text{m/s}^2]$$

so that
$$F = 515{\cdot}5 \text{ N.}$$

(*b*) If $F$ be the upward pull, in newtons, when the mass is being lowered with an acceleration of 0·5 m/s², the force available for downward acceleration is $(490{\cdot}5 - F)$ newtons. Hence,

$$(490{\cdot}5 - F)\ [\text{N}] = 50\ [\text{kg}] \times 0{\cdot}5\ [\text{m/s}^2]$$

so that
$$F = 465{\cdot}5 \text{ N.}$$

**Example 7.5** *What force will a man having a mass of* 70 *kg exert on the floor of a lift* (a) *ascending and* (b) *descending with an acceleration of* 0·6 *m/s²?*

When the lift is *stationary* or moving at a *uniform* velocity, force exerted by the man on the floor of the lift

$$= \text{weight of the man}$$
$$\simeq 70 \times 9{\cdot}81 = 686{\cdot}7 \text{ N.}$$

Hence, by Newton's Third Law,

force exerted by the floor of the lift on the man $= 686{\cdot}7$ N.

(*a*) When the lift is ascending with an acceleration of $0{\cdot}6 \text{ m/s}^2$,

force required to accelerate the man upwards
$$= 70 \text{ [kg]} \times 0{\cdot}6 \text{ [m/s}^2\text{]} = 42 \text{ N.}$$

This upward force is provided by an increase in the force exerted by the floor of the lift on the man,

hence total force exerted on the man
$$= 686{\cdot}7 + 42 = 728{\cdot}7 \text{ N}$$
= total force exerted by the man on the floor of the lift.

(*b*) When the lift is descending with an acceleration of $0{\cdot}6 \text{ m/s}^2$,

force required to accelerate the man downwards
$$= 70 \text{ [kg]} \times 0{\cdot}6 \text{ [m/s}^2\text{]} = 42 \text{ N.}$$

This downward force is derived from the weight of the man; consequently the net force exerted by the man on the floor of the lift is 42 N less than the value when the lift is stationary or moving at a constant velocity.

Hence, the downward force exerted by the man on the floor of the lift $= 686{\cdot}7 - 42 = 644{\cdot}7$ N.

### 7.6 Fletcher's trolley

The relationship between force, mass and acceleration (section 7.4) can be easily confirmed experimentally by using a Fletcher's trolley. This apparatus usually consists of a block of wood A (fig. 7.1) mounted on wheels, with transverse holes which can be loaded with iron cylinders to vary the mass of the trolley. The latter rests on a board B, supported on a bench C in such a way that the level of B can be altered slightly.

Attached to A is a cord passing over a pulley and carrying a pan D; and supported above A is a long blade of thin steel, fixed at one end and carrying a pencil or an inked brush at the other end. The tip of the pencil or brush is in contact with a piece of paper fixed to the top surface of A. The level of board B is adjusted so that with pan D empty, the trolley, once started, moves at a uniform speed down the slight gradient. The effect of any friction at the wheels and pulley is thereby compensated.

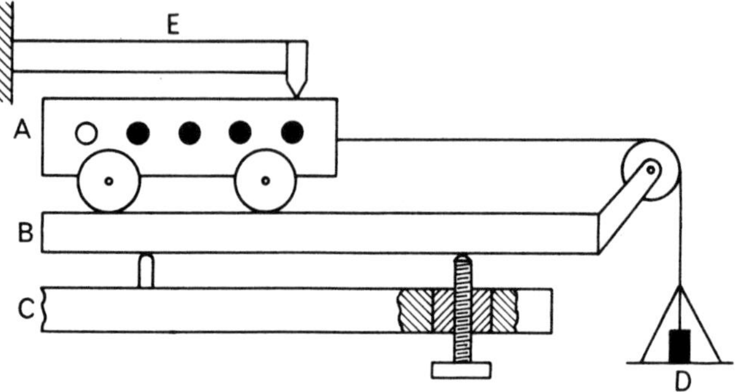


Fig. 7.1 Fletcher's trolley.

One of the iron cylinders is then transferred from A on to pan D, thus causing the trolley to accelerate towards the right in fig. 7.1. During this movement, strip E is stationary, thus enabling the pencil or brush to trace the horizontal axis shown in fig. 7.2. The test is repeated with strip E vibrating, and the pencil or brush will now trace the wavy line shown in fig. 7.2.

Fig. 7.2 Graph obtained with Fletcher's trolley.

During successive oscillations of strip E, the trolley moves through distances PQ, QR, RS, etc. Hence, if $t$ be the time of a complete oscillation of E,

$$\left.\begin{array}{r}\text{average speed of trolley while}\\ \text{travelling distance PQ}\end{array}\right\} = \mathrm{PQ}/t \qquad (7.3)$$

$$\left.\begin{array}{r}\text{average speed of trolley while}\\ \text{travelling distance QR}\end{array}\right\} = \mathrm{QR}/t \qquad (7.4)$$

$$\text{and} \quad \left.\begin{array}{r}\text{average speed of trolley while}\\ \text{travelling distance RS}\end{array}\right\} = \text{RS}/t \qquad (7.5)$$

From expressions (7.3) and (7.4),

$$\left.\begin{array}{r}\text{increase in average speed}\\ \text{during one oscillation}\end{array}\right\} = \text{QR}/t - \text{PQ}/t$$
$$= (\text{QR} - \text{PQ})/t$$

Since this increase takes place in time $t$,

$$\text{corresponding acceleration} = (\text{QR} - \text{PQ})/t^2$$

Similarly, from expressions (7.4) and (7.5),

$$\left.\begin{array}{r}\text{increase in average speed}\\ \text{during one oscillation}\end{array}\right\} = (\text{RS} - \text{QR})/t$$

$$\text{and} \quad \text{corresponding acceleration} = (\text{RS} - \text{QR})/t^2.$$

Hence the acceleration is proportional to the difference in length of the base line between zero values of the wavy line for consecutive oscillations of blade E.

By measuring the lengths of PQ, QR, RS, etc., we can show that within the limits of experimental error,

$$\text{QR} - \text{PQ} = \text{RS} - \text{QR} = \text{etc.}$$

i.e. we can show that the acceleration due to a constant force acting on a given mass remains constant.

The test can be repeated with various numbers of iron cylinders transferred from trolley A to pan D to show that for a given mass the acceleration is proportional to the accelerating force, namely the weight of the cylinders carried by pan D.

## *Summary of Chapter 7*

*First Law of Motion.* A body continues in its state of rest or of uniform motion in a straight line unless it is compelled by an external force to change that state.

*Second Law of Motion.* When a body is acted upon by an external force, the rate of change of momentum is proportional to the force and takes place in the direction of the force.

*Third Law of Motion.* To every force there is an equal and opposite force.

$$\text{Momentum} = \text{mass} \times \text{velocity}.$$

If a force $F$, in newtons, applied to a body of mass $m$, in kilograms, produces an acceleration $a$, in metres/second$^2$, then:

$$F = ma \tag{7.2}$$

## EXAMPLES 7

1. Calculate the acceleration produced when a force of 30 N acts on a mass of 50 kg.
2. What is the force required to give a mass of 10 kg an acceleration of 5 m/s$^2$?
3. Calculate the force, in newtons, to give a mass of 500 g an acceleration of 600 mm/s$^2$.
4. A body having a mass of 10 kg is travelling in a straight line at 10 m/s. What is its momentum?

   What is the average force required to increase its velocity from zero to 20 m/s in 4 s?
5. A body travelling in a straight line at 20 m/s has a momentum of 300 kg m/s. What is its mass?

   Calculate the force required to reduce the velocity at a uniform rate to 12 m/s in 10 s.
6. A body of mass 20 Mg is acted upon by a force of 15 kN. Assuming the body to be initially at rest, calculate the time taken for the body to acquire a velocity of 500 km/h.
7. A mass of 200 kg is at rest on a perfectly smooth horizontal surface. A constant horizontal force, applied for 5 s, causes it to move 18 m in that time. Calculate (*a*) the value of the force and (*b*) the final speed.
8. A 3-Mg lorry is travelling at 40 km/h on a level road. It is brought to rest with uniform retardation in 10 s. Calculate (*a*) the retarding force in kilonewtons and (*b*) the distance travelled during retardation.
9. Calculate the resistive force required to reduce the speed of a 200-kg mass at a uniform rate from 15 m/s to 5 m/s in 2 min.
10. Calculate the force exerted on the floor of a lift by a person having a mass of 80 kg when the lift is (*a*) ascending and (*b*) descending with an acceleration of 1·5 m/s$^2$ in each case.
11. A mass of 25 kg, initially at rest, is lifted 30 m by means of a rope of negligible mass at an acceleration of 2 m/s$^2$. Calculate (*a*) the tension in the rope and (*b*) the time taken to lift the load.
12. A block of iron hangs from a spring balance. When the block is stationary, the reading on the balance is 65 N. When the block is hauled vertically upwards with a uniform acceleration, the reading on the balance is 72 N. What is the value of the acceleration?
13. A body has a mass of 1·5 kg. When weighed by means of a spring balance in a moving lift, the balance reading was 13 N. If the lift was accelerating, in what direction was it travelling and what was the value of the acceleration?
14. A pit cage having a mass of 2 Mg is being lowered from rest down a shaft and accelerates uniformly until, when 20 m from the top, the velocity is

400 m/min. Calculate the force in the rope just above the cage. If the cage is then brought to rest uniformly from the speed of 400 m/min in 20 m, what is now the force in the rope?

15. A vehicle is travelling at 100 km/h along a straight level road when the brakes are applied, locking all the wheels. Calculate (*a*) the retardation in metres per second squared, (*b*) the velocity, in kilometres/hour, after 3 s. Assume the coefficient of friction between the wheels and the road to be constant at 0·6.
16. A railway waggon having a mass of 20 Mg is travelling at 25 km/h along a straight level track. Its wheels are then locked and friction provides a uniform retardation of 1·5 m/s². Calculate (*a*) the distance travelled by the waggon before coming to rest, (*b*) the retarding force due to friction and (*c*) the coefficient of friction between the wheels and the rails.

## ANSWERS TO EXAMPLES 7

1. 0·6 m/s²
2. 50 N.
3. 0·3 N.
4. 100 kg m/s, 50 N.
5. 15 kg, 12 N.
6. 185 s.
7. 288 N, 7·2 m/s.
8. 3·33 kN, 55·5 m.
9. 16·67 N.
10. 904·8 N, 664·8 N.
11. 295 N, 5·48 s.
12. 1·055 m/s².
13. 1·14 m/s², downward.
14. 17·4 kN, 21·84 kN.
15. 5·89 m/s², 36·4 km/h.
16. 16·1 m, 30 kN, 0·153.

CHAPTER 8

# Work, power and energy

## 8.1 Work

It was stated in section 1.9 that when a force is exerted against some form of resistance through a distance in the direction of the force, *work* is done, and that the SI unit of work is the *joule*, namely *the work done when a force of* 1 *newton is exerted through a distance of* 1 *metre in the direction of the force*. Hence, if a force $F$, in newtons, is exerted through a distance $s$, in metres, in the direction of the force,

$$\text{work done} = F\,[\text{newtons}] \times s\,[\text{metres}]$$
$$= Fs \text{ joules} \qquad (8.1)$$

An alternative unit of work is the *kilowatt hour* (kW h).

$$1\text{ kW h} = 1000 \text{ watt hours}$$
$$= 1000 \times 3600 \text{ watt seconds or joules}$$
$$= 3\,600\,000\text{ J} = 3{\cdot}6\text{ MJ.}$$

**Example 8.1** *A mass of* 40 *kg rests on a horizontal surface. If the coefficient of friction is* 0·2, *calculate the work done in moving this mass through a distance of* 10 *m.*

$$\text{Vertical force on surface} \simeq 40 \times 9{\cdot}81 = 392{\cdot}4\text{ N.}$$

From section 5.1,

$$\left.\begin{array}{r}\text{horizontal force required}\\ \text{to overcome friction}\end{array}\right\} = 392{\cdot}4\,[\text{N}] \times 0{\cdot}2 = 78{\cdot}48\text{ N,}$$

$$\therefore \qquad \text{work done} = 78{\cdot}48\,[\text{N}] \times 10\,[\text{m}] = 784{\cdot}8\text{ J.}$$

## 8.2 Work represented by an area of diagram of work

The amount of work done by a force exerted through a distance in its own direction may be represented by a rectangular area or diagram of work, one side of which represents the force to scale and the other the distance to scale. For example, in fig. 8.1, if a force of 50 N is represented on a scale of 1 mm to 1 N and moves through a distance of 12 m represented on a scale of 1 mm to 0·2 m, then 1 mm$^2$ of area represents

$$1\,[\text{N}] \times 0{\cdot}2\,[\text{m}] = 0{\cdot}2\ \text{J}.$$

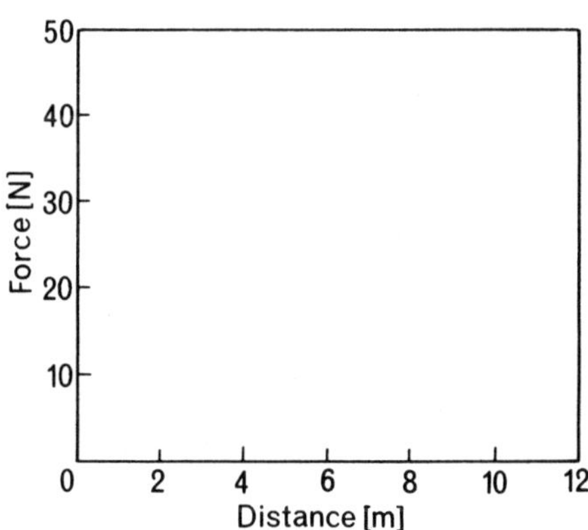


Fig. 8.1 Work represented by an area.

The sides of the rectangle are 50 mm by 60 mm so that the total area is 3000 mm$^2$.

Hence the work done is represented by an area of 3000 mm$^2$ on a scale of 1 mm$^2$ to 0·2 J,

$$\therefore \qquad \text{work done} = 0{\cdot}2\,[\text{J/mm}^2] \times 3000\,[\text{mm}^2] = 600\ \text{J}.$$

For calculating the work done by a force of constant value, the diagram is of no advantage; for instance, in the case just considered, we could have calculated the work done by merely multiplying the force by the distance thus:

$$\text{work done} = 50\,[\text{N}] \times 12\,[\text{m}] = 600\ \text{J}.$$

When the force varies in some known but perhaps rather complicated manner, calculation of the area of the work diagram may be a convenient way of determining the work done.

**Example 2** *A spring, initially in a state of ease (neither stretched nor compressed), is extended* 50 *mm. Calculate the work done if the spring requires a force of* 0·8 *N per millimetre of stretch.*

$$\text{Force for 50-mm extension} = 50\,[\text{mm}] \times 0{\cdot}8\,[\text{N/mm}] = 40\ \text{N}.$$

Fig. 8.2 shows how the extension increases in proportion to the force as the latter is increased from zero to 40 N, and the work done is represented by the area of the shaded triangle.

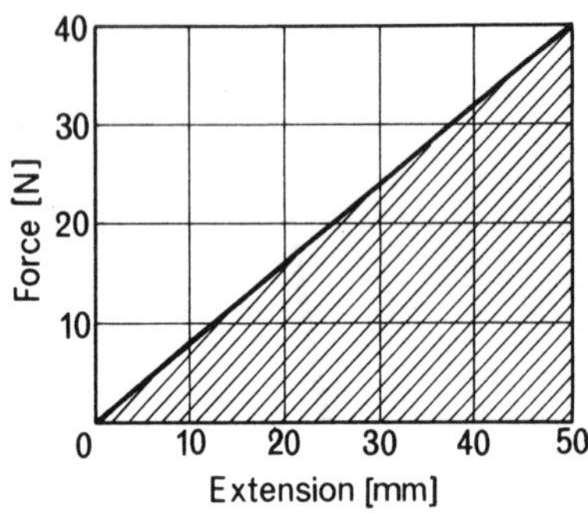


Fig. 8.2 Force/extension diagram for Example 8.2.

Alternatively, the work done can be calculated thus:

average force during extension $= 40/2 = 20$ N

and total extension $= 50$ mm $= 0{\cdot}05$ m,

$\therefore$ work done $=$ average force $\times$ extension

$= 20$ [N] $\times$ $0{\cdot}05$ [m] $= 1$ J.

**Example 8.3** *A chain,* 20 *m long, has a mass of* 12 *kg/m length and hangs vertically. Draw the work diagram and determine* (a) *the work done in winding the chain on to a drum at the top and* (b) *the work done in winding the chain up the first* 10 *m.*

(*a*) Weight of chain per metre length $\simeq 12 \times 9{\cdot}81 = 117{\cdot}7$ N

$\therefore$ total weight of chain $= 117{\cdot}7 \times 20 = 2354$ N.

Hence the total lifting force required at first is 2354 N and the lifting force diminishes uniformly to zero as the chain is being wound on the drum.

Fig. 8.3 is the diagram of work and shows how the lifting force varies with the length through which this force is being exerted at all stages of the winding process.

If the scales are 1 mm to 50 N for the lifting force and 1 mm to $0{\cdot}5$ m for the distance through which the chain is raised, then 1 mm$^2$ represents 50 N $\times$ $0{\cdot}5$ m, namely 25 J. The total area of the diagram, i.e. triangle AOB, is OB × mean height EC. But OB = 40 mm and EC = 23·5 mm,

$$\therefore \quad \text{area of triangle AOB} = 40 \times 23{\cdot}5 = 940\ \text{mm}^2$$

$$\text{and} \quad \text{work done} = 25\ [\text{J/mm}^2] \times 940\ [\text{mm}^2]$$
$$= 23\,500\ \text{J} = 23{\cdot}5\ \text{kJ}.$$

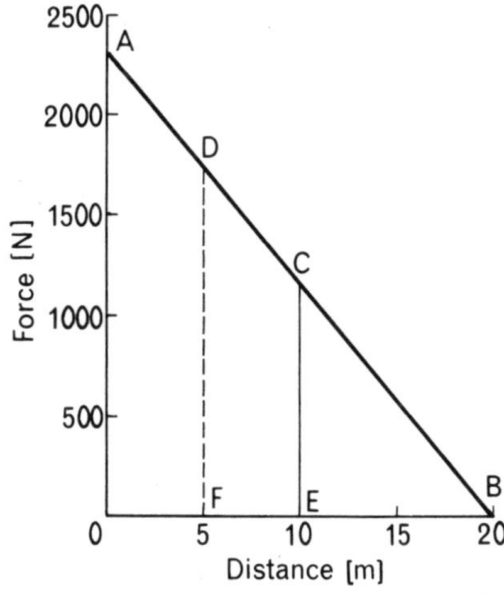


Fig. 8.3 Work done by a uniformly varying force.

In such a simple case, it is not necessary to calculate the area and the number of joules per square millimetre. The diagram shows that the *average* force, represented by EC, is half the maximum force OA which is 2354 N; hence,

$$\text{work done} = \text{average lifting force} \times \text{distance}$$
$$= \tfrac{1}{2} \times 2354\ [\text{N}] \times 20\ [\text{m}]$$
$$= 23\,540\ \text{J} = 23{\cdot}54\ \text{kJ}.$$

The discrepancy between the two values is due to the approximation in reading the value of EC from the graph.

(*b*) Similarly, though the work done in the first 10 m of lift is represented by the area ACEO, it is not necessary to calculate the area. It is evident from fig. 8.3 that:

$$\text{FD} = \tfrac{3}{4} \times \text{OA} = \tfrac{3}{4} \times 2354 = 1765\ \text{N}$$
$$= \text{average force during first 10 m of lift,}$$

$$\therefore \quad \left.\begin{array}{r}\text{work done during first} \\ \text{10 m of lift}\end{array}\right\} = 1765\ [\text{N}] \times 10\ [\text{m}]$$

$$= 17\,650\ \text{J} = 17{\cdot}65\ \text{kJ}.$$

**Example 8.4** *A load is hauled with a tractive effort F, in newtons, which varies with the distance x, in metres, moved as shown in the following table:*

| $x$ [m] | 0 | 20 | 50 | 80 | 110 | 130 | 160 | 190 | 200 |
|---|---|---|---|---|---|---|---|---|---|
| $F$ [N] | 1280 | 1270 | 1220 | 1110 | 905 | 800 | 720 | 670 | 660 |

*Determine the total work done* (a) *in kilojoules,* (b) *in watt hours.*

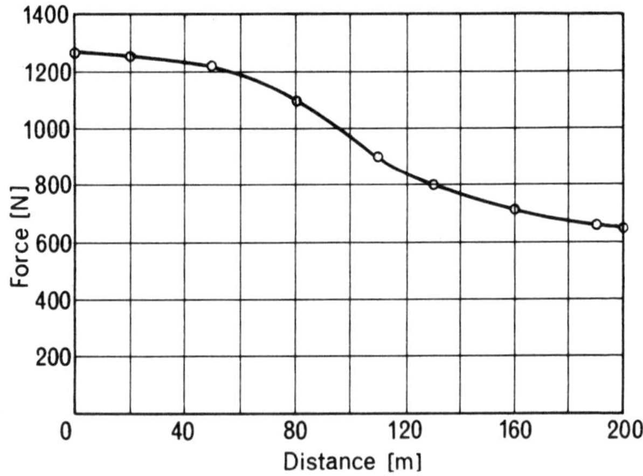


Fig. 8.4 Work done by a variable force.

First plot the graph showing the values of $F$ for different values of $x$ (as in fig. 8.4), using scales of, say, 1 mm to 2 m for $x$ and 1 mm to 20 N for $F$.

The average tractive effort can be determined by means of the mid-ordinate rule, i.e. the height is measured at the middle of each of 10 vertical strips into which the diagram is divided. The sum of these average heights divided by 10 gives the average height of the whole graph. The following table gives the tractive effort, derived from the graph of fig. 8.4, at the mid-point of each 20-m strip:

| $x$ [m] | 10 | 30 | 50 | 70 | 90 | 110 | 130 | 150 | 170 | 190 |
|---|---|---|---|---|---|---|---|---|---|---|
| $F$ [N] | 1275 | 1260 | 1220 | 1150 | 1050 | 905 | 800 | 740 | 695 | 670 |

The sum of the 10 values of $F$ is 9765 N,

hence average tractive effort $= 9765/10 = 976{\cdot}5$ N

and work done $= 976{\cdot}5\ [\text{N}] \times 200\ [\text{m}]$

$= 195\,300\ \text{J} = 195{\cdot}3\ \text{kJ}.$

Since 1 watt hour = 3600 watt seconds or joules,

$\therefore$ work done $= 195\,300/3600 = 54{\cdot}3$ W h.

## 8.3 Work done by an oblique force

In section 2.9 it was shown that when a force $F$, acting on a body, is inclined at an angle $\theta$ to the direction of motion (fig. 8.5), the force can be resolved into two components, namely $F \cos \theta$ acting in the direction of motion and $F \sin \theta$ acting at right angles to the direction

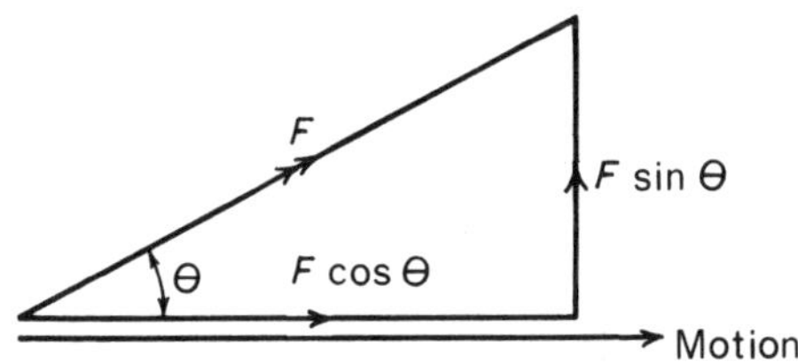


Fig. 8.5 Work done by an oblique force.

of motion. Since there is no movement of the body in the direction of component $F \sin \theta$, the latter does not do any work. Hence the work done by the oblique force $F$ is the product of the first component, $F \cos \theta$, and the distance $s$ through which the body moves:

i.e. work done $= (F \cos \theta) \times s.$

**Example 8.5** *A barge is being towed along a canal at 4 km/h. The tow rope is inclined at an angle of 25° to the direction of motion of the barge, and the pull on the rope is 300 N. Calculate the work done in 2 minutes.*

$4\ \text{km/h} = 4000\ [\text{m}]/3600\ [\text{s}] = 1{\cdot}111\ \text{m/s}$

$\therefore$ distance travelled in 2 minutes $= 1{\cdot}111\ [\text{m/s}] \times (2 \times 60)\ [\text{s}] = 133{\cdot}3\ \text{m}.$

$$\left.\begin{array}{r}\text{Component of pull in}\\ \text{direction of force}\end{array}\right\} = 300 \cos 25° = 300 \times 0{\cdot}9063$$

$$= 271{\cdot}9 \text{ N,}$$

$$\therefore \quad \text{work done} = 271{\cdot}9 \text{ [N]} \times 133{\cdot}3 \text{ [m]}$$

$$= 36\ 250 \text{ J} = 36{\cdot}25 \text{ kJ.}$$

## 8.4 Work done in rotation

Let us consider the case of a crank handle or a pulley attached to a shaft, as in fig. 8.6, and suppose a force of 70 N to be exerted at right angles to the crank arm, 200 mm long, or at the circumference of the pulley of 200 mm radius, in order to turn the shaft against some resistance. Also, suppose the force to be always in the direction of the motion of the point to which it is applied.

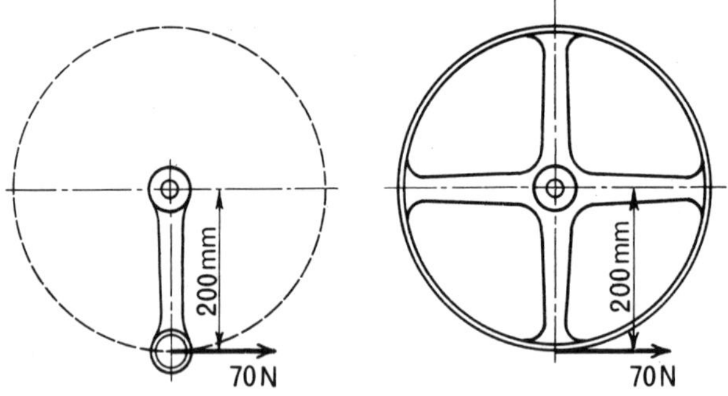


Fig. 8.6 Work done in rotation.

The distance through which the point of application of the force travels in 1 revolution is $2\pi \times 0{\cdot}2$ [m], namely 1·257 m,

$$\therefore \quad \text{work done in 1 revolution} = 70 \text{ [N]} \times 1{\cdot}257 \text{ [m]} = 88 \text{ J.}$$

In general, if a force $F$ newtons acts at a radius $r$ metres,

$$\text{work done in 1 revolution} = F \times 2\pi r \text{ joules} \qquad (8.2)$$

The product of a turning force and the radius of the circle at which it acts is termed the *torque* or *turning moment* about the axis of rotation. Hence, for a force $F$ newtons acting at a radius $r$ metres,

$$\text{torque (or turning moment)} = T = Fr \text{ newton metres}$$

and from expression (8.2),

$$\left.\begin{array}{r}\text{work done, in joules,}\\ \text{in } n \text{ revolutions}\end{array}\right\} = Fr \times 2\pi n$$

$$= \text{torque in newton metres} \times \text{angle in radians}$$

$$= T\theta \qquad (8.3)$$

**Example 8.6** *An electric motor has to lift a body having a mass of* 50 *kg by means of a rope wound round a drum having a diameter of* 1·2 *m. Calculate* (a) *the torque to be exerted and* (b) *the work done when the drum makes* 20 *revolutions.*

(*a*) Force on rope ≃ 50 × 9·81 = 490·5 N

and radius of drum = 0·6 m,

∴ torque = 490·5 [N] × 0·6 [m]

= 294·3 N m.

(*b*) From expression (8.3),

work done = 294·3 [N m] × (2π × 20) [rad]

= 37 000 J = 37 kJ.

**Example 8.7** *A pulley is* 800 *mm in diameter and the difference in tension on the two sides of the driving belt is* 2000 *N. If the speed of the pulley is* 300 *rev/min, what is the work done in* 5 *min?*

Radius of pulley = 400 mm = 0·4 m,

∴ torque = 2000 [N] × 0·4 [m]

= 800 N m.

No. of revolutions in 5 minutes = 300 × 5 = 1500.

From expression (8.3),

work done = 800 [N m] × (2π × 1500) [rad]

= 7 540 000 J = 7·54 MJ.

## 8.5 Power

*Power is the rate of doing work* and the SI unit of power is the *watt*, namely 1 joule per second. In practice, the watt is often found to be inconveniently small; consequently the *kilowatt* (kW) is frequently used, the kilowatt being 1000 watts. For still larger powers, the *megawatt* (MW) is used, where:

$$1 \text{ MW} = 1000 \text{ kW} = 1\,000\,000 \text{ W}.$$

Similarly, when we are dealing with a large amount of work (or energy), it is often convenient to express the latter in *kilowatt hours* (see section 8.1).

If $T$ be the torque or turning moment, in newton metres, and if $n$ be the speed, in revolutions per second, then from expression (8.3) we have:

$$\begin{aligned} \text{work done per second} &= \text{torque in newton metres} \\ &\quad \times \text{speed in radians per second} \\ &= T\,[\text{newton metres}] \\ &\quad \times 2\pi n\,[\text{radians/second}] \\ &= 2\pi nT \text{ joules/second or watts} \end{aligned}$$

i.e.

$$\begin{aligned} \text{power} &= 2\pi nT \text{ watts} \\ &= \omega T \text{ watts} \qquad (8.4) \end{aligned}$$

where

$$\begin{aligned} \omega &= \text{angular velocity in radians/second} \\ &= 2\pi n \text{ radians/second.} \end{aligned}$$

If the rotational speed be $N$ revolutions per minute,

$$\text{power} = 2\pi TN/60 \text{ watts} \qquad (8.5)$$

**Example 8.8** *A motor vehicle hauls a trailer at* 75 *km/h when exerting a steady pull of* 800 *N. Calculate* (a) *the work done in* 20 *min* (i) *in megajoules and* (ii) *in kilowatt hours,* (b) *the power required.*

(*a*) (*i*)

$$\begin{aligned} \text{Distance travelled} &= 75\,[\text{km/h}] \times (20/60)\,[\text{h}] \\ &= 25 \text{ km,} \end{aligned}$$

$\therefore$

$$\begin{aligned} \text{work done} &= 800\,[\text{N}] \times (25 \times 1000)\,[\text{m}] \\ &= 20\,000\,000 \text{ J} = 20 \text{ MJ.} \end{aligned}$$

(*ii*)

$$\begin{aligned} \text{Since } 1 \text{ kW h} &= 3{\cdot}6 \text{ MJ} \\ \text{work done} &= 20/3{\cdot}6 = 5{\cdot}56 \text{ kW h.} \end{aligned}$$

(*b*)

$$\begin{aligned} \text{Power} &= \frac{\text{work done in joules}}{\text{time in seconds}} \\ &= \frac{20\,000\,000\,[\text{J}]}{(20 \times 60)\,[\text{s}]} = 16\,670 \text{ W} \\ &= 16{\cdot}67 \text{ kW.} \end{aligned}$$

Alternatively,

$$\text{power} = \frac{\text{work done in kilowatt hours}}{\text{time in hours}}$$

$$= \frac{5{\cdot}56\ [\text{kW h}]}{(20/60)\ [\text{h}]} = 16{\cdot}68\ \text{kW}.$$

**Example 8.9** *An electric motor is developing* 8 *kW at a speed of* 1200 *rev/min. Calculate* (a) *the work done in* 45 *min* (i) *in kilowatt hours,* (ii) *in megajoules, and* (b) *the torque in newton metres.*

(*a*) (*i*)

$$\text{Work done} = 8\ [\text{kW}] \times (45/60)\ [\text{h}] = 6\ \text{kW h}.$$

(*ii*)

$$\text{Since } 1\ \text{kW h} = 3{\cdot}6\ \text{MJ},$$

$$\therefore \quad \text{work done} = 6 \times 3{\cdot}6 = 21{\cdot}6\ \text{MJ}.$$

(*b*) From expression (8.5),

$$8000\ [\text{W}] = 2\pi T \times (1200/60)\ [\text{rev/s}]$$

$$\therefore \quad T = 63{\cdot}7\ \text{N m}.$$

**Example 8.10** *The tensions on the two sides of a belt passing round a pulley are* 2200 *N and* 460 *N respectively. The effective diameter of the pulley is* 400 *mm and the speed is* 700 *rev/min. Calculate* (*a*) *the power transmitted and* (b) *the work done in* 10 *min* (i) *in megajoules and* (ii) *in kilowatt hours.*

(*a*) The effective driving force of a belt is the difference between the tensions on the tight and the slack sides,

$$\therefore \quad \text{net driving force} = 2200 - 460 = 1740\ \text{N}.$$

$$\text{Effective radius of pulley} = 200\ \text{mm} = 0{\cdot}2\ \text{m},$$

$$\therefore \quad \text{torque} = 1740\ [\text{N}] \times 0{\cdot}2\ [\text{m}] = 348\ \text{N m}.$$

From expression (8.5), we have:

$$\text{power} = 2\pi \times 348\ [\text{N m}] \times (700/60)\ [\text{rev/s}] = 25\,500\ \text{W} = 25{\cdot}5\ \text{kW}.$$

(*b*) (*i*)

$$\text{Work done in 10 min} = 25\,500\ [\text{W}] \times (10 \times 60)\ [\text{s}] = 15\,300\,000\ \text{J} = 15{\cdot}3\ \text{MJ}.$$

(*ii*) Work done in 10 min $= 25{\cdot}5\ [\text{kW}] \times (10/60)\ [\text{h}]$
$= 4{\cdot}25$ kW h.

Alternatively, since 1 kW h $= 3{\cdot}6$ MJ

$\therefore$ work done in 10 min $= 15{\cdot}3/3{\cdot}6 = 4{\cdot}25$ kW h.

## 8.6 Determination of the output power of a machine by means of a brake

In the case of a comparatively small machine, the output power can be measured by some form of mechanical brake such as that shown in fig. 8.7, where a belt (or rope) on an air or water-cooled pulley has its ends attached to spring balances $S_1$ and $S_2$, calibrated in newtons. The balances are supported by a rigid horizontal beam B, and the tension on the belt can be controlled by wing-nuts W.

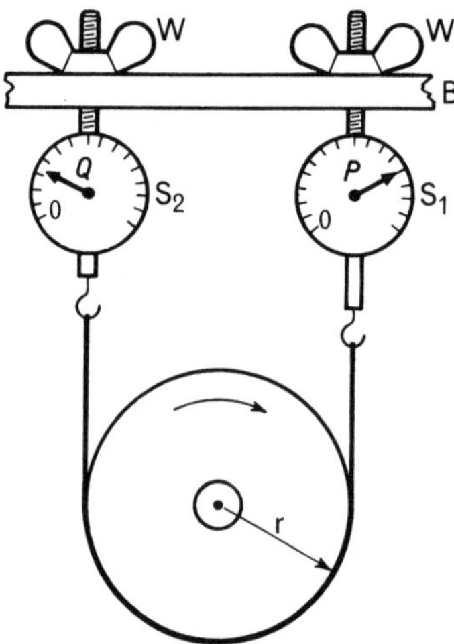


Fig. 8.7 Brake test.

Suppose the brake pulley to be rotating clockwise and the tension on the belt to be adjusted to give readings of $P$ and $Q$ newtons on $S_1$ and $S_2$ respectively. The pull $P$ exerted by $S_1$ has to balance the pull $Q$ exerted by $S_2$ and the friction force $F$ between the belt and the pulley,

i.e. $$P = Q + F$$

$\therefore$ $$F = (P - Q) \text{ newtons.}$$

If $r$ be the effective radius of the brake, in metres,

$$\text{torque due to brake friction} = Fr$$
$$= (P - Q)r \text{ newton metres.}$$

If $N$ be the speed of the pulley in revolutions/minute, then from expression (8.5),

$$\text{output power} = 2\pi(P - Q)rN/60 \qquad (8.6)$$

The output power of the machine is converted into heat at the brake, and the size of the machine that can be tested by this method is limited by the difficulty of dissipating this heat.

### 8.7 Efficiency of a Machine

In all machines, some of the power supplied to the machine is lost in overcoming friction, etc., so that the useful power available is less than the input power. The ratio of the output power to the input power is termed the *efficiency* of the machine,

i.e.
$$\text{efficiency} = \frac{\text{output power}}{\text{input power}}$$

and is expressed as a 'per unit' value or as a percentage; thus, if the input and output powers are 75 kW and 60 kW respectively,

$$\text{efficiency} = 60/75 = 0{\cdot}8 \text{ per unit}$$
$$= 0{\cdot}8 \times 100 = 80 \text{ per cent.}$$

The input and output powers must obviously be expressed in the same unit.

**Example 8.11** *In a brake test on an electric motor, the readings on balances $S_1$ and $S_2$ (fig.* 8.7) *were* 440 *N and* 87 *N respectively. The effective diameter of the pulley was* 500 *mm and the speed was* 800 *rev/min. The power supplied to the motor was* 8·5 *kW. Calculate* (a) *the output power of the motor and* (b) *the efficiency.*

(*a*) Effective radius of pulley = 250 mm = 0·25 m.

$$\text{Net pull due to friction} = 440 - 87 = 353 \text{ N},$$
$$\therefore \quad \text{torque} = 353\,[\text{N}] \times 0{\cdot}25\,[\text{m}] = 88{\cdot}25 \text{ N m}.$$

Substituting in expression (8.5), we have:

$$\text{output power} = 2\pi \times 88{\cdot}25\ [\text{N m}] \times (800/60)\ [\text{rev/s}]$$
$$= 7390\ \text{W} = 7{\cdot}39\ \text{kW}.$$

(*b*)
$$\text{Efficiency} = 7{\cdot}39\ [\text{kW}]/8{\cdot}5\ [\text{kW}] = 0{\cdot}87 \text{ per unit}$$
$$= 87 \text{ per cent.}$$

## 8.8 Energy

When a body is capable of doing work, it is said to possess *energy* which may take various forms such as mechanical energy, thermal energy, chemical energy and electrical energy. In mechanics we are concerned only with mechanical energy which is of two kinds, namely *kinetic energy* and *potential energy.*

*The kinetic energy of a body is the energy it possesses by virtue of its motion.* Thus a body, set in motion by a force doing work upon it, acquires kinetic energy which enables it to do work against resisting forces. An important engineering application is the flywheel. Work is done on the flywheel as its speed is increased; and later, when the machine to which it is attached slows down, some of the stored kinetic energy is given out by the flywheel and helps to drive the machine. This is the reason why a machine driving a fluctuating load, such as a stamping press, is usually fitted with a flywheel to maintain a more constant speed than would otherwise be the case.

If a force is exerted on a body and there is no opposing resistance except the inertia of the body, the whole of the work done becomes the kinetic energy of the body. Thus, if a force $F$, acting on a mass $m$, gives it a uniform acceleration $a$, then from expression (7.2):

$$F = ma$$

If $s$ is the distance travelled by the body while it accelerates from standstill to a velocity $v$, then from expression (6.4), we have:

$$s = \tfrac{1}{2}\, v^2/a$$

$$\therefore \quad \text{work done} = F\ [\text{newtons}] \times s\ [\text{metres}]$$
$$= ma \times \tfrac{1}{2}v^2/a$$
$$= \tfrac{1}{2}mv^2 \text{ joules} \tag{8.7}$$
$$= \text{kinetic energy of body.}$$

*The potential energy of a body is the energy it possesses by virtue of its position or state of strain.* For instance, a body raised to a height above the ground has potential energy since its weight can do work as the body returns to the ground.

If a body having a mass $m$, in kilograms, is lifted vertically through a height $h$, in metres, and if $g$ is the gravitational acceleration in metres/second$^2$ at that point,

$$\begin{aligned}\text{force required} &= \text{weight of the body}\\ &= mg \text{ newtons}\end{aligned}$$

and

$$\begin{aligned}\text{work done} &= \text{weight of the body} \times \text{height}\\ &= mgh \text{ joules} \qquad (8.8)\\ &= \text{potential energy of body.}\end{aligned}$$

In section (1.8) it was stated that, for latitudes extending from 34° to 69°, the value of $g$ is 9·81 m/s$^2$ $\pm$ 0·1 per cent. Hence, for this region,

$$\text{potential energy} = 9{\cdot}81\, mh \text{ joules, within } \pm\, 0{\cdot}1 \text{ per cent.}$$

Another illustration of the storage of potential energy is given in Example 8.2, where a spring, initially at ease, i.e. in its zero position, is extended 50 mm, thereby storing 1 joule of potential energy (sometimes referred to as *elastic strain energy*) in the spring.

The pendulum of a clock is an example of energy being changed backwards and forwards between the kinetic form and the potential form. Thus the oscillating mass has its maximum kinetic energy at the lowest point of its travel, its potential energy being then zero. On the other hand, the potential energy of the pendulum is a maximum at the end of each swing, its speed and therefore its kinetic energy being then zero. The small loss of energy due to friction is supplied by the impulse given regularly through the escapement mechanism from the mainspring.

One of the natural sources of potential energy is water lifted by evaporation from sea-level to lakes and rivers at higher levels into which it is deposited as rain or snow. With the aid of pipes supplying water turbines at a lower level, the potential energy of the water stored in a reservoir at a high level can be converted into kinetic energy and thereby used to drive the turbines which in turn drive electrical generators or other machinery. Thus the potential energy of the water in the reservoir is converted into useful work.

**Example 8.12** *A body having a mass of* 30 *kg is supported* 50 *m above* *the earth's surface. What is its potential energy?*

*If the body is allowed to fall freely, calculate its potential and kinetic energies* (a) *when the body is 20 m above the ground and* (b) *just before it touches the ground.*

$$\text{Weight of body} \simeq 30 \times 9{\cdot}81 = 294{\cdot}3 \text{ N},$$

$$\therefore \quad \left.\begin{array}{r}\text{work done in lifting}\\ \text{the body 50 m}\end{array}\right\} = 294{\cdot}3\ [\text{N}] \times 50\ [\text{m}]$$
$$= 14\,715 \text{ J} = 14{\cdot}715 \text{ kJ},$$

i.e. potential energy of the body when it is 50 m above ground is 14·715 kJ. Since the body is stationary, its kinetic energy is zero. (*a*) When the body is 20 m above ground,

$$\text{its potential energy} = 294{\cdot}3\ [\text{N}] \times 20\ [\text{m}]$$
$$= 5886 \text{ J} = 5{\cdot}886 \text{ kJ}.$$

$$\text{Vertical distance travelled by body} = 50 - 20 = 30 \text{ m}.$$

Since $\qquad v = \sqrt{(2gs)} \qquad (6.7)$

and assuming $\qquad g = 9{\cdot}81 \text{ m/s}^2$, we have:

$$v = \sqrt{(2 \times 9{\cdot}81 \times 30)} = 24{\cdot}26 \text{ m/s}.$$

From expression (8.7),

$$\text{kinetic energy} = \tfrac{1}{2} \times 30\ [\text{kg}] \times (24{\cdot}26)^2\ [\text{m/s}]^2$$
$$= 8829 \text{ J} = 8{\cdot}829 \text{ kJ}.$$

Alternatively, from the principle of the conservation of energy (section 8.10), the sum of the potential and kinetic energies must remain constant since no energy is being converted into any other form of energy; hence:

$$\text{kinetic energy} = \text{potential energy at 50 m} - \text{potential energy at 20 m}$$
$$= 14{\cdot}715 - 5{\cdot}886 = 8{\cdot}829 \text{ kJ}.$$

(*b*) $\left.\begin{array}{r}\text{Velocity of body just}\\ \text{before it touches ground}\end{array}\right\} = \sqrt{(2 \times 9{\cdot}81 \times 50)} = 31{\cdot}32 \text{ m/s}$

$$\therefore \quad \text{kinetic energy} = \tfrac{1}{2} \times 30\ [\text{kg}] \times (31{\cdot}32)^2\ [\text{m/s}]^2$$
$$= 14\,715 \text{ J} = 14{\cdot}715 \text{ kJ},$$

i.e. the whole of the initial potential energy has been converted into kinetic energy. When the body is finally brought to rest by the resistive force of the ground, practically the whole of this energy is converted into heat.

**Example 8.13** *A motor vehicle having a mass of* 2 *Mg is travelling along a straight level road at* 60 *km/h. What is its kinetic energy in kilojoules?*

*If the resistance to motion remains constant at* 14 *newtons per kilonewton weight of the vehicle, how far will it travel, without driving force or brakes, before coming to rest?*

$$60 \text{ km/h} = 60\,000 \text{ [m]}/3600 \text{ [s]} = 16{\cdot}67 \text{ m/s}.$$

$$\text{Mass of vehicle} = 2 \text{ Mg} = 2000 \text{ kg},$$

$$\therefore \quad \text{kinetic energy} = \tfrac{1}{2} \times 2000 \text{ [kg]} \times (16{\cdot}67)^2 \text{ [m/s]}^2$$

$$= 278\,000 \text{ J} = 278 \text{ kJ}.$$

$$\text{Weight of vehicle} \simeq 2000 \times 9{\cdot}81 = 19\,620 \text{ N}$$

$$= 19{\cdot}62 \text{ kN},$$

$$\therefore \quad \text{resistance to motion} = 19{\cdot}62 \text{ [kN]} \times 14 \text{ [N/kN]}$$

$$= 275 \text{ N}.$$

Since the whole of the kinetic energy of 278 000 J has to be expended in bringing the vehicle to rest,

$$\therefore \quad 278\,000 \text{ [J]} = 275 \text{ [N]} \times \text{distance in metres}$$

so that $\text{distance} = 1010 \text{ m} = 1{\cdot}01 \text{ km}.$

## 8.9 Other forms of energy

There are other forms of energy such as *thermal* energy, *electrical* energy and *chemical* energy. Thermal energy can be produced by doing mechanical work; for example, when a piece of metal is hammered vigorously, the gain of thermal energy may be sufficient to be detected by the sense of touch. Thermal energy is also produced from mechanical work when the surface of one body rubs or slides over the surface of another body, particularly when the surfaces are rough.

Electrical energy can be generated from the mechanical energy developed by an engine or a turbine driving an electric generator. Conversely, electrical energy supplied to an electric motor enables the latter to supply mechanical energy to a machine coupled to its shaft.

Chemical energy can be produced from electrical energy as, for example, when a secondary cell (section 19.4) is charged. Conversely, when the cell is discharged, chemical energy is converted into electrical energy. The latter can be converted into thermal and light energies when a suitable lamp is connected across the cell, or into mechanical energy if the cell is used to drive an electric motor.

## 8.10 Principle of the conservation of energy

This important principle states that whenever energy is converted from one form to another, no energy is lost. All the energy involved in the conversion can be accounted for in some form or another. This constancy of the total energy is referred to as the *Principle of the Conservation of Energy.*

### *Summary of Chapter* 8

Work is done when force is exerted through a distance in its own direction. Energy is thus converted to or from some form of mechanical energy including kinetic energy and potential energy; the latter may be either gravitational or elastic strain energy. Or it may be converted into thermal energy, electrical energy or chemical energy. But in any transformation, no energy is destroyed (the Principle of Conservation of Energy).

Work is the product of force and the distance moved in the direction of the force, or the product of the torque and the angle of rotation in radians. The SI unit of work is the *joule.*

Power is the rate of doing work and the SI unit of power is the *watt*, namely 1 joule/second.

The symbol for work and energy is $W$ and that for power is $P$.

$$W\,[\text{joules}] = F\,[\text{newtons}] \times s\,[\text{metres}] \tag{8.1}$$

$$= T\,[\text{newton metres}] \times \theta\,[\text{radians}] \tag{8.3}$$

$$1\ \text{kW h} = 3\,600\,000\ \text{J} = 3{\cdot}6\ \text{MJ}.$$

Work done by an oblique force $= Fs\cos\theta$

$$\text{Power [watts]} = \omega\,[\text{rad/s}] \times T\,[\text{N m}] \tag{8.4}$$

$$= 2\pi T\,[\text{N m}] \times (N/60)\,[\text{rev/s}] \tag{8.5}$$

The energy of a body is its capacity for doing work.

Kinetic energy is the energy possessed by a body by virtue of its motion.

Kinetic energy, in joules $= \frac{1}{2}\, m\,[\text{kg}] \times v^2\,[\text{m/s}]^2$ (8.7)

Potential energy is the energy possessed by a body by virtue of its position or state of strain.

For a body of mass $m$ lifted through height $h$,

$$\text{potential energy, in joules} = m\,[\text{kg}] \times g\,[\text{m/s}^2] \times h\,[\text{m}] \quad (8.8)$$
$$\simeq 9{\cdot}81\, mh.$$

$$\text{Efficiency} = \frac{\text{output power}}{\text{input power}}$$

## EXAMPLES 8

1. The work done in moving a body through a distance of 100 m in the direction of the force is 750 J. Calculate the value of the force.
2. Calculate the work done, in kilojoules, in lifting a mass of 800 kg through a vertical height of 40 m.
3. A force of 250 N is applied to a body. If the work done is 30 kJ, what is the distance through which the body is moved?
4. The work done in lifting a body vertically through a distance of 120 m is 5 W h. Calculate the mass of the body.

   If the time taken is 2 min, what is the average value of the power?
5. A force $F$ newtons acting on a body in the direction of its motion varies as follows for different distances $s$ metres from the initial position:

| $F$ | 200 | 360 | 450 | 480 | 400 | 300 |
|---|---|---|---|---|---|---|
| $s$ | 0 | 10 | 20 | 30 | 40 | 50 |

   Draw the work diagram to scale and determine (*a*) the average force and (*b*) the work done, in kilojoules, when the body is moved through a distance of 50 m. Assume the points on the graph to be joined by straight lines.
6. A car has the following tractive forces, in kilonewtons, exerted on it after it has travelled distances $s$ metres from rest:

| $F$ | 6·4 | 6·2 | 5·6 | 4·8 | 4·1 | 3·6 | 3·3 |
|---|---|---|---|---|---|---|---|
| $s$ | 0 | 5 | 10 | 15 | 20 | 25 | 30 |

   Determine the work done (*a*) in kilojoules and (*b*) in watt hours when the car has moved through 30 m.

   If the time taken to travel 30 m is 20 s, what is the average value of the power in kilowatts?
7. A helical spring is extended by a force which increases uniformly from zero to 600 N. The corresponding extension of the spring is 200 mm. Draw a work diagram and determine the total work done.

8. A compression spring is 150 mm long when unloaded and 120 mm long when subjected to a force of 20 N. It is 80 mm long when fully compressed. Draw a diagram showing the relationship between load and length of spring between zero and full load. Determine (*a*) the force to produce full compression and (*b*) the work done in compressing the spring between 150 mm and 80 mm and (*c*) the work done in compressing the spring between 120 mm and 80 mm.
9. A chain, having a mass of 18 kg/m of length, is 30 m long and hangs vertically. How much work is done in upwinding the chain on to a drum?
10. A chain, 200 m long, having a mass of 12 kg/m and hanging vertically, is wound up on to a drum. Draw to scale a graph showing the lifting forces as ordinates and the lengths drawn up, from 0 to 200 m, as abscissae. From the graph calculate the work done in winding up (*a*) the first 80 m of the chain and (*b*) the whole chain.
11. Calculate the work done (*a*) in megajoules and (*b*) in kilowatt hours, in emptying a circular shaft, 4 m diameter and 60 m deep, which is full of water, assuming the water flows away at the top of the shaft. Neglecting losses, calculate the time taken by a pump, developing 4 kW, to empty the shaft. Assume 1 $m^3$ of water to have a mass of 1000 kg.
12. If the work done in moving a body is 1500 J and the time taken is 12 s, what is the mean value of the power?
13. A body having a mass of 300 kg is lifted through a distance of 200 m in 15 s. What is the average value of the power?
14. The output power of an electric motor is 8 kW and is maintained constant for 6 h. Calculate the work done (*a*) in kilowatt hours and (*b*) in megajoules.
15. If the speed of the motor in Q. 14 is 500 rev/min, what is the value of the torque in newton metres?
16. An engine, running at 1800 rev/min, develops a torque of 4 kN m. Calculate (*a*) the power developed and (*b*) the work done, in kilowatt hours, in 3 h.
17. A motor develops a torque of 3 kN m. At what speed must it run in order that it may develop 200 kW?
18. A locomotive, hauling a train at 110 km/h, exerts a pull of 10 kN. Calculate (*a*) the work done in 30 min (*i*) in megajoules, (*ii*) in kilowatt hours and (*b*) the power developed by the locomotive.
19. A bus having a mass of 3 Mg is travelling along a level road at 60 km/h. The resistance to motion is 10 N per kilonewton of weight. Calculate (*a*) the tractive effort required, (*b*) the power and (*c*) the work done in 40 min (*i*) in megajoules and (*ii*) in kilowatt hours.
20. A train has a mass of 150 Mg and is hauled at a constant speed of 90 km/h along a straight horizontal track. The track resistance is 6 mN per newton of weight. Calculate (*a*) the tractive effort in kilonewtons, (*b*) the work done in 20 min (*i*) in megajoules and (*ii*) in kilowatt hours and (*c*) the power in kilowatts.
21. A barge is towed along a canal at 5 km/h. The tow rope is inclined at an angle of 30° to the direction of motion of the barge and the pull on the rope is 400 N. Calculate (*a*) the work done, in kilojoules, in 15 min and (*b*) the power required.

22. The rope used to haul a sledge along level ground is inclined at an angle 20° with the ground. The tension in the rope is 300 N. Calculate the work done when the sledge is hauled a distance of 80 m.

    If this work is done in 3 min, what is the average value of the power?
23. A belt friction brake applied at the circumference of a pulley, 400 mm diameter, exerts a backward drag of 220 N. If the speed of the pulley is 660 rev/min, calculate (*a*) the torque on the pulley and (*b*) the power absorbed by the brake.

    If the pulley is driven by an electric motor and if the input power to the motor is 3·6 kW, what is the efficiency of the motor?
24. In a brake test on an internal-combustion engine, the readings on the spring balances (fig. 8.7) were 490 N and 27 N. The effective diameter of the brake wheel was 0·9 m, and the speed of the engine was 500 rev/min. Calculate the output power of the engine.
25. A 150-mm diameter cylinder is being turned on a lathe driven at 180 rev/min. The tangential force exerted by the cutting tool is 800 N. If the efficiency of the lathe is 80 per cent, calculate the output power of the driving motor.

    If the corresponding efficiency of the motor is 75 per cent, calculate the input power to the motor.
26. A force of 2 kN acts on a plunger having a diameter of 200 mm. What is the average pressure on the plunger, in kilonewtons per square metre? If this plunger is used to force oil under a piston having a diameter of 500 mm, what is the mass of a body that the piston can support? What power is required to lift this mass through 100 mm in 4 s? Neglect all losses.
27. Calculate the kinetic energy, in kilojoules, possessed by a car having a mass of 900 kg when travelling at 100 km/h.

    What constant braking force would be required to bring the car to rest in 6 s?
28. Calculate the kinetic energy (*i*) in megajoules, (*ii*) in kilowatt hours, possessed by a body having a mass of 500 kg when travelling at 120 m/s.

    What constant force would be required to bring it to rest in 2 km?
29. A body having a mass of 50 kg is supported 40 m above ground. What is its potential energy?

    If the body is allowed to fall freely, calculate its potential and kinetic energies (*a*) when the body is 30 m above ground and (*b*) just before it touches the ground. Assume $g = 9{\cdot}81\ \text{m/s}^2$.
30. The energy expended in lifting a 30-kg mass is 6 kJ. Calculate the height through which it has been lifted.

    If this mass is then allowed to fall freely, what is its kinetic energy after it has fallen a distance of 10 m?
31. A locomotive, hauling a train at 90 km/h, exerts a pull of 20 kN. The total mass of the locomotive and carriages is 300 Mg. Calculate (*a*) the work done, in megajoules, in 10 min, (*b*) the power developed by the locomotive and (*c*) the kinetic energy of the locomotive and carriages, in megajoules (neglecting rotational inertia).

## ANSWERS TO EXAMPLES 8

1. 7·5 N.
2. 314 kJ.
3. 120 m.
4. 15·3 kg, 150 W.
5. 388 N, 19·4 kJ.
6. 145·5 kJ, 40·4 W h; 7·275 kW.
7. 60 J.
8. 46·7 N; 1·633 J, 1·333 J.
9. 79·5 kJ.
10. 1·507 MJ, 2·354 MJ.
11. 222 MJ, 61·7 kW h; 15·4 h.
12. 125 W.
13. 39·24 kW.
14. 48 kW h, 172·8 MJ.
15. 153 N m.
16. 754 kW, 2262 kW h.
17. 637 rev/min.
18. 550 MJ, 153 kW h; 306 kW.
19. 294·3 N, 4·9 kW; 11·77 MJ, 3·27 kW h.
20. 8·83 kN; 265 MJ, 73·6 kW h; 221 kW.
21. 433 kJ, 482 W.
22. 22·56 kJ, 125·3 W.
23. 44 N m, 3·04 kW, 84·4 per cent.
24. 10·9 kW.
25. 1·413 kW, 1·884 kW.
26. 63·7 kN/m$^2$, 1275 kg, 312·5 W,
27. 347 kJ, 4·17 kN.
28. 3·6 MJ, 1 kW h, 1·8 kN.
29. 19·62 kJ; 14·715 kJ, 4·905 kJ; 0, 19·62 kJ.
30. 20·4 m, 2943 J.
31. 300 MJ, 500 kW, 93·75 MJ.

CHAPTER 9

# Simple machines

## 9.1 Machines

A machine is a mechanical device for transmitting motion, force and energy. Lifting machines are usually arranged to enable a small *effort* (or driving force) to raise a much larger *load* (or resisting force). To achieve this, the effort must move through a much greater distance than that through which the load is raised, the work done by the effort being equal to the useful work in lifting the load together with the work required to overcome friction. This relationship is known as the *Principle of Conservation of Energy* and has already been referred to in section 8.10.

## 9.2 Mechanical advantage, velocity (or movement) ratio and efficiency

It was mentioned in section 9.1 that the purpose of most lifting machines is to enable a large load $W$ to be moved by the application of a relatively small effort $P$. The ratio of the load to the effort is termed the *mechanical advantage* of the machine,

i.e.
$$\text{mechanical advantage} = \frac{\text{load}}{\text{effort}} = \frac{W}{P} \tag{9.1}$$

It is evident that for a given load $W$, the smaller the value of the effort $P$, the greater is the mechanical advantage.

The ratio of the distance moved by the effort to that moved by the load is termed the *movement ratio* or more commonly the *velocity ratio*,

i.e.
$$\text{velocity ratio} = \frac{\text{distance moved by effort}}{\text{distance moved by load}} \tag{9.2}$$

The value of the velocity ratio is dependent upon the arrangement of the machine and is *constant* for a given machine, whereas the value of the mechanical advantage, in general, decreases as the load is reduced, being zero at no load.

From the Principle of the Conservation of Energy,

work done by effort = work done on load
+ work done to overcome friction.

The work done in overcoming friction is converted into heat and is thus wasted as far as the machine is concerned. The ratio of the work done on the load to that done by the effort is termed the *efficiency* of the machine,

i.e.
$$\text{efficiency} = \frac{\text{useful work done}}{\text{work done by effort}} \tag{9.3}$$

Since work done by effort $P = P \times$ distance moved by effort and useful work done on load $W = W \times$ distance moved by load,

$$\therefore \qquad \text{efficiency} = \frac{W \times \text{distance moved by load}}{P \times \text{distance moved by effort}}$$

$$= \frac{W}{P} \times \frac{\text{distance moved by load}}{\text{distance moved by effort}}$$

$$= \frac{\text{mechanical advantage}}{\text{velocity ratio}} \tag{9.4}$$

For an *ideal* machine, i.e. a machine having no friction, the efficiency is unity (= 100 per cent), so that:

work done by effort = work done on load,

∴ *ideal* effort × distance moved by effort
= $W$ × distance moved by load

so that
$$\textit{ideal}\text{ effort} = \frac{W}{\text{velocity ratio}} \tag{9.5}$$

and
$$\textit{ideal}\text{ mechanical advantage} = \frac{W}{\text{ideal effort}}$$

$$= \text{velocity ratio} \tag{9.6}$$

Let us now consider some of the simplest types of machines and determine the velocity ratio of each type.

### 9.3 The lever

The lever was probably the earliest device used by man to enable him to move a large load by means of the limited physical effort he could exert. For instance, guide books on Stonehenge have illustrations showing how men used wooden poles as levers to erect the heaviest stones, each having a mass of about 50 Mg.

Fig. 9.1 shows a straight lever, pivotted at C. A mass, having weight $W$, is suspended at B and a downward effort $P$ is applied at A to balance weight $W$.

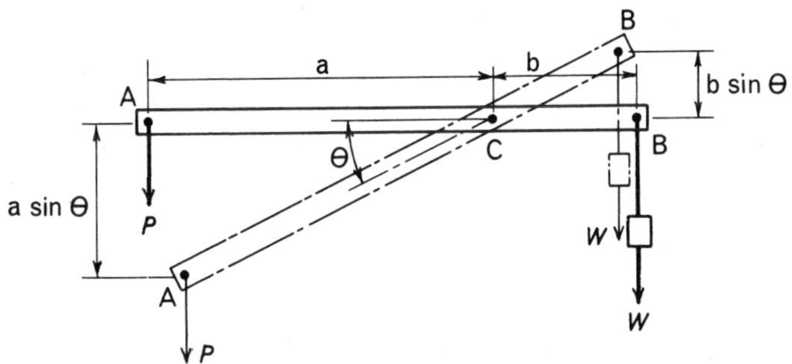


Fig. 9.1 A lever.

Suppose distances AC and BC to be $a$ and $b$ respectively, and the weight of the lever to be negligible. If the lever is tilted anticlockwise through an angle $\theta$,

$$\left.\begin{array}{r}\text{distance through which effort } P\\ \text{moves in its own direction}\end{array}\right\} = a \sin \theta$$

so that $$\text{work done by effort} = P \times a \sin \theta.$$

$$\left.\begin{array}{r}\text{Similarly, distance through which}\\ \text{load } W \text{ is moved in its own direction}\end{array}\right\} = b \sin \theta$$

so that $$\text{work done on load} = W \times b \sin \theta$$

Hence, $$\text{velocity ratio} = \frac{a \sin \theta}{b \sin \theta} = \frac{a}{b}.$$

## 9.4 The wheel and axle

The wheel and axle may be regarded as an adaptation of the lever to allow continuous rotation of the device about the pivot.

Fig. 9.2 shows a wheel A, of radius $a$, and an axle B, of radius $b$, carried by a shaft C. A body having weight $W$ is attached to a cord E, wound around axle B, and a downward effort $P$ is applied to cord D wound around the rim of wheel A.

When the wheel and axle make one revolution in an *anticlockwise* direction, a length, $2\pi b$, of cord E is wound on axle B, thereby lifting the load a distance $2\pi b$. At the same time, a length, $2\pi a$, of cord D is unwound from wheel A, enabling the effort to move a distance $2\pi a$.

Hence,
$$\text{velocity ratio} = \frac{\text{distance moved by effort}}{\text{distance moved by load}}$$
$$= \frac{2\pi a}{2\pi b} = \frac{a}{b}$$

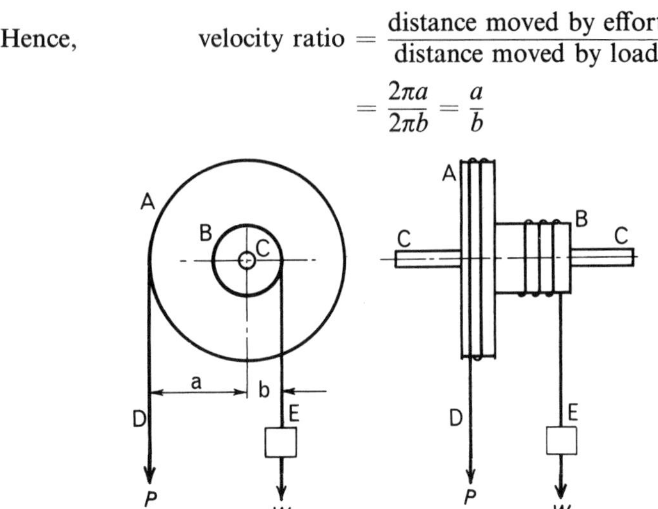


Fig. 9.2 The wheel and axle.

**Example 9.1** *In a certain wheel and axle machine, the diameters of the wheel and axle are* 450 *mm and* 60 *mm respectively. The efficiency is* 97 *per cent (or* 0·97 *per unit) when a body having a mass of* 40 *kg is being lifted. Calculate* (a) *the velocity ratio,* (b) *the ideal effort,* (c) *the actual effort and* (d) *the mechanical advantage.*

$$\text{Weight of body lifted} = W \simeq 40 \times 9{\cdot}81$$
$$= 392{\cdot}4 \text{ N.}$$

(*a*)
$$\text{Velocity ratio} = \frac{\text{diameter of wheel}}{\text{diameter of axle}}$$
$$= \frac{450\ [\text{mm}]}{60\ [\text{mm}]} = 7{\cdot}5.$$

(*b*) From expression (9.5), we have:

$$\text{ideal effort} = \frac{W}{\text{velocity ratio}}$$
$$= \frac{392{\cdot}4\ [\text{N}]}{7{\cdot}5} = 52{\cdot}3 \text{ N.}$$

(*c*) From expression (9.4), we have:

$$\text{efficiency} = \frac{\text{load}}{\text{effort}} \times \frac{1}{\text{velocity ratio}}$$

$$\therefore \qquad 0{\cdot}97 = \frac{392{\cdot}4\ [\text{N}]}{P} \times \frac{1}{7{\cdot}5}$$

$$\text{so that actual effort} = P = 54\ \text{N}.$$

Alternatively,

$$\text{actual effort} = \frac{\text{ideal effort}}{\text{efficiency}}$$

$$= 52{\cdot}3\ [\text{N}]/0{\cdot}97 = 54\ \text{N}.$$

(*d*)

$$\text{Mechanical advantage} = \frac{W}{P} = \frac{392{\cdot}4\ [\text{N}]}{54\ [\text{N}]} = 7{\cdot}275.$$

Alternatively, from expression (9.4), we have:

$$\text{mechanical advantage} = \text{velocity ratio} \times \text{efficiency}$$

$$= 7{\cdot}5 \times 0{\cdot}97 = 7{\cdot}275.$$

## 9.5 Inclined plane

As mentioned in section 4.1, the principle of the inclined plane was known at least 5000 years ago when it was employed in the construction of the Pyramids. The smooth inclined plane was discussed in Chapter 4, but in this section we shall consider the general case where friction may not be negligible.

Suppose a body having weight $W$ to be hauled up the whole of the inclined surface (fig. 9.3) by a force $P$ acting parallel to the plane.

Hence, work done by effort $= P \times \text{AB}$.

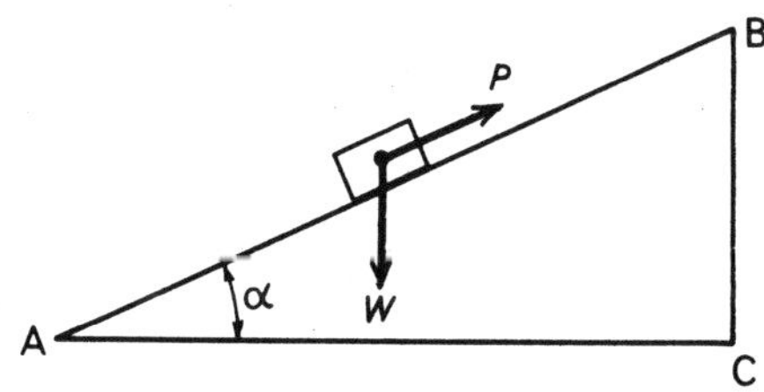


Fig. 9.3 The inclined plane.

While the load has been hauled a distance AB up the inclined plane, it has also been lifted through the vertical distance BC, so that work done on load $= W \times \mathrm{BC}$.

$$\text{Velocity ratio} = \frac{\mathrm{AB}}{\mathrm{BC}} = \frac{1}{\sin\alpha}$$

where $\alpha$ is the angle between the plane and the horizontal.

## 9.6 The screw and the screw-jack

The screw is really an inclined plane converted into a helical inclined path around the bolt on which the screw has been cut, thereby enabling the circular motion of a nut to be converted into a linear motion along the screwed portion of the bolt. This means that a small effort applied tangentially at the end of, say, a spanner moves through a distance equal to ($2\pi \times$ radius of effort) while the nut travels a distance equal to the lead of the thread, i.e. the distance between the centres of adjacent threads for single-start thread.

Hence,

$$\text{velocity ratio} = \frac{2\pi \times \text{radius of effort}}{\text{lead of thread}}$$

e.g. if the radius at which the effort is applied is, say, 200 mm and the lead of the thread is 1·8 mm,

$$\text{velocity ratio} = \frac{2\pi \times 200\ [\text{mm}]}{1{\cdot}8\ [\text{mm}]} = 698.$$

Allowing for an efficiency of, say, 15 per cent, we have from expression (9.4):

$$\text{mechanical advantage} = \text{velocity ratio} \times \text{efficiency}$$
$$= 698 \times 0{\cdot}15 = 105.$$

A device, using the principle of the screw, that is commonly employed for lifting heavy objects is the screw-jack shown in fig. 9.4.

An effort $P$ is applied tangentially at the end of an arm of radius $r$. For a screw having a right-hand thread, one turn of the effort in an anticlockwise direction, viewed from above the jack, lifts the load $W$ through a distance equal to the *lead*, $l$, of the screw. (In a single-start screw, the lead is equal to the pitch of the thread.)

For one revolution of the effort,

$$\text{distance moved by effort} = 2\pi r$$

and

$$\text{distance moved by load} = l$$

$\therefore$

$$\text{velocity ratio} = 2\pi r/l.$$

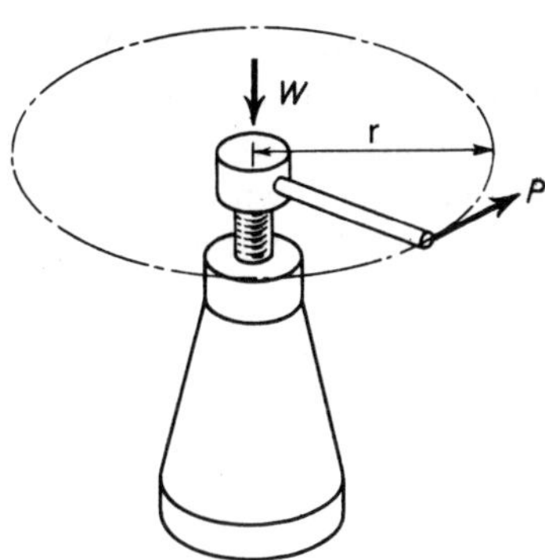


Fig. 9.4 The screw-jack.

**Example 9.2** *A body having a mass of* 200 *kg is resting on top of a screw-jack. The screw has a lead of* 8 *mm, and an effort of* 36 *N has to be applied tangentially at a radius of* 250 *mm to lift the load. Calculate* (a) *the velocity ratio,* (b) *the mechanical advantage and* (c) *the efficiency.*

$$\text{Weight of body supported on screw-jack} \simeq 200 \times 9{\cdot}81 = 1962 \text{ N.}$$

(*a*)
$$\text{Velocity ratio} = 2\pi r/l = 2\pi \times 250 \text{ [mm]}/8 \text{ [mm]} = 196.$$

(*b*)
$$\text{Mechanical advantage} = W/P = 1962 \text{ [N]}/36 \text{ [N]} = 54{\cdot}5.$$

(*c*)
$$\text{Efficiency} = \frac{\text{mechanical advantage}}{\text{velocity ratio}} = 54{\cdot}5/196 = 0{\cdot}278 \text{ per unit} = 27{\cdot}8 \text{ per cent.}$$

## 9.7 Pulleys

*A single-pulley system.* Fig. 9.5(a) shows a pulley block fitted with one pulley (or sheave), with a rope passing over the pulley and supporting at one end a body having weight $W$. This load can be lifted by applying an effort $P$ to the other end of the rope. With this simple arrangement, the velocity ratio is unity since the distance through which the effort is applied is exactly the same as that through which the load is raised.

The effort $P$ has to be greater than the load $W$ to allow for friction at the pulley, so that the mechanical advantage ($= W/P$) is less than unity. The only advantage of the single-pulley system is that a person, when hauling a rope downwards, is able to make use of his own weight when exerting the pull and thus finds it easier than to lift the load upwards.

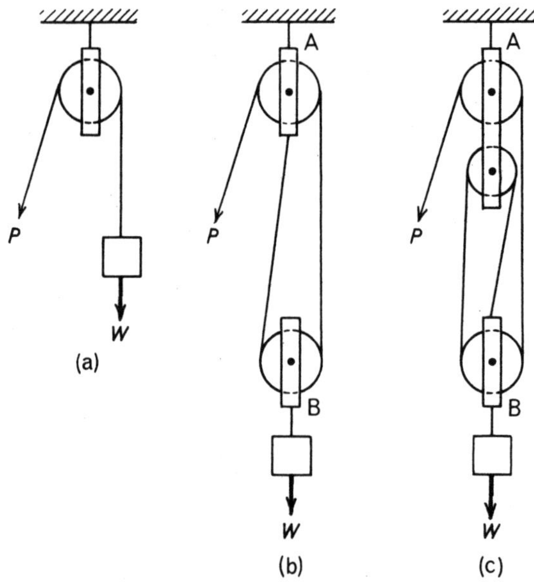


Fig. 9.5 Arrangements of pulley-blocks.

*A two-pulley system.* The velocity ratio can be increased by using more pulleys; e.g. in fig. 9.5(b) there are two pulley-blocks A and B, each with one pulley. A rope has one end attached to pulley-block A and passes round the pulleys of B and A, as shown. The body to be lifted is attached to pulley-block B.

If the load $W$ were raised 1 m by means of effort $P$, then *each* of the lengths of rope between the pulley-blocks would be shortened by 1 m and the effort $P$ would move 2 m. Hence,

$$\text{velocity ratio} = \frac{\text{distance moved by effort}}{\text{distance moved by load}}$$

$$= \frac{2\,[\mathrm{m}]}{1\,[\mathrm{m}]} = 2.$$

*A three-pulley system.* Fig. 9.5(c) shows pulley-block A fitted with two pulleys. These pulleys are usually on the same spindle, but for convenience of explanation, they are shown one above the other. With this arrangement, effort $P$ moves a distance of 3 m when the load is lifted 1 m, so that the velocity ratio is now 3.

In general, if $n$ be the *total* number of pulleys (or sheaves) on the two pulley-blocks,

$$\text{velocity ratio} = n.$$

**Example 9.3** *A body having a mass of* 30 *kg was lifted by an effort of* 95 *N by means of a pulley-block arrangement. The upper and lower blocks had* 3 *pulleys and* 2 *pulleys respectively. Calculate* (a) *the velocity ratio,* (b) *the mechanical advantage and* (c) *the efficiency.*

(*a*) $\text{Total number of pulleys} = 3 + 2 = 5,$

$\therefore$ $\text{velocity ratio} = 5.$

(*b*) $\text{Weight of body} \simeq 30 \times 9{\cdot}81 = 294{\cdot}3\ \text{N},$

$\therefore$ $\text{mechanical advantage} = \dfrac{W}{P} = \dfrac{294{\cdot}3\ [\text{N}]}{95\ [\text{N}]} = 3{\cdot}1.$

(*c*)
$$\text{Efficiency} = \frac{\text{mechanical advantage}}{\text{velocity ratio}}$$
$$= 3{\cdot}1/5 = 0{\cdot}62 \text{ per unit}$$
$$= 62 \text{ per cent.}$$

## 9.8 Belt and chain drives

A belt or a chain is used when a shaft has to be driven from a parallel shaft that is too far away for the use of gear wheels (section 9.9). Fig. 9.6 shows a belt drive in which A is the *driver* pulley and B the *driven* pulley. The transfer of motion from pulley A to the belt and again from the belt to pulley B is dependent upon friction at each area of contact between belt and pulley.

If $d_A$ and $d_B$ = diameters of pulleys A and B respectively and $n_A$ and $n_B$ = speeds, in revolutions/second, of A and B respectively,

then linear speed of rim of pulley A $= \pi d_A n_A$

and linear speed of rim of pulley B $= \pi d_B n_B$.

If there is no slipping, the linear speed of the rim of each pulley is the same as the speed of the belt, hence $$\pi d_A n_A = \pi d_B n_B$$

$$\therefore \quad \frac{\text{speed of driver pulley A}}{\text{speed of driven pulley B}} = \frac{\text{diameter of driven pulley B}}{\text{diameter of driver pulley A}} \qquad (9.7)$$

i.e. the speeds of the pulleys are inversely proportional to their diameters.

When a driving torque is applied to driver pulley A, the side of the belt approaching the pulley tightens and that leaving the pulley slackens. Thus, with the driver pulley rotating clockwise as in fig. 9.6, the lower side of the belt has a larger tension than the upper side.

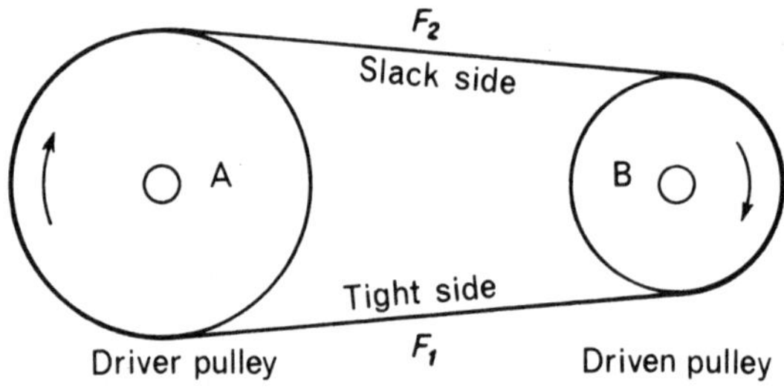


Fig. 9.6 A belt drive.

If $F_1$ and $F_2$ be the tensions in the tight and slack sides respectively of the belt,

effective force due to friction $= F_1 - F_2$

and power transmitted $=$ net force [newtons] $\times$ speed of belt [metres/second]

$$= (F_1 - F_2) \times \pi d_A n_A \text{ watts} \qquad (9.8)$$

$$\text{or} = (F_1 - F_2) \times \pi d_B n_B \text{ watts} \qquad (9.9)$$

These expressions are similar to expression (8.6) deduced for a brake test in section 8.6.

One of the main disadvantages of a belt is its liability to slip. This disadvantage can be avoided by the use of a chain whose links engage with teeth on the driver and driven wheels. One of the best-known examples is the use of a chain to transmit power from the pedals to the rear wheel of a bicycle.

Since the number of teeth on each wheel is proportional to the diameter of the wheel,

$$\frac{\text{speed of driver wheel}}{\text{speed of driven wheel}} = \frac{\text{diameter of driven wheel}}{\text{diameter of driver wheel}}$$
$$= \frac{\text{number of teeth on driven wheel}}{\text{number of teeth on driver wheel}}.$$

**Example 9.4** *A belt-driven pulley has a diameter of* 500 *mm and its speed is* 300 *rev/min. The tensions in the two sides of the belt are* 1800 *N and* 400 *N respectively. Calculate the power transmitted by the belt.*

$$\text{Effective friction force} = 1800 - 400 = 1400\ \text{N}.$$

$$\begin{aligned}\text{Linear speed of belt} &= \pi \times 0{\cdot}5\ [\text{m}] \times (300/60)\ [\text{rev/s}]\\ &= 7{\cdot}85\ \text{m/s},\\ \therefore\quad \text{power transmitted} &= 1400\ [\text{N}] \times 7{\cdot}85\ [\text{m/s}]\\ &= 11\,000\ \text{W} = 11\ \text{kW}.\end{aligned}$$

Alternatively,

$$\begin{aligned}\text{speed} &= 300/60 = 5\ \text{rev/s}\\ \text{and torque} &= 1400\ [\text{N}] \times 0{\cdot}25\ [\text{m}] = 350\ \text{N m},\end{aligned}$$

hence, from expression (8.4), we have:

$$\begin{aligned}\text{power transmitted} &= 2\pi nT\\ &= 2\pi \times 5\ [\text{rev/s}] \times 350\ [\text{N m}]\\ &= 11\,000\ \text{W} = 11\ \text{kW}.\end{aligned}$$

## 9.9 Gear wheels

Gear wheels are used to transmit motion and power from one shaft to a parallel shaft in close proximity, and to enable the speed of rotation to be stepped up or down to suit specific requirements, e.g. to reduce the high speed of an electric motor to the relatively low speed of a lathe.

A gear wheel has a number of specially-shaped teeth around its periphery. These teeth mesh with similar teeth on a second wheel as shown in fig. 9.7, where wheel A (the *driver*) drives wheel B (the *follower*).

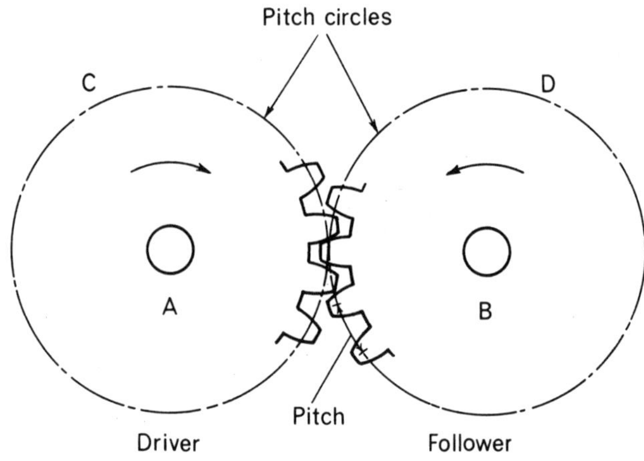


Fig. 9.7 Gear wheels.

In order that the teeth may mesh correctly, the pitch of the teeth, i.e. the peripheral distance between the centres of adjacent teeth, measured on the pitch circles C and D, must be the same for the two wheels. Hence, for a given pitch, the number of teeth is proportional to the diameter of the pitch circle;

$$\therefore \quad \frac{\text{number of teeth on wheel A}}{\text{number of teeth on wheel B}} = \frac{\text{circumference of wheel A}}{\text{circumference of wheel B}}.$$

If $n_A$ and $n_B$ be the speeds of wheels A and B respectively, in revolutions per second,

$$\text{peripheral speed of driver A, measured at pitch circle} = \text{circumference of A} \times n_A$$

$$\text{and peripheral speed of follower B, measured at pitch circle} = \text{circumference of B} \times n_B.$$

Since there can be no slip, the peripheral speed at the pitch circles is the same for the two wheels,

i.e. $\quad \text{circumference of A} \times n_A = \text{circumference of B} \times n_B$

$$\therefore \quad \frac{\text{speed of driver A}}{\text{speed of follower B}} = \frac{\text{circumference of follower}}{\text{circumference of driver}}$$

$$= \frac{\text{number of teeth on follower}}{\text{number of teeth on driver}} \qquad (9.10)$$

i.e. the speeds of the gear wheels are inversely proportional to the number of teeth on the wheels.

It will be noted that gear wheels, A and B, in fig. 9.7 rotate in opposite directions. The directions of rotation of A and B can be arranged to be the same by introducing an idle wheel C between A and B as shown in fig. 9·8. If idler C rotates at $n_C$ revolutions per second,

$$\frac{n_A}{n_C} = \frac{\text{number of teeth on C}}{\text{number of teeth on A}}.$$

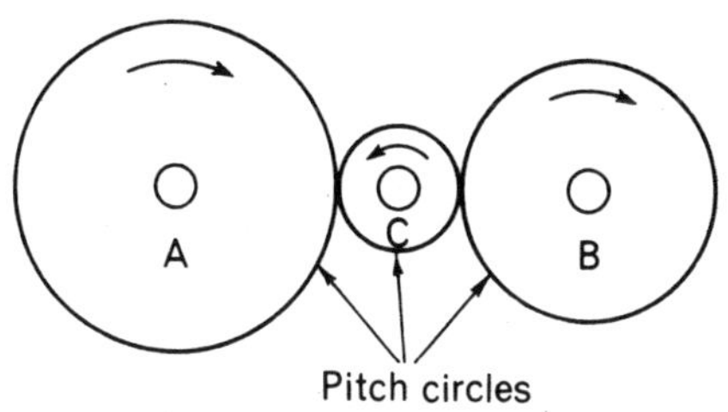


Fig. 9.8 Effect of an idle wheel.

Similarly, $$\frac{n_C}{n_B} = \frac{\text{number of teeth on B}}{\text{number of teeth on C}}$$

$$\therefore \quad \frac{n_A}{n_C} \times \frac{n_C}{n_B} = \frac{\text{number of teeth on C}}{\text{number of teeth on A}} \times \frac{\text{number of teeth on B}}{\text{number of teeth on C}}$$

hence $$\frac{n_A}{n_B} = \frac{\text{number of teeth on B}}{\text{number of teeth on A}}$$

i.e. the gear ratio of the gear train ABC is independent of the number of teeth on the idler.

**Example 9.5** *In the compound gear train shown in fig. 9.9, gear wheels B and C are rigidly attached to an idle shaft. Wheels A, B and C have* 60, 20 *and* 50 *teeth respectively, and the speed of A is* 200 *rev/min. Calculate the number of teeth on wheel D so that its speed may be* 1000 *rev/min.*

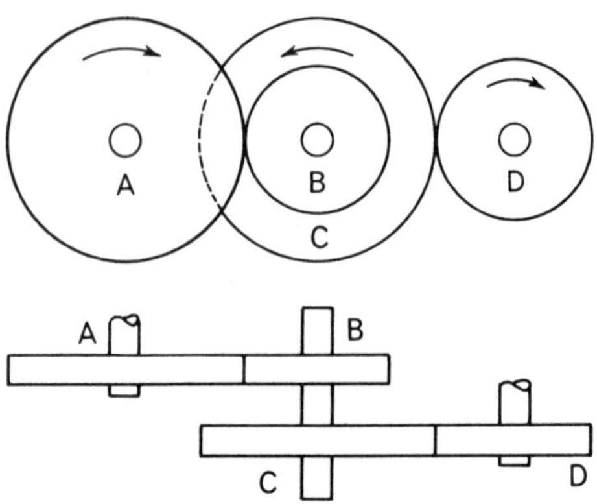


Fig. 9.9 A compound gear train.

$$\frac{\text{Speed of B}}{\text{Speed of A}} = \frac{\text{number of teeth on A}}{\text{number of teeth on B}} = \frac{60}{20} = 3.$$

$\therefore$ speed of B $= 3 \times 200$ [rev/min] $= 600$ rev/min.

Since gear wheels B and C are fixed to the same shaft, their speeds must be the same,

$\therefore$ speed of C $= 600$ rev/min.

Also,
$$\frac{\text{speed of C}}{\text{speed of D}} = \frac{\text{number of teeth on D}}{\text{number of teeth on C}}$$

i.e.
$$\frac{600}{1000} = \frac{\text{number of teeth on D}}{50}$$

$\therefore$ number of teeth on D $= 30$.

## *Summary of Chapter* 9

Several simple machines have been discussed which enable a small effort to lift a relatively large load. Also, transmission of motion and power by belt, chain and gear wheels have been described.

By the Principle of Conservation of Energy,

work done by effort = work done on load + work done in overcoming friction

$$\text{Mechanical advantage} = \frac{\text{load}}{\text{effort}} \tag{9.1}$$

$$\text{Velocity ratio} = \frac{\text{distance moved by effort}}{\text{distance moved by load}} \qquad (9.2)$$

$$\text{Efficiency} = \frac{\text{work done on load}}{\text{work done by effort}} \qquad (9.3)$$

$$= \frac{\text{mechanical advantage}}{\text{velocity ratio}} \qquad (9.4)$$

$$\text{Ideal effort} = \frac{\text{load}}{\text{velocity ratio}} \qquad (9.5)$$

$$\text{Ideal mechanical advantage} = \frac{\text{load}}{\text{ideal effort}}$$

$$= \text{velocity ratio} \qquad (9.6)$$

For belt drive,

$$\frac{\text{speed of driver pulley}}{\text{speed of driven pulley}} = \frac{\text{diameter of driven pulley}}{\text{diameter of driver pulley}} \qquad (9.7)$$

and

$$\text{power transmitted} = (F_1 - F_2) \times \pi dn \qquad (9.8)$$

For gear wheels,

$$\frac{\text{speed of driver}}{\text{speed of follower}} = \frac{\text{number of teeth on follower}}{\text{number of teeth on driver}} \qquad (9.10)$$

## EXAMPLES 9

1. In a certain lifting machine, the effort moves 0·3 m for every millimetre lift of the load. If the efficiency of the machine is 45 per cent when the load is 6 kN, calculate (*a*) the velocity ratio, (*b*) the effort required and (*c*) the mechanical advantage.
2. In a certain machine, an effort of 40 N was required to raise a body having a mass of 50 kg. When the body was raised a distance of 30 mm, the effort moved through a distance of 540 mm. Calculate (*a*) the velocity ratio, (*b*) the mechanical advantage, (*c*) the efficiency, (*d*) the ideal effort and (*e*) the ideal mechanical advantage.
3. A lever AB, 3 m long, is pivotted at a point 0·4 m from end A. Calculate the value of the load, in kilonewtons, at end A that can be raised by a downward effort of 500 N applied at right angles to the lever at end B. Assume the weight of the lever and the friction at the pivot to be negligible. Also calculate the mechanical advantage of the arrangement.
4. A crowbar, 2 m long, is used to move an object at one end of the bar. If the force required to move the object is 800 N and the maximum effort that can be exerted at the other end of the lever is 120 N, calculate the maximum distance of the fulcrum from the end of the bar at which the load is located.

5. The diameters of the wheel and axle (fig. 9.2) are 300 mm and 75 mm respectively. When a body having a weight of 300 N is being lifted, the efficiency is 80 per cent. Calculate (*a*) the velocity ratio, (*b*) the effort required, (*c*) the mechanical advantage, (*d*) the useful work done when the load is raised 20 mm and (*e*) the corresponding work done by the effort.
6. In a wheel-and-axle machine, the wheel has a diameter of 200 mm and the axle has a diameter of 40 mm. Calculate (*a*) the effort required to lift a mass of 15 kg, if the efficiency is 0·9 per unit, (*b*) the velocity ratio, (*c*) the mechanical advantage, (*d*) the ideal effort, (*e*) the effort to overcome friction and (*f*) the ideal mechanical advantage.
7. A simple winch has a drum, 250 mm diameter, around which a rope is wound. The handle, which operates at a radius of 400 mm, is connected directly to the drum. The winch is used to raise a body having a mass of 50 kg. Calculate (*a*) the velocity ratio, (*b*) the tangential effort required at the handle if the efficiency of the winch is 80 per cent.
8. A block of concrete having a mass of 200 kg is hauled up an inclined plane having a gradient of 1 in 5. If the efficiency is 35 per cent, calculate (*a*) the effort required if its direction is parallel to the plane, (*b*) the velocity ratio and (*c*) the mechanical advantage.
9. A body having a weight of 3 kN is dragged up a surface inclined at an angle of 20° to the horizontal. The effort, acting parallel to the surface, is 1·3 kN. Calculate (*a*) the velocity ratio, (*b*) the mechanical advantage, (*c*) the efficiency, (*d*) the work done in hauling the body a distance of 3 m and (*e*) the corresponding useful work done.
10. The screw of a certain screw-jack has a lead of 10 mm. An effort of 120 N is applied tangentially at the end of a bar having a radius of 250 mm. If the efficiency is 0·3 per unit, calculate (*a*) the load, in kilonewtons, (*b*) the velocity ratio and (*c*) the mechanical advantage.
11. A screw-jack is used to lift a mass of 500 kg. The lead of the screw is 12 mm. Calculate the force required at the end of a 300-mm arm, assuming the efficiency to be 28 per cent. Also determine the velocity ratio and the mechanical advantage.
12. A lifting tackle consists of two pulley blocks. Each block has two pulleys (or sheaths). One end of the cord is attached to the upper pulley block. Sketch the arrangement and calculate (*a*) the velocity ratio, (*b*) the effort required to lift a mass of 120 kg, assuming the efficiency to be 85 per cent and (*c*) the mechanical advantage.
13. A system of pulleys consists of an upper block fitted with three pulleys and a lower block fitted with two pulleys. An effort of 240 N is required to raise a body having a mass of 85 kg. Sketch the arrangement and calculate (*a*) the velocity ratio, (*b*) the mechanical advantage, (*c*) the efficiency, (*d*) the work done in lifting the body 50 mm and (*e*) the corresponding work done by the effort.
14. A pulley, 150 mm in diameter and rotating at 900 rev/min, is belt-coupled to another pulley. What should be the diameter of the latter pulley in order that it may rotate at 280 rev/min?
15. A shaft A has a pulley 120 mm in diameter from which shaft B is to be driven by a belt. Shaft B has to drive a shaft C, the pulley on the latter

being 400 mm diameter. The speeds of shafts A, B and C are to be 900, 500 and 180 rev/min respectively. Calculate the diameters of the two pulleys on shaft B.

16. Energy is transmitted from the shaft of an engine by a belt passing over a pulley having a diameter of 600 mm. The effective pull on the belt is 2·8 kN. Calculate (*a*) the torque, in newton metres, and (*b*) the power transmitted when the speed is 4 rev/s.
17. The pull on the tight side of a belt is 720 N and that on the slack side is 80 N. The driving pulley has a diameter of 460 mm and is rotating at 480 rev/min. Calculate (*a*) the torque and (*b*) the power transmitted.
18. A motor drives a line of shafting by means of a chain drive. The wheel on the motor has 23 teeth and that on the shafting 82 teeth. If the speed of the motor is 900 rev/min, what is the speed of the shafting?
19. By applying a force of 200 N to the pedal of his bicycle, a cyclist exerts a turning moment of 35 N m on the crank. The bicycle has 48 teeth on the front chain-wheel and 18 on the rear sprocket-wheel. The wheels are 700 mm in diameter. Calculate (*a*) the effective length of the crank, (*b*) the speed of the rear wheel, in revolutions/minute, when the pedals are being driven at 180 rev/min, (*c*) the velocity ratio of the machine, i.e. the ratio of the linear speed of the pedals to the linear speed of the outermost surface of the rear wheel, and (*d*) the force, parallel to the road, acting on the machine. Neglect any losses.
20. Two gear wheels, A and B, have 80 and 30 teeth respectively. If the speed of A is 600 rev/min, what is the speed of B?
21. Three gear wheels, A, B and C, are on three parallel shafts. A has 14 teeth and meshes with B, which has 37 teeth and meshes with C which has 49 teeth. If the speed of A is 1300 rev/min, what is the speed of C?
22. In a double-reduction gear train (fig. 9.9), wheel A has 24 teeth and rotates at 800 rev/min. Wheels B and C have 96 and 36 teeth respectively and are rigidly attached to each other. If wheel D has 60 teeth, calculate its speed.

## ANSWERS TO EXAMPLES 9

1. 300, 44·4 N, 135.
2. 490·5 N, 18, 12·26, 68·1 per cent, 27·25 N, 18.
3. 3·25 kN, 6·5.
4. 0·261 m.
5. 4, 93·75 N, 3·2, 6 J, 7·5 J.
6. 32·7 N, 5, 4·5, 29·43 N, 3·27 N, 5.
7. 3·2, 192 N.
8. 1120 N, 5, 1·75.
9. 2·92, 2·31, 79 per cent, 3900 J, 3080 J.
10. 5·65 kN, 157, 47·1.
11. 111·8 N, 157, 44.
12. 4, 346 N, 3·4.
13. 5, 3·48, 69·6 per cent, 41·7 J, 60 J.
14. 482 mm.
15. 216 mm, 144 mm.
16. 840 N m, 21·1 kW.
17. 147·2 N m, 7·4 kW.
18. 252·5 rev/min.
19. 175 mm, 480 rev/min, 0·1875, 37·5 N.
20. 1600 rev/min.
21. 371·4 rev/min.
22. 120 rev/min.

CHAPTER 10

# Stress and strain

## 10.1 Stress

When a material has a force exerted on it, the material is said to be *stressed* or in a state of *stress*. If a rod is subjected to a *tension*, the force per unit area of cross-section of the rod is referred to as *tensile stress*. If the rod is subjected to compression, the force per unit area of cross-section is termed a *compressive stress*. Hence, in general:

$$\text{stress} = \frac{\text{force}}{\text{cross-sectional area}} \tag{10.1}$$

Tensile and compressive stresses are sometimes referred to as *normal stresses* since they act at right angles to the cross-sectional area which is used for calculating the value of the stress.

The *SI unit of stress* is the *newton per square metre* (symbol, $N/m^2$). At the 1971 meeting of the CGPM, it was decided to adopt the term *pascal* (symbol, Pa) for this unit, in memory of the French philosopher Blaise Pascal (1623–62) who carried out many brilliant experiments in hydrostatics and pneumatics.

It is often more convenient to express stress in kilonewtons per square metre ($kN/m^2$) or in meganewtons per square metre ($MN/m^2$), where

$$1\ \text{kN/m}^2 \text{ or } 1\ \text{kPa} = 10^3\ \text{N/m}^2 \text{ or } 10^3\ \text{Pa}$$

and

$$1\ \text{MN/m}^2 \text{ or } 1\ \text{MPa} = 10^6\ \text{N/m}^2 \text{ or } 10^6\ \text{Pa}.$$

**Example 10.1** *A tie-bar has a cross-sectional area of* 125 $mm^2$ *and is subjected to a pull of* 10 *kN. Calculate the stress in meganewtons per square metre* (*or megapascals*).

$$\text{Force} = 10\ \text{kN} = 10\,000\ \text{N}.$$

$$\text{Cross-sectional area} = 125\ \text{mm}^2 = 125 \times 10^{-6}\ \text{m}^2,$$

$$\therefore \quad \text{stress} = \frac{10\,000\ [\text{N}]}{125 \times 10^{-6}\ [\text{m}^2]} = 80 \times 10^6\ \text{N/m}^2$$

$$= 80\ \text{MN/m}^2 \text{ or MPa}.$$

**Example 10.2** *A wire,* 1·5 *mm diameter, supports a mass of* 60 *kg. Calculate the stress.*

$$\text{Tension in wire} = \text{weight of the mass}$$
$$\simeq 60 \times 9{\cdot}81 = 588{\cdot}6 \text{ N.}$$

$$\text{Cross-sectional area} = (\pi/4) \times (1{\cdot}5)^2 = 1{\cdot}767 \text{ mm}^2$$
$$= 1{\cdot}767 \times 10^{-6} \text{ m}^2$$

$$\therefore \quad \text{stress} = \frac{588{\cdot}6 \text{ [N]}}{1{\cdot}767 \times 10^{-6} \text{ [m}^2\text{]}} = 333{\cdot}5 \times 10^6 \text{ N/m}^2$$
$$= 333{\cdot}5 \text{ MN/m}^2 \text{ or MPa.}$$

## 10.2 Pressure

When the tyre of a bicycle is inflated, the compressed air exerts an outward thrust on the inner surface of the tyre, and the thrust per unit area is termed the *pressure*. The SI unit of pressure is the *newton per square metre* or *pascal*, exactly the same as for stress.

Pressure is sometimes expressed in *bars*, where 1 bar $= 10^5$ N/m$^2$. A pressure of 1 bar is approximately the same as the atmospheric pressure at sea level. The atmospheric pressure known as the *standard atmospheric pressure* is 101 325 N/m$^2$ and can therefore be expressed as 1·013 25 bars or 1013·25 millibars. The millibar is the unit generally used in the measurement of meteorological pressures.

The term *pressure* generally refers to the thrust per unit area exerted by *fluids* such as water, air, steam, etc., whereas the term *stress* refers to the force per unit area exerted in a solid body. Sometimes the term 'pressure' is used to denote the total thrust on a surface; but to avoid confusion, it is better in such a case to use the term 'thrust' or 'force'.

**Example 10.3** *Steam at a pressure of* 1·5 *MPa is applied to a piston having a diameter of* 200 *mm. Calculate* (a) *the thrust on the piston and* (b) *the work done when the piston moves through a distance of* 350 *mm, assuming the steam pressure to remain constant.*

(*a*)

$$\text{Area of piston} = (\pi/4) \times (200)^2 = 31\,400 \text{ mm}^2$$
$$= 0{\cdot}0314 \text{ m}^2.$$

$$\text{Pressure on piston} = 1{\cdot}5 \text{ MPa} = 1{\cdot}5 \times 10^6 \text{ Pa,}$$
$$\therefore \quad \text{thrust on piston} = 1{\cdot}5 \times 10^6 \text{ [Pa]} \times 0{\cdot}0314 \text{ [m}^2\text{]}$$
$$= 47\,100 \text{ N} = 47{\cdot}1 \text{ kN.}$$

(*b*) Displacement of piston $= 350$ mm $= 0{\cdot}35$ m,

$\therefore$ work done $= 47\,100\ [\text{N}] \times 0{\cdot}35\ [\text{m}]$

$= 16\,500$ J $= 16{\cdot}5$ kJ.

## 10.3 Strain

When a rod or a wire is pulled, it stretches; and the total stretch or elongation, expressed as a fraction of the unstretched length, is termed *direct strain* or merely *strain* when it is obvious that the change in length is due to tension or compression. Thus, if the pull on a rod having an unstretched length of 1 m produces an extension of 2·5 mm,

$$\text{strain} = \frac{\text{extension}}{\text{original length}} \tag{10.2}$$

$$= \frac{2{\cdot}5\ [\text{mm}]}{(1 \times 1000)\ [\text{mm}]} = 0{\cdot}0025.$$

It will be noticed that the extension and the original length must be expressed in the same unit. In the above expression, they are both expressed in millimetres, but the result would be exactly the same if they were both expressed in, say, metres, thus:

$$\text{strain} = \frac{(2{\cdot}5/1000)\ [\text{m}]}{1{\cdot}0\ [\text{m}]} = 0{\cdot}0025.$$

Hence strain is merely a ratio and possesses no units.

## 10.4 Tensile test on a steel wire

The behaviour of various materials, when stretched by gradually increasing forces, may be studied experimentally in the laboratory.

The results of a test performed on a steel wire are given in the following table. The initial length of the wire was 907 mm and the initial cross-sectional area was 0·34 mm².

| Load [N] | 0 | 10 | 20 | 30 | 40 | 50 | 60 | 70 | 80 | 90 | 100 |
|---|---|---|---|---|---|---|---|---|---|---|---|
| Extension [mm] | 0 | 0·10 | 0·22 | 0·35 | 0·50 | 0·62 | 0·75 | 1·1 | 2·1 | 4·0 | 7·5 |

| Load [N] | 105 | 110 | 118 | 128 | 135 | 144 | 149 | 154 |
|---|---|---|---|---|---|---|---|---|
| Extension [mm] | 10·4 | 12·9 | 17·8 | 21·8 | 29·1 | 42·7 | 54 | 70 (broke) |

An examination of the tabulated figures shows that there are two stages in the stretching of the wire:

(*a*) *The Elastic Stage.* Up to a load of about 60 N, the extension is small and is proportional to the load, as shown in fig. 10.1. During this stage of the test, it is found that if the load is removed, the stretch disappears and the wire returns to its original length. The fact that during this elastic stage the extension is proportional to the load is characteristic of the elastic straining of materials.

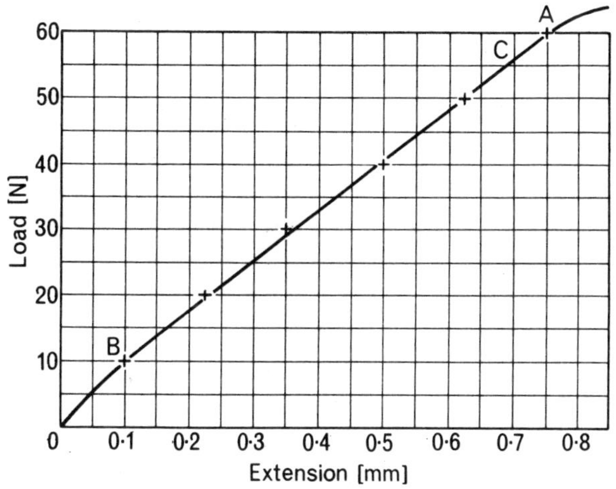


Fig. 10.1 Elastic extension of a wire.

(*b*) *The Plastic Stage.* The results for loads of 80 N upwards are plotted in fig. 10.2, where the extension is shown to a much smaller scale than in fig. 10.1.

It is evident from the table and from fig. 10.2 that for loads exceeding about 60 N, the extension for a given increase in load is much greater than for loads below 60 N. For instance, when the load is increased from 60 N to 149 N, the extension increases from 0·75 mm to 54 mm, i.e. when the load is increased about 2·5 times, the extension increases more than 70 times.

During this plastic stage of the test, it is found that when the load is removed, the wire does not return to its original length and is said to have a *plastic deformation* or *permanent set*. A metal that is able to undergo cold plastic deformation, usually as the result of tension, is said to be *ductile* or to possess *ductility*.

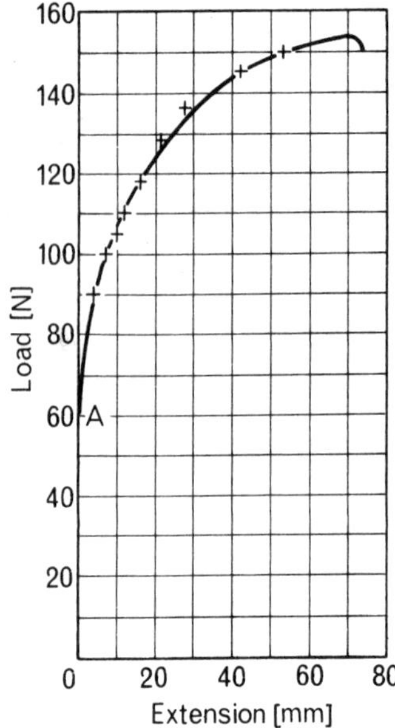


Fig. 10.2 Tensile test on a steel wire.

## 10.5 Hooke's law

Robert Hooke (1635–1703), a famous English physicist, architect and inventor, carried out a series of experiments with springs and wires and was the first to publish, in 1676, a clear statement to the effect that when a material is worked within its elastic range, *the strain is proportional to the stress.*

## 10.6 Modulus of elasticity (direct) or Young's modulus

For a material worked within its elastic range, we have from Hooke's law that:

$$\text{strain} \propto \text{stress},$$

i.e.

$$\frac{\text{stress}}{\text{strain}} = \text{a constant}.$$

This constant is termed *modulus of elasticity (direct)* or *Young's modulus*, after Thomas Young* who was the first to determine this constant for some materials.

* Thomas Young (1773-1829) was one of the most versatile geniuses in history. He studied languages and medicine and was a Professor of Physics at the Royal Institution, London. He was the first to translate the Egyptian hieroglyphics on the Rosetta Stone at the British Museum and was also famous for his share in establishing the undulatory theory of light.

The symbol for modulus of elasticity (direct) or Young's modulus is $E$,

$$E = \frac{\text{tensile or compressive stress}}{\text{strain}} \qquad (10.3)$$

i.e.

$$= \frac{\text{force/original cross-sectional area}}{\text{change of length/original length}}$$

Young's modulus may be regarded as a measure of the resistance which a material offers to extension or compression, i.e. the larger the value of $E$, the smaller is the extension or compression produced by a given stress.

When the value of Young's modulus is being determined from experimental results such as those shown in fig. 10.1, it is usually better not to use figures for load and extension at any one point on the graph but to take the *difference* in extension and the corresponding *difference* in load between two points, such as B and C in fig. 10.1, on a straight line drawn through the points determined experimentally. This is due to the difficulty in determining accurately the point representing zero load and zero extension.

Let us now proceed to calculate the value of $E$ from the graph of fig. 10.1:

$$\text{load at B} = 10 \text{ N},$$

$$\text{load at C} = 55 \text{ N},$$

$$\therefore \quad \text{increase in load} = 45 \text{ N}.$$

$$\text{Initial cross-sectional area} = 0{\cdot}34 \text{ mm}^2 \text{ (see section 10.4)}$$

$$= 0{\cdot}34 \times 10^{-6} \text{ m}^2,$$

$$\therefore \quad \text{increase in stress} = 45 \text{ [N]}/(0{\cdot}34 \times 10^{-6}) \text{ [m}^2\text{]}$$

$$= 132{\cdot}4 \times 10^6 \text{ N/m}^2 \text{ or Pa}.$$

$$\text{Extension at B} = 0{\cdot}10 \text{ mm}$$

and

$$\text{extension at C} = 0{\cdot}69 \text{ mm},$$

$$\therefore \quad \text{increase in extension} = 0{\cdot}59 \text{ mm}.$$

$$\text{Initial length of wire} = 907 \text{ mm (see section 10.4)}$$

$$\therefore \quad \text{increase in strain} = 0{\cdot}59 \text{ [mm]}/907 \text{ [mm]}$$

$$= 0{\cdot}65 \times 10^{-3}.$$

Hence $E = \dfrac{\text{increase in stress}}{\text{increase in strain}} = \dfrac{132{\cdot}4 \times 10^6\ [\text{N/m}^2]}{0{\cdot}65 \times 10^{-3}}$

$= 204 \times 10^9$ N/m² or Pa

$= 204\,000$ MN/m² or MPa

$= 204$ giganewtons*/metre² (or GN/m² or GPa).

**Example 10.4** *A mild-steel rod,* 4 *m long and* 30 *mm in diameter, carries a tensile load of* 100 *kN. Calculate the extension, assuming* $E = 200\ GPa$.

$$\text{Cross-sectional area} = (\pi/4) \times (30)^2 = 706{\cdot}5\ \text{mm}^2$$
$$= 706{\cdot}5 \times 10^{-6}\ \text{m}^2,$$
$$\text{tensile load} = 100\ \text{kN} = 100\,000\ \text{N},$$
$$\therefore \quad \text{stress} = 100\,000\ [\text{N}]/(706{\cdot}5 \times 10^{-6})\ [\text{m}^2]$$
$$= 1{\cdot}415 \times 10^8\ \text{Pa}.$$
$$E = 200\ \text{GPa} = 200 \times 10^9\ \text{Pa}.$$

Substituting in expression (10.3), we have:

$$200 \times 10^9\ [\text{Pa}] = \frac{1{\cdot}415 \times 10^8\ [\text{Pa}]}{\text{strain}}$$
$$\therefore \quad \text{strain} = 7{\cdot}075 \times 10^{-4}$$
$$\text{and} \quad \text{extension} = 7{\cdot}075 \times 10^{-4} \times 4\ \text{m} = 2{\cdot}83 \times 10^{-3}\ \text{m}$$
$$= 2{\cdot}83\ \text{mm}.$$

**Example 10.5** *A member of a structure is* 5 *m long and has a cross-sectional area of* 900 *mm². What is the greatest pull which can be applied if the elongation is not to exceed* 2·8 *mm? Assume* $E = 220\ GPa$.

$$\text{Maximum allowable strain} = \frac{2{\cdot}8\ [\text{mm}]}{(5 \times 1000)\ [\text{mm}]}$$
$$= 0{\cdot}56 \times 10^{-3}$$
$$\text{and} \quad E = 220\ \text{GPa} = 220 \times 10^9\ \text{Pa}.$$

Substituting in expression (10.3), we have:

$$220 \times 10^9\ [\text{Pa}] = \frac{\text{maximum allowable stress}}{0{\cdot}56 \times 10^{-3}}$$
$$\therefore \quad \text{maximum allowable stress} = 123{\cdot}2 \times 10^6\ \text{Pa}.$$

* *Giga* is derived from a Greek word meaning giant, and dictionaries state that its pronunciation should be the same as 'giga' in gigantic.

$$\text{Cross-sectional area} = 900\ \text{mm}^2 = 900 \times 10^{-6}\ \text{m}^2,$$

$$\therefore \quad \text{maximum allowable pull} = (123{\cdot}2 \times 10^6)\ [\text{Pa}] \times (900 \times 10^{-6})\ [\text{m}^2]$$
$$= 110\,900\ \text{N} = 110{\cdot}9\ \text{kN}.$$

**Example 10.6** *An extensometer used on a tie-bar of a steel bridge shows an extension of* 0·11 *mm on a length of* 200 *mm. Assuming* $E = 210$ *GPa, calculate the stress in the tie-bar.*

$$\text{Strain} = \frac{0{\cdot}11\ [\text{mm}]}{200\ [\text{mm}]} = 0{\cdot}55 \times 10^{-3}$$

and $$E = 210\ \text{GPa} = 210 \times 10^9\ \text{Pa}.$$

Substituting in expression (10.3), we have:

$$210 \times 10^9\ [\text{Pa}] = \frac{\text{stress}}{0{\cdot}55 \times 10^{-3}}$$

$$\therefore \quad \text{stress} = 115{\cdot}5 \times 10^6\ \text{Pa}$$
$$= 115{\cdot}5\ \text{MPa}.$$

## 10.7 Stress-strain graph for mild steel

Fig. 10.3 shows a typical graph for a tensile test on a mild-steel specimen. The stresses have been calculated by dividing the loads by the *original* cross-sectional area of the test piece, and the strains by dividing the extensions by the *original* length selected for the test. The graph differs from a load-extension graph, such as fig. 10.2, only in the matter of scales.

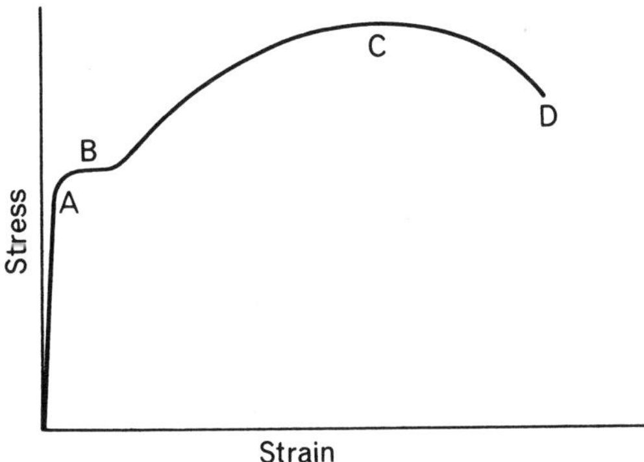


Fig. 10.3 Stress-strain graph for mild steel.

Point A in fig. 10.1 and 10.3 marks approximately the *limit of proportionality* of stress to strain, and is practically the *elastic limit,* i.e. the highest stress that can be applied without producing permanent deformation. For stresses up to the elastic limit, the material returns to its original length on removal of the load.

Portion B of the graph in fig. 10.3 gives the *yield stress*, namely the stress at which, in a tensile test, elongation of the test piece first occurs without increase of load. This plastic yielding is not found in all ductile materials but is a marked feature of the softer irons and steels.

The maximum stress, represented by C in fig. 10.3 is termed the *tensile strength*,* and is obtained by dividing the maximum load applied during a tensile test by the *original* cross-sectional area of the specimen and not by the reduced area of section after plastic extension. At about point C, the cross-sectional area at some point along the length of the test specimen begins to decrease rapidly and the load required to increase the extension beyond C consequently decreases until the specimen ultimately fractures at a load corresponding to point D.

**Example 10.7** *Taking the elastic limit in fig.* 10.1 *as* 60 *N and the maximum load as* 154 *N, calculate* (a) *the stress at the elastic limit,* (b) *the tensile strength and* (c) *the percentage final elongation. The wire had an initial length of* 907 *mm and an initial cross-sectional area of* 0·34 *mm*².

(*a*) Cross-sectional area $= 0{\cdot}34\ \text{mm}^2 = 0{\cdot}34 \times 10^{-6}\ \text{m}^2$,

$\therefore$ stress at elastic limit $= 60\ [\text{N}]/(0{\cdot}34 \times 10^{-6})\ [\text{m}^2]$

$= 176{\cdot}5 \times 10^6\ \text{Pa} = 176{\cdot}5\ \text{MPa}.$

(*b*) Tensile strength $= 154\ [\text{N}]/(0{\cdot}34 \times 10^{-6})\ [\text{m}^2]$

$= 453 \times 10^6\ \text{Pa} = 453\ \text{MPa}.$

(*c*) In the table given in section 10.4, it is stated that the elongation was 70 mm when the wire broke,

$$\therefore \quad \text{percentage final elongation} = \frac{70\ [\text{mm}]}{907\ [\text{mm}]} \times 100$$

$$= 7{\cdot}7 \text{ per cent.}$$

* This term replaces *ultimate tensile stress*, the use of which is not now recommended by the British Standards Institution.

**10.8 Proof stress**

The graph of fig. 10.4 is typical of the stress-strain relationship for many materials such as copper, etc. In this diagram, the strain has been plotted as a *percentage* of the original length, i.e. percentage strain = 100 × elongation/original length.

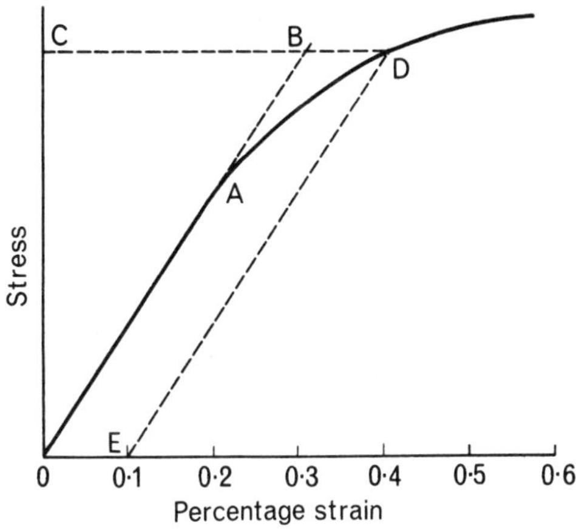


Fig. 10.4 Proof stress.

Point A represents the limit of proportionality for this specimen. Had the strain remained proportional to the stress, the stress-strain relationship would have been represented by the dotted line OB, which is OA extended beyond point A. If we draw a horizontal line CD at a stress OC, then CB represents the percentage strain if the strain had remained proportional to the stress, and CD represents the *actual* percentage strain. Consequently, BD represents the amount by which the percentage strain has departed from proportionality, i.e. BD is the *non-proportional* elongation of the specimen, expressed as a percentage of the original length.

Draw the dotted line DE parallel to OB, then:

$$OE = BD.$$

If OE is, say, 0·1 per cent, then OC is termed the *proof stress* to give a non-proportional elongation equal to 0·1 per cent of the original length of the specimen.

It is usual to express the proof stress as the stresses required to give a non-proportional elongation of 0·1, 0·2 and 0·5 per cent of the original length.

**Example 10.8** *A specimen has an initial gauge length of* 55 *mm and a cross-sectional area of* 150 *mm*$^2$. *A test on the specimen gave the following results:*

| Load [kN] | 0 | 10 | 20 | 30 | 35 | 38 | 40 |
|---|---|---|---|---|---|---|---|
| Extension [mm] | 0 | 0·075 | 0·15 | 0·23 | 0·30 | 0·38 | 0·6 |

*Determine the* 0·2 *per cent proof stress in meganewtons per square metre* (*or megapascals*).

The graph representing the above table is given in fig. 10.5.

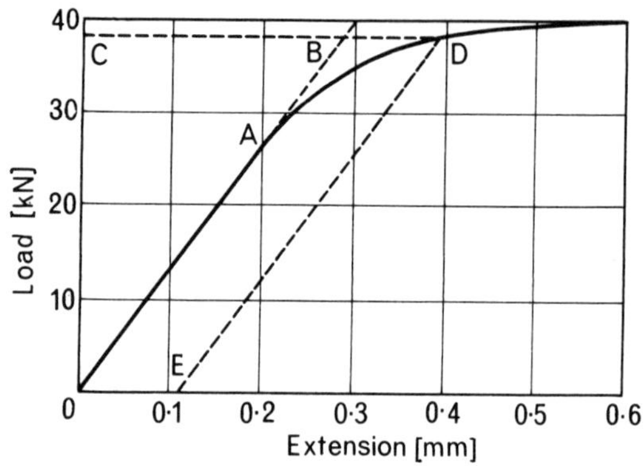


Fig. 10.5 Load-extension graph for Example 10.8

For 0·2 per cent non-proportional extension,

$$\text{actual non-proportional extension} = 55\ [\text{mm}] \times 0{\cdot}2/100 = 0{\cdot}11\ \text{mm}.$$

Hence, in fig. 10.5, from a point E corresponding to an extension of 0·11 mm, draw ED parallel to OA; and from point D, draw a horizontal line to cut the vertical axis at C. Then the *proof load* to give 0·2 per cent non-proportional extension is OC.

From the graph in fig. 10.5, OC = 38 kN = 38 000 N.

$$\text{Cross-sectional area} = 150\ \text{mm}^2 = 150 \times 10^{-6}\ \text{m}^2$$

and

$$\text{corresponding proof stress} = \frac{38\ 000\ [\text{N}]}{150 \times 10^{-6}\ [\text{m}^2]} = 253 \times 10^6\ \text{N/m}^2 = 253\ \text{MN/m}^2 \text{ or MPa.}$$

## 10.9 Tensile strength and factor of safety

The tensile strength of a material has already been referred to as the maximum tensile load (in a test) divided by the original cross-sectional area of the specimen. It is obvious that no such stress must be allowed in any part of a machine or structure. The *permissible stress* should, in fact, always be far below the tensile strength and also almost always below the stress corresponding to the elastic limit. The ratio of the tensile strength to the permissible tensile stress is termed the *factor of safety*.

i.e.

$$\text{factor of safety} = \frac{\text{tensile strength}}{\text{permissible tensile stress}} \qquad (10.4)$$

Hence,

$$\text{permissible tensile stress} = \frac{\text{tensile strength}}{\text{factor of safety}}$$

The factor of safety to be adopted depends upon many circumstances and it often covers many contingencies very imperfectly known.

The following table gives approximate values of the tensile strength and Young's modulus for several of the most commonly used materials, but materials having a particular name vary widely in their properties and often more precise information is required about the actual materials to be employed. This is particularly true in regard to cast irons and steels of widely differing kinds.

TENSILE STRENGTH AND YOUNG'S MODULUS

| Material | Tensile strength in MPa | Young's modulus in GPa |
|---|---|---|
| Wrought iron | 300 to 400 | 190 |
| Mild steel | 450 to 600 | 210 |
| Cast iron | 120 to 160 | 120 |
| Copper (hard drawn) | 300 to 350 | 110 |
| Aluminium (hard drawn) | 150 to 190 | 70 |

Compressive strength of cast iron, 600 to 900 MPa

## 10.10 Compression

Fig. 10.6 shows a short thick bar AB, with opposing forces *F* pushing or thrusting axially at its ends, i.e. in the opposite directions to the forces which cause tension. In such a case, the material is said to be under *compression* or *compressive stress*. The transmission of forces with action and reaction perpendicular to any cross-section at right angles to the axis is similar to that in the case of

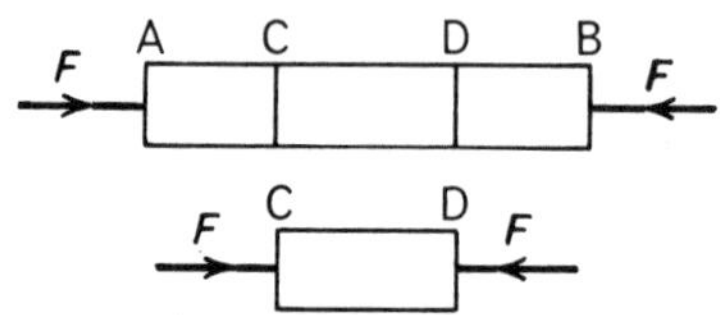


Fig. 10.6 Compressive forces.

tension, but the directions of the forces are reversed. For instance, a short length CD is acted upon by normal *inward* thrusts $F$ across the faces C and D as shown in fig. 10.6, and itself exerts *outward* thrusts, each equal to $F$, on the parts CA and DB.

The average compressive stress across a section is found by dividing force $F$ by the area of cross-section,

i.e. $$\text{compressive stress} = \frac{\text{compressive force}}{\text{cross-sectional area}}$$

The compressive strain is measured as a fraction of the original length,

i.e. $$\text{compressive strain} = \frac{\text{decrease in length}}{\text{original length}}$$

and the modulus of elasticity (direct) or Young's modulus

$$= E = \frac{\text{compressive stress}}{\text{compressive strain}}.$$

The value of Young's modulus for a given material is practically the same for compression as it is for tension.

A tensile force tends to straighten out any kink or curvature in a long rod, but a compressive force tends to increase such imperfections and to buckle the rod. Consequently a structural element used to resist compression, generally called a strut or a column, has to be rather broad in relation to its length, as in the case of tubes and many steel sections such as I, H, L and T shapes.

## 10.11 Shearing

A material is under the action of *shear stress* across a surface within it when there are forces tending to make the part on one side of the surface slide past the part on the other side of the surface. Thus a pin or rivet shown in fig. 10.7 is subject to shear stress across the

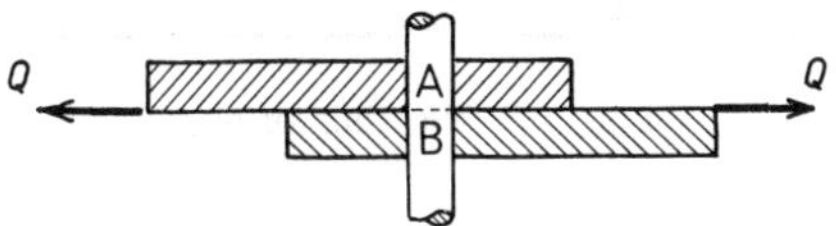


Fig. 10.7 Shear stress.

plane surface (dotted) AB. The upper part A of the pin tends to slide to the left relative to the lower part B. The tendency of the stress is to *shear* the pin into two parts, A and B. When shears are used for cutting metal or vegetation, actual shearing fracture takes place; but in parts of machines and structures, the *shear stress* is much below the *shear strength*. The average shear stress in fig. 10.7 is:

$$\frac{\text{shear force}}{\text{area of cross-section between A and B}}$$

A common case of the use of metal to resist shearing is to be found in rivetted joints and in instances where one part of a machine is capable of turning relative to another about a pin. For example, in the knuckle joint shown in fig. 10.8, the pin resists the pull at *two*

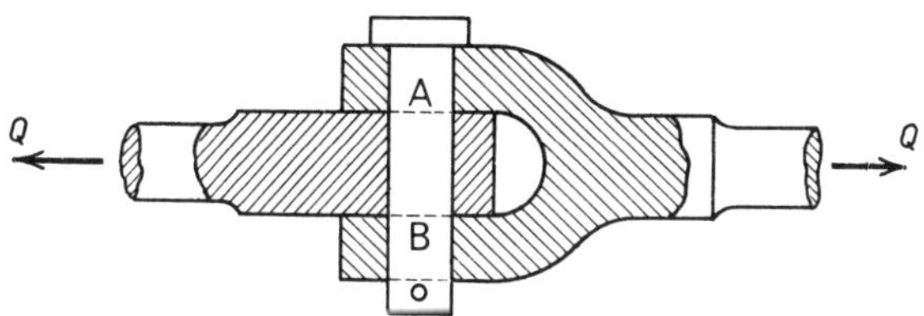


Fig. 10.8 Pin in double shear.

sections, A and B, so that the average shear stress in the pin is equal to force $Q$ divided by *twice* the area of cross-section of the pin. The pin is said to be subjected to *double shear*.

*Shear Strain.* The form of shear strain which results from a shear stress is illustrated in fig. 10.9. The left-hand view represents a number of piled metal plates which in the right-hand view have been pulled to the right. Suppose these plates were separated one

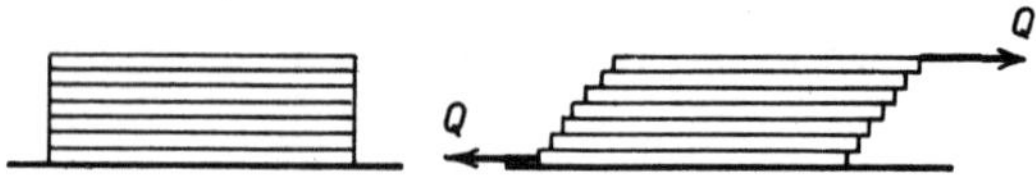


Fig. 10.9 'Shearing' of plates

from another by layers or films of thick oil. Then the movement to the right due to a pull $Q$ would be resisted by each film of oil and the films would be subject to shear stress. In a continuous material,

Fig. 10.10 Form of shear strain.

the form will be as illustrated in fig. 10.10, which might represent the visible shear strain produced in rubber. Simple shear strains in metals cannot be easily demonstrated.

## *Summary of Chapter* 10

Stress is the force per unit area of a solid body.
Pressure is the thrust per unit area exerted by a fluid on a surface.
Elastic strains are those which disappear on the removal of stress.
For most materials worked within the elastic limit, the strain is proportional to the stress (Hooke's law).

$$\text{Tensile (or compressive) stress} = \frac{\text{tensile (or compressive) force}}{\text{cross-sectional area}} \quad (10.1)$$

$$\text{Tensile (or compressive) strain} = \frac{\text{change in length}}{\text{original length}} \quad (10.2)$$

Modulus of elasticity (direct) or Young's modulus

$$= E = \frac{\text{tensile (or compressive) stress}}{\text{tensile (or compressive) strain}} \quad (10.3)$$

Proof stress is the tensile stress that produces a non-proportional elongation equal to a specified percentage of the original length.

$$\text{Tensile strength} = \frac{\text{maximum tensile force}}{\text{original cross-sectional area}}$$

$$\text{Factor of safety} = \frac{\text{tensile strength}}{\text{permissible tensile stress}} \qquad (10.4)$$

$$\text{Shear stress} = \frac{\text{shear force}}{\text{cross-sectional area}}$$

$$\text{Shear strength} = \frac{\text{maximum shear force}}{\text{cross sectional area}}$$

## EXAMPLES 10

1. What is the stress, in megapascals, in a wire, 1 mm diameter, under a load of 120 N?
2. Calculate the diameter of a steel rod (to the nearest millimetre) to carry a tensile load of 90 kN, if the stress is not to exceed 100 MPa.
3. What pull, in kilonewtons, can be borne by a tie rod, 40 mm diameter, if the stress is limited to 120 MPa?
4. A cylindrical concrete vertical pillar, having a diameter of 250 mm, supports a load of 10 kN. Calculate the stress in megapascals.
5. Calculate the area, in square millimetres, of a surface over which a thrust of 20 kN must be distributed in order that the pressure may be 2·8 MPa.
6. The floor of a tank is 1·2 m × 0·9 m. If the tank is 0·6 m high and full of water, calculate (*a*) the thrust on the floor surface and (*b*) the pressure on the floor. Assume 1 litre of water to have a mass of 1 kg.
7. Steam at a pressure of 2 MPa is applied to a piston having a diameter of 100 mm. Calculate (*a*) the thrust, in kilonewtons, on the piston and (*b*) the work done, in kilojoules, when the piston moves through a distance of 200 mm, assuming the pressure to remain constant.
8. Convert:
   (*a*) 20 N/mm$^2$ into megapascals;
   (*b*) 500 kPa into millinewtons per square millimetre;
   (*c*) 80 kN/mm$^2$ into gigapascals;
   (*d*) 50 GPa into kilonewtons per square millimetre;
   (*e*) 400 Pa into micronewtons per square millimetre.
9. If a spring is elastically stretched 60 mm by a load of 72 N, how much will it be stretched by a load of 40 N?
10. A helical spring carrying a load of 600 N is compressed by 20 mm. What would be the load required to compress the spring by 8 mm?
11. A steel rod, 10 mm diameter and 3 m long, stretches 1·8 mm for a pull of 10 kN. Calculate the value of Young's modulus for the steel.

12. A steel rod, 20 mm diameter, is subjected to a pull of 40 kN. Calculate (*a*) the stress, (*b*) the strain and (*c*) the total extension on a length of 200 mm. Assume $E = 200$ GPa.
13. A body having a mass of 1 Mg (or 1 t) is attached to the lower end of a vertical aluminium rod, 30 mm diameter and 2 m long. Calculate (*a*) the stress in the rod and (*b*) the extension, assuming $E = 70$ GPa.
14. When a metal block is suspended by a steel wire, 5 m long and 2·5 mm diameter, the elongation is 3 mm. Calculate the mass of the block, assuming $E = 200$ GPa.
15. Calculate the diameter of a steel rod to carry a load of 8 kN if the extension is not to exceed 0·04 per cent. Assume $E = 210$ GPa.
16. The following results were obtained on a mild-steel specimen having an initial gauge length of 50 mm and an initial cross-sectional area of 160 mm²:

| Load [kN] | 5 | 10 | 15 | 20 | 25 | 28 | 29·3 |
|---|---|---|---|---|---|---|---|
| Extension [mm] | 0·0066 | 0·0135 | 0·020 | 0·0266 | 0·0335 | 0·0505 | 0·075 |

Draw a load-extension graph and calculate the modulus of elasticity (direct) of the material.

17. In an experiment on a 2-m length of 1-mm diameter steel wire, the following results were obtained:

| Load [N] | 50 | 75 | 100 | 125 | 150 |
|---|---|---|---|---|---|
| Extension [mm] | 0·48 | 0·71 | 0·99 | 1·18 | 1·44 |

Calculate Young's modulus for the steel.

If a tie-bar in a bridge is made of the same material and is 4 m long, what will be the extension when the stress is 100 MPa?

18. The following results were obtained on a rod having a diameter of 12·5 mm and a length of 250 mm:

| Load [kN] | 12·5 | 25 | 37·5 | 40 | 41 | 42·5 | 45 |
|---|---|---|---|---|---|---|---|
| Extension [mm] | 0·12 | 0·24 | 0·36 | 0·55 | 1·8 | 3·85 | 5·1 |

| Load [kN] | 47·5 | 50 | 52·5 |
|---|---|---|---|
| Extension [mm] | 6·3 | 7·8 | 9·5 |

Plot load-extension graphs, using a scale of 1 mm to 0·5 kN and 1 mm to 0·02 mm for extensions up to 2 mm and 1 mm to 0·1 mm for extensions up to 10 mm. From these graphs, determine (*a*) the approximate value of the elastic limit, (*b*) Young's modulus and (*c*) the 0·2 per cent proof stress.

19. The following results were obtained from a tensile test on a copper wire having a diameter of 1·3 mm and a length of 2·3 m:

| Load [N] | 25 | 50 | 75 | 100 | 125 | 150 | 160 | 170 |
|---|---|---|---|---|---|---|---|---|
| Extension [mm] | 0·38 | 0·75 | 1·15 | 1·53 | 2·8 | 8·6 | 15·0 | 28·2 |

Plot the load-extension graph and determine (*a*) Young's modulus and (*b*) the 0·5 per cent proof stress.

20. A steel tie-rod, 10 m long and 3 mm diameter, is subjected to a pull of 600 N. Calculate (*a*) the stress, (*b*) the elongation if the modulus of

elasticity (direct) is 210 GPa and (*c*) the factor of safety if the tensile strength is 500 MPa.

21. The overhead conductor of an electric transmission line is a hard-drawn copper wire, 10 mm diameter, having a tensile strength of 320 MPa and a Young's modulus of 110 GPa. Assuming a factor of safety of 6, calculate (*a*) the maximum allowable pull on the wire and (*b*) the corresponding elongation on a span of 40 m.
22. A wrought-iron vertical pillar, 1·3 m long, has a cross-sectional area of 2000 mm$^2$. The pillar supports a load of 80 kN. Assuming $E = 180$ GPa, calculate the compression.
23. A hollow cast-iron vertical cylinder, 3 m long when unloaded, has an outer diameter of 150 mm and an inner diameter of 130 mm. Calculate (*a*) the maximum load that can be supported by the cylinder if the stress is not to exceed 80 MPa and (*b*) the corresponding decrease in length. Assume $E = 100$ GPa.
24. Two lengths of flat tie-bar (fig. 10·7) are connected together by a lap-riveted joint with three rivets, each 10 mm diameter. Calculate the maximum pull that can be applied to the tie-bars if the shear stress in the rivets is not to exceed 60 MPa.
25. If a knuckle-joint (fig. 10.8) is subjected to a pull of 100 kN, calculate the diameter of pin necessary in order that the average shear stress shall not exceed 80 MPa.

## ANSWERS TO EXAMPLES 10

1. 153 MPa.
2. 34 mm.
3. 150·8 kN.
4. 0·204 MPa.
5. 7140 mm$^2$.
6. 6·35 kN, 5·88 kPa.
7. 15·7 kN, 3·14 kJ.
8. 20 MPa; 500 mN/mm$^2$; 80 GPa; 50 kN/mm$^2$; 400 μN/mm$^2$.
9. 33·3 mm.
10. 240 N.
11. 212 GPa.
12. 127·5 MPa, $6{\cdot}375 \times 10^{-4}$, 0·1275 mm.
13. 13·9 MPa, 0·397 mm.
14. 60 kg.
15. 11 mm.
16. 234 GPa.
17. 265·5 GPa, 1·506 mm.
18. About 306 MPa, 212 GPa, about 330 MPa.
19. 114 GPa, about 119 MPa.
20. 84·9 MPa, 4·04 mm, 5·9.
21. 4·19 kN, 19·4 mm.
22. 0·289 mm.
23. 352 kN, 2·4 mm.
24. 14·1 kN.
25. 28·2 mm.

CHAPTER 11

# Heat and temperature

## 11.1 Heat

References have been made in previous chapters to the fact that work done in overcoming friction is converted into heat. There are numerous everyday actions that involve the same principle, e.g. rubbing one's hands together, striking a match, applying the brake on a vehicle, etc. In each case, the mechanical energy expended reappears in the form of thermal energy and produces an increase of temperature.

In order to appreciate what is meant by thermal energy, we have to consider the molecular structure of matter. All substances consist of molecules, the molecule of an element (e.g. hydrogen or copper) or of a compound (e.g. water or common salt) being the smallest particle of the element or compound that can normally exist alone. In a solid, the molecules are arranged in an orderly manner. Their mean positions relative to one another are fixed, but they oscillate about their mean positions at a speed that increases with increase in temperature. Hence molecules of solids possess kinetic energy owing to their vibration about their mean positions.

In a liquid, the molecules are able to move about at random. In a gas, the number of molecules in a given space is far less than in a liquid; consequently they are much freer to move about at random than they are in a liquid. The higher the temperature of the liquid or the gas, the greater is the speed of movement of the molecules. It follows that thermal energy is stored in a substance in the form of kinetic energy of the molecules of which the substance is composed. The gain of heat by a material is usually accompanied by the following effects:

(*a*) an increase in the temperature of the material,
(*b*) an increase in the volume of the material.

There may also be a change in the state of the material; for instance, a solid such as ice may be melted, or a liquid such as water may be vaporized.

## 11.2 Heat transfer

Heat may be transferred from a hot body to a cooler body in one or more of the following ways:

(*a*) by conduction,
(*b*) by convection,
(*c*) by radiation.

(*a*) *Conduction.* Conduction of heat takes place between bodies in actual contact if they are at different temperatures. It also takes place between parts of the same body if there is a difference of temperature between these parts.

The thermal conductivity of different materials varies widely; for instance, if one end of a glass rod, say 10 cm long, was heated to a high temperature, the other end could be held in the hand without any discomfort; but this would not be possible with a metal rod of the same length. In other words, metals are good conductors of heat, whereas materials such as glass, cork, cotton wool and gases are very poor conductors.

(*b*) *Convection.* Convection is the conveyance of heat by the actual movement of a hot fluid which may be a liquid or a gas. In most cases, convection takes place automatically; for example, in a hot-water heating system, water heated by a boiler expands so that its density becomes lower than that of the cold water. Consequently the hot water rises and thereby produces circulation of water in the pipes, enabling cold water to move towards the source of heat. Thus heat is transmitted by convection from the boiler to the place where that heat can be utilized, e.g. for warming a room.

Similarly, when air is heated by, say, a lamp or a radiator, it expands, and its consequent lower density causes it to rise, thereby producing an upward draught in the vicinity of the source of heat and a downward movement of air elsewhere in the room.

(*c*) *Radiation.* Radiation is the transmission of heat by wave or vibratory motion in the space between the source and the body on which the waves impinge. The nature of heat waves is the same as that of light waves and of radio waves—the only difference is the frequency of the waves, i.e. the number of vibrations or oscillations per second. Thermal radiation can be transmitted through a vacuum; i.e. the transference of energy by radiation is not dependent upon a material medium, as is the case with conduction and convection. For example, thermal energy from the sun reaches the earth's atmosphere by radiation through empty space.

Thermal radiation, when not impeded by any object, travels in straight lines, obeys the same laws as light and is therefore reflected from a polished metal surface. For a given surface temperature, the heat radiated per unit area is much less if that surface is polished than if it is black. As the temperature of a body is raised, the frequency range of the radiated energy increases, and a point is reached at which the body becomes luminous, i.e. it radiates visible (or light) energy as well as thermal energy.

## 11.3 Temperature

The temperature of a body relates to its degree of hotness or coldness and is independent of the size and physical nature of the body. Two bodies are said to be at the same temperature if no heat flows from one body to the other when they are brought into contact with each other.

The sense of touch is a very unreliable guide as to the temperature of a body. For instance, if we were to touch two plates—one of iron and the other of wood—which had been immersed for a few minutes in hot water, the iron plate would *feel* hotter than the wooden plate, though they were actually at the same temperature. This is due to the fact that heat is conducted far better by iron than by wood. It is therefore necessary to use an instrument, called a *thermometer*, to measure the temperature of a body.

Of the various types of thermometer, the mercury thermometer is by far the most commonly used for measuring temperatures ranging between the freezing and boiling points of water. The mercury thermometer consists of a glass rod G (fig. 11.1), having a small uniform bore C. At one end of the rod is a bulb B having a very thin glass wall so that heat can pass easily to and from the mercury contained in the bulb. For a given rise of temperature, mercury expands more than glass; consequently, when the mercury thermometer is placed in contact with a hot body, the expansion of the mercury in bulb B causes the level to rise in bore C, and a scale alongside C can be calibrated to read the temperature.

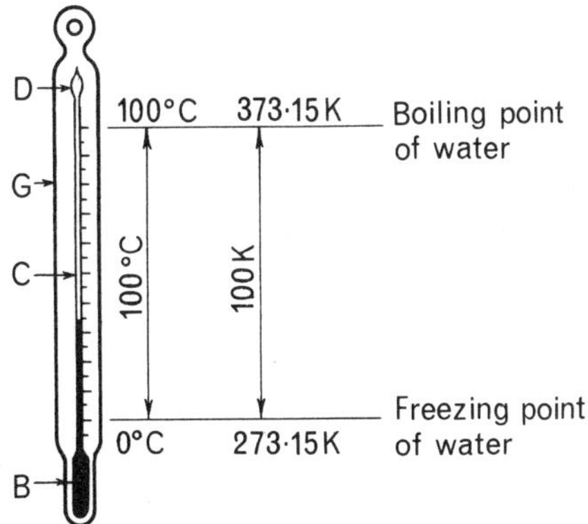


Fig. 11.1 Mercury thermometer.

The upper end of G is sealed when the mercury is heated sufficiently to fill the bore, so that when the mercury cools, the space above the mercury is a vacuum. A small bulb D is provided to prevent breakage of the thermometer due to the temperature rising a little above that for which the thermometer is calibrated.

## 11.4 Temperature scales

The SI unit of temperature is the *kelvin** (K) and the corresponding scale of temperature is referred to as the *thermodynamic temperature scale*. This scale is based upon thermodynamic considerations that are well beyond the scope of this book. All we need to state here is that, on this scale, the temperature at which ice melts under standard atmospheric pressure is exactly 273·15 K, as shown in fig. 11.1.

The thermodynamic scale suggests that if the temperature of a body could be reduced to zero, the body would possess no thermal energy. Such a condition can never be attained in practice.

* The SI unit for thermodynamic temperature has been changed from 'degree Kelvin' to 'kelvin' and the unit symbol from '°K' to 'K'. The name 'kelvin' commemorates Lord Kelvin (1824–1907), a famous British scientist and inventor who introduced the dynamical theory of heat.

A temperature change of one kelvin is precisely the same as the temperature change of one degree Celsius which is the unit of the *Celsius (or centigrade) scale* suggested by Anders Celsius, a Swedish physicist and astronomer, in 1742. On this scale, the temperature at which ice melts is taken as zero (0°C) and that at which water boils under standard atmospheric pressure is taken as 100°C, as shown in fig. 11.1. In other words, one degree Celsius is one-hundredth of the difference between the temperature at which ice melts (or water freezes) and that at which water boils under standard atmospheric pressure.

The temperature of melting ice under standard atmospheric pressure can therefore be stated as 0°C or 273·15 K and that of boiling water under standard atmospheric pressure can be stated as 100°C or 373·15 K. In general, if $t$ be the temperature of a body in degrees Celsius and $T$ be the same temperature in kelvins, then:

$$t = T - 273{\cdot}15 \qquad (11.1)$$

It should be noted that °C or K, can be used to denote temperature interval* and also the number of temperature units above or below a specified datum. For example, a temperature rise from, say, 20 degrees Celsius to 50 degrees Celsius can be expressed as 30°C or 30 K. If the temperature of a material is, say, 10 degrees Celsius, this can be expressed as 10°C or 283·15 K.

## 11.5 Quantity of heat

Since heat is a form of energy, the SI unit of heat is the *joule*, a quantity already discussed in sections 1.9 and 8.1. In practice, however, it is often more convnient to use the *kilojoule* (kJ) or the *megajoule* (MJ) as the unit of thermal energy.

Prof. James P. Joule, after whom the unit of energy is named, took an important part in establishing the exact relationship between the number of work units expended and the number of calorimetric units of heat produced, the calorimetric unit being the heat required to raise the temperature of unit mass of water by 1°C. If the unit of mass is the gram, the calorimetric unit is referred to

* The abbreviation 'deg', commonly used to express a temperature interval, is now regarded as obsolescent by the international General Conference of Weights and Measures.

as a *calorie*; but the calorie is *not* an SI unit and should no longer be used in thermal calculations.

Unfortunately, the amount of heat required to raise the temperature of a given mass of water by 1°C varies slightly over the 0–100°C range. The following table gives the heat required to raise the temperature of 1 kg of water by 20°C over this range:

| *Temperature range* | *Heat required* |
|---|---|
| 0–20°C | 83 900 J/kg |
| 20–40°C | 83 600 J/kg |
| 40–60°C | 83 600 J/kg |
| 60–80°C | 83 800 J/kg |
| 80–100°C | 84 200 J/kg |

It follows from the above data that, for most purposes, we can assume the heat required to raise the temperature of 1 kg of water by 1°C to be 4190 J or 4·19 kJ.

The relationship between the number of work units and the number of calorimetric units used to be referred to as the *mechanical equivalent of heat* or *Joule's equivalent*; but since both heat and work are *now* expressed in joules, the term 'mechanical equivalent of heat' has no significance. In other words, work and heat are interchangeable; and when work is transformed into heat or heat into work, the quantity of work, in joules, is equal to the quantity of heat, in joules. This relationship is known as the *First Law of Thermodynamics.*

## 11.6 Specific heat capacity

Different substances absorb different amounts of heat to raise a given mass of the substance by one degree. The quantity of heat required to raise the temperature of 1 kg of the substance by 1°C is termed the *specific heat capacity* of that substance. For instance, it was stated in section 11.5 that the quantity of heat required to raise the temperature of 1 kg of water by 1°C is approximately 4190 J. Hence we can say that the specific heat capacity of water is 4190 J/kg °C. It follows that if the temperature of $m$ kilograms of water is raised by $t$ degrees,

$$\text{quantity of heat required} = 4190\, mt \text{ joules} \qquad (11.2)$$

In general, if the specific heat capacity of a substance is $c$ joules per kilogram degree Celsius, the heat required to raise the temperature of $m$ kilograms of the substance by $t$ degrees

$$= mct \text{ joules} \tag{11.3}$$

The following table gives approximate values of the specific heat capacity for some well-known substances for a range of temperature between 0°C and 100°C:

| *Substance* | *Specific heat capacity* |
|---|---|
| Water | 4190 J/kg °C |
| Ice (0°C to —20°C) | 2100 J/kg °C |
| Copper | 390 J/kg °C |
| Iron | 500 J/kg °C |
| Aluminium | 950 J/kg °C |
| Brass | 370 J/kg °C |
| Dry air at standard atmospheric pressure | 1015 J/kg °C |

**Example 11.1** *Calculate the quantity of heat required to raise the temperature of* 6 *kg of water from* 10°*C to* 25°*C.*

Increase of temperature = 25 — 10 = 15°C.

Substituting in expression (11.2), we have:

$$\begin{aligned}\text{heat required} &= 4190\ [\text{J/kg °C}] \times 6\ [\text{kg}] \times 15\ [\text{°C}]\\ &= 377\,000\ \text{J} = 377\ \text{kJ}.\end{aligned}$$

**Example 11.2** *How many kilograms of copper can be raised from* 15°*C to* 60°*C by the absorption of* 80 *kJ of heat? Assume the specific heat capacity of copper to be* 390 *J/kg °C.*

Increase of temperature = 60 — 15 = 45°C.

Quantity of heat = 80 kJ = 80 000 J.

Substituting in expression (11.3), we have:

$$80\,000\ [\text{J}] = m \times 390\ [\text{J/kg °C}] \times 45\ [\text{°C}]$$

$$\therefore \quad m = 4{\cdot}56\ \text{kg}.$$

**Example 11.3** *Calculate the mass of aluminium that can be raised in temperature from* 10°C *to* 150°C *by the heat released from burning* 1 $m^3$ *of fuel gas. Assume that the heat obtainable from* 1 $m^3$ *of the gas is* 17 *MJ, and that* 75 *per cent of this heat can be made effective. The specific heat capacity of aluminium is* 950 *J/kg °C.*

Increase of temperature = 150 — 10 = 140°C.

$$\left.\begin{array}{r}\text{Effective heat per cubic}\\ \text{metre of gas}\end{array}\right\} = 17 \times 0{\cdot}75 = 12{\cdot}75\ \text{MJ}$$

$$= 12{\cdot}75 \times 10^6\ \text{J}.$$

Let $m$ be the mass of the aluminium, in kilograms.

Substituting in expression (11.3), we have:

$$12{\cdot}75 \times 10^6\ [\text{J}] = m \times 950\ [\text{J/kg °C}] \times 140\ [\text{°C}]$$

$$\therefore \qquad m = 95{\cdot}9\ \text{kg}.$$

## 11.7 Heat transfer in mixtures

When a hot substance is mixed or brought into contact with a cooler one, heat is transferred from the hotter to the cooler substance until the two are ultimately at the same temperature, which is below the original temperature of the hotter substance and above that of the cooler one. One of the substances is frequently a liquid; sometimes both are. It is then necessary to use some sort of containing vessel, and this vessel as well as the liquid will be heated or cooled. The heat thus gained or lost by the vessel should be taken into account.

Calculations on mixtures of hot and cold substances are based on the principle that the heat lost by the hot substance is equal to that gained by the cold substance; i.e. it is assumed that no loss of heat occurs to any external substance such as the surrounding air.

**Example 11.4** *A metal vessel and its water contents are together equivalent to* 6 *kg of water and their temperature is* 8°C. *If* 5 *kg of water at* 35°C *are added, what will be the resulting temperature of the mixture?*

If $t$ be the resulting temperature, then from expression (11.2),
total *gain* of heat by the cold water and vessel

$$= 4190\ [\text{J/kg °C}] \times 6\ [\text{kg}] \times (t - 8)\ [\text{°C}]$$

and total *loss* of heat by the hot water

$$= 4190\ [\text{J/kg °C}] \times 5\ [\text{kg}] \times (35 - t)\ [\text{°C}]$$

Since the loss of heat to external substances is being assumed negligible, the loss of heat by the hot water is equal to the gain of heat by the cold water and vessel; hence,

$$4190 \times 6 \times (t - 8) = 4190 \times 5 \times (35 - t)$$

$$\therefore \quad t = 20{\cdot}27\text{°C}.$$

**Example 11.5** *A hot copper ball having a mass of* 6 *kg is lowered into a vessel containing water which, with the vessel, is equivalent to a mass of* 5 *kg of water at* 12*°C. The temperature rises to* 28*°C. Assuming no loss of heat to the surroundings, calculate the initial temperature of the copper ball. Assume the specific heat capacity of copper to be* 390 *J/kg °C.*

Heat gained by water and vessel

$$= 4190\ [\text{J/kg °C}] \times 5\ [\text{kg}] \times (28 - 12)\ [\text{°C}]$$
$$= 335\ 200\ \text{J}.$$

If the initial temperature of the copper was $t$,
heat lost by the copper

$$= 390\ [\text{J/kg °C}] \times 6\ [\text{kg}] \times (t - 28)\ [\text{°C}]$$
$$= (2340\ t - 65\ 500)\ \text{J}.$$

Since heat lost by copper = heat gained by water and vessel

$$\therefore \quad 2340\ t - 65\ 500 = 335\ 200$$

so that $t = 171$°C.

## 11.8 Water equivalent

In the preceding examples, the vessel containing water was taken, with the water, as being together equivalent to a certain mass of water. This is a convenient practice, so it is important to know how the 'equivalent' can be determined. If the vessel is made of metal having a mass $m$ kilograms and a specific heat capacity $c$ joules per kilogram degree Celsius,

heat required to raise its temperature 1°C = $mc$ joules.

If $m_W$ be the water equivalent of the vessel in kilograms, i.e. the mass of water requiring the same amount of heat to raise its temperature 1°C: then, from expression (11.2),

$$\left.\begin{array}{r}\text{heat required to raise the temperature}\\ \text{of } m_W \text{ kilograms of water } 1^\circ\text{C}\end{array}\right\} = 4190\, m_W \text{ joules.}$$

Hence, $$4190\, m_W = mc$$

$\therefore$ water equivalent of vessel $= m_W = mc/4190$ kilograms (11.4)

For example, if a copper vessel has a mass of 0·8 kg and the copper has a specific heat capacity of 390 J/kg °C,

$$\left.\begin{array}{r}\text{water equivalent}\\ \text{of the vessel}\end{array}\right\} = 0{\cdot}8\text{ [kg]} \times 390\text{ [J/kg °C]}/4190\text{ [J/kg °C]}$$
$$= 0{\cdot}0745\text{ kg.}$$

The vessel and the water content are together equivalent to the mass of water plus the water equivalent of the vessel.

**Example 11.6** *An iron bucket has a mass of* 2 *kg and contains* 8 *kg of water. Calculate the water equivalent of* (a) *the bucket alone,* (b) *the bucket and water contained in it, assuming the specific heat capacity of the iron to be* 500 *J/kg °C.*

*If the initial temperature is 6°C, to what temperature will the bucket and water be raised by the addition of* 3 *kg of water at* 95°*C?*

From expression (11.4),

$$\left.\begin{array}{r}\text{water equivalent of}\\ \text{empty bucket}\end{array}\right\} = 2\text{ [kg]} \times 500\text{ [J/kg °C]}/4190\text{ [J/kg °C]}$$
$$= 0{\cdot}238\text{ kg,}$$

$\therefore$ water equivalent of bucket and water $= 8 + 0{\cdot}238 = 8{\cdot}238$ kg.

If $t$ be the final temperature after addition of the hot water,

heat gained by bucket and cold water

$$= 4190\text{ [J/kg °C]} \times 8{\cdot}238\text{ [kg]} \times (t - 6)\text{ [°C]}$$

and heat lost by the hot water

$$= 4190\text{ [J/kg °C]} \times 3\text{ [kg]} \times (95 - t)\text{ [°C].}$$

Since this loss of heat is equal to the heat gained by the bucket and cold water,

$$4190 \times 8{\cdot}238 \times (t - 6) = 4190 \times 3 \times (95 - t)$$

$$\therefore \qquad t = 29{\cdot}8^\circ\text{C.}$$

**Example 11.7** *A mass of 40 g of aluminium is heated to 200°C and then quickly immersed in 160 g of water contained in a copper vessel having a mass of 24 g, the initial temperature of the water being 12°C. If the final temperature of the water is 21·8°C, calculate the specific heat capacity of the aluminium. Assume the specific heat capacity of copper to be 390 J/kg °C and the loss of heat to be negligible.*

From expression (11.4),

water equivalent of copper vessel

$$= (24/1000)\ [\text{kg}] \times 390\ [\text{J/kg °C}]/4190\ [\text{J/kg °C}]$$
$$= 0{\cdot}002\,23\ \text{kg},$$

$\therefore$ water equivalent of vessel and water $= 0{\cdot}16 + 0{\cdot}002\,23$
$= 0{\cdot}162\,23$ kg.

Heat absorbed by vessel and water

$$= 4190\ [\text{J/kg °C}] \times 0{\cdot}162\,23\ [\text{kg}] \times (21{\cdot}8 - 12)\ [\text{°C}]$$
$$= 6660\ \text{J}.$$

If $c$ be the specific heat capacity of the aluminium, then from expression (11.3),

heat given out by the aluminium

$$= c \times (40/1000)\ [\text{kg}] \times (200 - 21{\cdot}8)\ [\text{°C}]$$
$$= 7{\cdot}128\,c\ \text{joules}.$$

$$\left.\begin{array}{r}\text{Since heat given}\\ \text{out by aluminium}\end{array}\right\} = \text{heat absorbed by water and vessel}$$

$\therefore$ $7{\cdot}128\,c = 6660$

so that $c = 935$ J/kg °C.

## 11.9 Calorimeters

Any vessel which can be used to contain a liquid, generally water, for the purpose of measuring heat by measuring the rise of temperature of the vessel and contents when a hot body is placed in it (as in the foregoing examples) is called a *calorimeter*. In order to reduce errors by loss of heat, the vessel is often placed within another container, from which it is 'insulated' by some material which is a bad conductor of heat. Acquaintance with simple calorimeters should be made in a laboratory.

## 11.10 Sources of thermal energy

In this book we first come across thermal energy as produced from mechanical energy when work is done in overcoming friction. But, of course, heat in this case is an unwanted by-product and mechanical work wasted in that way is not an important source of heat. There is a large store of thermal energy in the hot interior of the earth, which is at a much higher temperature than the surface. The most important source of energy is the sun. The heat from the sun produces vegetation which is the foundation of all coal, wood and other solid fuels as well as of oils. The sun is also the source of energy which lifts water as vapour to high levels where it falls as rain or snow and thus provides water-power.

Another source of heat discovered and developed during recent years is the nuclear reactor used, for example, in nuclear power stations. The nuclear reactor is a structure in which a fission chain reaction can be maintained and controlled, *fission* being the splitting of a heavy nucleus (or core of an atom consisting of protons and neutrons, see section 13.2) into two approximately equal fragments. Fission is accompanied by the release of energy in the form of heat which may be absorbed by a gas such as carbon dioxide or helium. The gas is then passed through boilers to produce steam for driving turbo-generators. The reactor consists mainly of specially-treated (or enriched) uranium, with graphite introduced to limit the rate at which fission takes place.

When heat is produced in appliances such as electric heaters and lamps, it is generally derived from electrical energy generated from mechanical energy, which, in turn, has been derived from heat generated by the combustion of coal or oil or by a nuclear reactor.

*Fuel and calorific values.* The most common method of producing heat is the *combustion* or burning of fuels. The energy in fuels is in the form of chemical energy, and burning or combustion is a chemical action in which certain constituents in the fuel, say, coal or oil, unite with the *oxygen* in the atmosphere and, in doing so, give out energy as heat. In this process, the principle of conservation of energy holds good, and the quantity of heat given out is quite definite and can be calculated from a knowledge of the mass and composition of the fuel.

Carbon, which is a constituent of wood, coal, coke and oil, on burning completely gives out about 34 MJ per kilogram. Hydrogen, which is a constituent of coal, coal-gas and oil, releases more than 140 MJ per kilogram. Fuels such as wood, coal, oil and various mixtures of gas vary widely in their composition and in the amount of heat they can give out. A good coal will give out at least 32 MJ per kilogram. The quantity of heat released per kilogram of fuel completely burned is termed the *calorific value* of the fuel. The calorific value of a gas fuel is usually stated in megajoules per cubic metre; thus, the value for 'town' gas is about 18·6 MJ/m$^3$ and that for natural gas is about 37·9 MJ/m$^3$.

**Example 11.8** *It is required to raise the temperature of* 130 *litres of water from* 15°*C to* 45°*C. How much coal would provide the necessary amount of heat if its calorific value is* 32 *MJ/kg? How much gas at* 18·6 *MJ/m*$^3$ *would provide the same amount of heat? Neglect any losses.*

$$\text{Mass of 1 litre of water} = 1\ \text{kg}$$

$$\therefore\ \text{mass of 130 litres of water} = 130\ \text{kg.}$$

$$\text{Increase of temperature} = 45 - 15 = 30°\text{C.}$$

$$\begin{aligned}\text{Hence heat required} &= 4190\ [\text{J/kg °C}] \times 130\ [\text{kg}] \times 30\ [°\text{C}]\\ &= 16{\cdot}34 \times 10^6\ \text{J} = 16{\cdot}34\ \text{MJ.}\end{aligned}$$

For coal having a calorific value of 32 MJ/kg,

$$\text{mass of coal required} = 16{\cdot}34\ [\text{MJ}]/32\ [\text{MJ/kg}] = 0{\cdot}511\ \text{kg.}$$

For gas having a calorific value of 18·6 MJ/m$^3$,

$$\text{volume of gas required} = 16{\cdot}34\ [\text{MJ}]/18{\cdot}6\ [\text{MJ/m}^3] = 0{\cdot}878\ \text{m}^3.$$

**Example 11.9** *An engine uses* 11 *kg of oil per hour, the calorific value of the oil being* 44 *MJ/kg. If the engine has an overall efficiency of* 30 *per cent (i.e. it converts* 30 *per cent of the thermal energy into work), calculate the output power of the engine.*

Thermal energy per hour from oil

$$\begin{aligned}&= 11\ [\text{kg}] \times 44\ [\text{MJ/kg}] = 484\ \text{MJ}\\ &= 484 \times 10^6\ \text{J,}\end{aligned}$$

$\therefore$ thermal energy per second from oil

$$= (484 \times 10^6)/3600$$
$$= 134\,400 \text{ J.}$$

Thermal energy converted per second into work

$$= 134\,400 \text{ [J]} \times (30/100)$$
$$= 40\,320 \text{ J,}$$

$\therefore$ output power of engine $= 40\,320 \text{ W} = 40{\cdot}32 \text{ kW}$.

**Example 11.10** *A petrol engine developing 25 kW drives paddles within a hydraulic brake so that the energy transmitted is converted into heat which warms a stream of water passing through the brake. If the inlet temperature of this stream is 15°C and it is desired that the water shall not be heated above 85°C, calculate the flow of water that must be provided, in litres per minute, neglecting any heat losses.*

Since 1 kW = 1 kJ/s,

work done in 1 second = 25 kJ = 25 000 J.

Rise of temperature of water = 85 — 15 = 70°C.

If $m$ be the mass of water, in kilograms, required per second, then from expression (11.2) we have:

$$25\,000 \text{ [J]} = 4190 \text{ [J/kg °C]} \times m \times 70 \text{ [°C]}$$

$$\therefore \quad m = 0{\cdot}0852 \text{ kg.}$$

Since the volume of 1 kg of water is practically 1 litre,

$\therefore$ volume of water per second = 0·0852 litre

and volume of water per minute = 0·0852 × 60 = 5·11 litres.

## 11.11 Change of state: Latent heat

If heat is supplied at a constant rate to a lump of ice having an initial temperature of, say, —20°C, and if no heat is lost by conduction, convection or radiation, the temperature varies as shown by the graph in fig. 11.2. The temperature first increases at a uniform rate from —20°C to 0°C, as shown by line AB. It then remains constant at 0°C for the time BC required for the ice to melt into water. Further supply of heat raises the temperature uniformly to 100°C, as shown by line CD. If the pressure on the surface of the water is atmospheric, the water then begins to boil and the temperature remains constant at 100°C until the water is all evaporated. If the supply of heat is maintained, the temperature of the steam increases, as shown by line EF. The steam is then said to be *superheated*..

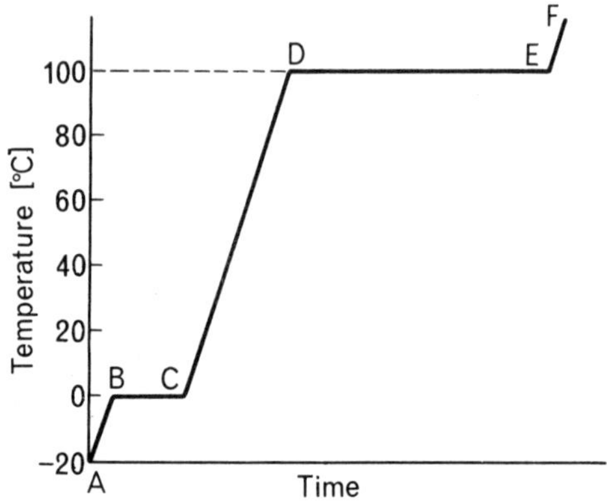


Fig. 11.2 Variation of temperature with time.

Another simple experiment to illustrate what happens when a substance changes from the liquid to the solid state is to heat paraffin wax in a test-tube to about 65 or 70°C, i.e. about 10 to 15°C above its melting point. A thermometer is then placed in the liquid wax and the temperature is observed at frequent intervals of time while the wax is cooling. When readings of temperature are plotted against time, the graph is as shown in fig. 11.3, from which it will be seen that during an interval AB, the temperature remains constant. This period represents the time during which the wax is solidifying.

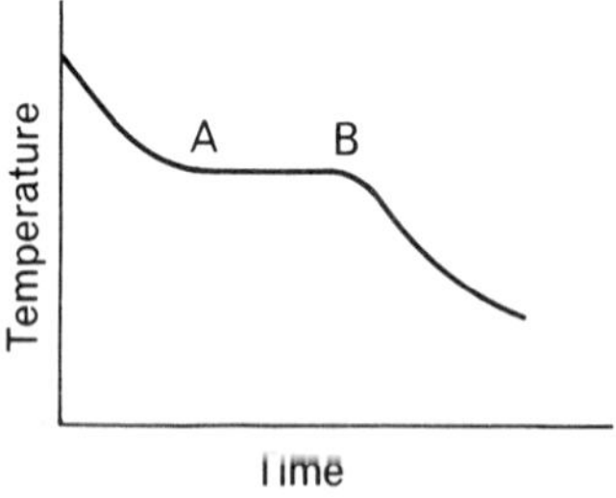


Fig. 11.3 Cooling curve for paraffin wax.

During periods BC and DE in fig. 11.2, when the water is changing from solid to liquid and from liquid to steam respectively, large quantities of heat are absorbed without any change of temperature. Similarly, during period AB in fig. 11.3, the wax loses a relatively large quantity of heat in changing from the liquid to the solid state without any change of temperature.

Let us consider what happens when water changes from solid to liquid and then from liquid to vapour. As mentioned in section 11.1, all materials consist of molecules; and in a solid, these molecules merely vibrate or oscillate about a constant mean position. In a liquid, a molecule can move from one position to another, being hindered only by frequent collisions with other molecules, i.e. they move about with a freedom they do not possess in the solid state. In a vapour, the molecules are much more scattered. There are fewer collisions between molecules though they move about at high speed in random directions. The volume occupied by a given mass of vapour is much greater than that of the same mass of the substance in the solid or liquid form.

The heat absorbed in effecting the change from ice to water and again from water to steam is not apparent as far as indications on a thermometer are concerned. This thermal energy is therefore said to be *latent*. Actually it is absorbed in two ways: (*a*) in giving the molecules more kinetic energy and (*b*) in the change from water to steam, work is done in pushing on the surrounding atmosphere or other expansible surrounding medium to make room for the increased volume occupied by the steam.

## 11.12 Specific latent heat of fusion

The heat required to change unit mass of a substance from the solid state to the liquid state at the same temperature is termed the *specific latent heat of fusion* of that substance. For example, when 1 kg of ice at 0°C is melted into water at 0°C, the heat absorbed is 335 kJ; i.e. the specific latent heat of fusion of ice is 335 kJ/kg.

**Example 11.11** *Two kilograms of dry ice at 0°C are placed in a vessel containing 5 kg of water at 38°C. The water equivalent of the vessel is 0·2 kg. Calculate the temperature of the water when the ice is completely melted, assuming no loss to or absorption of heat from outside the vessel. Take the specific latent heat of fusion of ice as 335 kJ/kg.*

Let $t$ be the final temperature of the mixture.

Water equivalent of original water and vessel

$$= 5{\cdot}0 + 0{\cdot}2 = 5{\cdot}2 \text{ kg.}$$

From expression (11.2), heat given out by water and vessel

$$= 4190 \,[\text{J/kg °C}] \times 5{\cdot}2 \,[\text{kg}] \times (38 - t) \,[\text{°C}]$$
$$= (828\,000 - 21\,800\,t) \text{ J.}$$

Since specific latent heat of fusion of ice $= 335$ kJ/kg $= 335\,000$ J/kg

$\therefore$ heat required to melt the ice

$$= 2 \,[\text{kg}] \times 335\,000 \,[\text{J/kg}] = 670\,000 \text{ J.}$$

Heat required to raise 2 kg of melted ice from 0°C to $t$

$$= 4190 \,[\text{J/kg °C}] \times 2 \,[\text{kg}] \times t$$
$$= 8380\,t \text{ joules.}$$

Since heat absorbed in melting the ice and heating to $t$ = heat given out by water and vessel

$\therefore$ $$670\,000 + 8380\,t = 828\,000 - 21\,800\,t$$

hence $$t = 5{\cdot}24\text{°C.}$$

## 11.13 Evaporation; Saturation temperature

In water at a constant temperature, some molecules will be moving at a higher velocity than the others; and some of the molecules near the surface of the water may possess sufficient kinetic energy to enable them to escape from the water to form vapour. If the evaporation takes place in an enclosed space, a state of equilibrium is reached when as many molecules return to the liquid in a given time as leave it in that time. The vapour is then said to be *saturated.*

If, however, the evaporation takes place in the open, for example, from a dish containing water, most of the molecules leaving the water do not return to it. If there is a draught blowing over the dish, the vapour is removed and evaporation takes place more rapidly. The heat required to convert the water into vapour comes from the water remaining in the dish; hence the temperature of this water falls. This is the reason why the contents of a jug can be kept cool by wrapping the jug with a wet cloth. The cooling effect is increased by exposing the jug and its wrapping to a draught.

When heat is continuously applied to the liquid, the temperature continues to rise until a value is reached at which the liquid vaporizes with *no further rise of temperature.* The liquid is then boiling and the temperature attained is termed the *boiling point* of the liquid.

The lower the pressure on the surface of the liquid, the lower is the temperature at which the liquid boils. For example, water at standard atmospheric pressure boils at 100°C; whereas at half the standard atmospheric pressure, the boiling point is about 82°C.

The boiling point of water at any particular pressure is also known as the *saturation temperature of steam* at that pressure. The term 'saturation' in this case implies that immediately the steam is subjected to a cooling influence, such as contact with a cooler material, it begins to condense. Steam at the saturation temperature is termed *saturated steam.*

Higher temperatures of steam are possible by the use of *superheated* or unsaturated steam produced by heating saturated steam which is not in contact with water. Superheated steam does not condense on being subjected to a cooling influence until it has cooled down to the saturation temperature corresponding to its pressure.

## 11.14 Specific latent heat of vaporization

The heat required to change 1 kg of a substance from the liquid state to vapour at the same temperature is termed the *specific latent heat of vaporization* of that substance. For example, when 1 kg of water at 100°C is converted into steam at 100°C, the heat absorbed is 2257 kJ, i.e. the specific latent heat of vaporization of water at standard atmospheric pressure is 2257 kJ/kg or 2·257 MJ/kg.

The latent heat absorbed in the formation of steam from water is given out in the condensation of steam into water.

**Example 11.12** *Calculate the heat required to convert* 2 *kg of water at* 60°*C into steam at* 100°*C at standard atmospheric pressure.*

From expression (11.2),

heat required to raise 2 kg of water from 60°C to 100°C

$$= 4190\ [\text{J/kg °C}] \times 2\ [\text{kg}] \times (100 - 60)\ [\text{°C}]$$
$$= 335\ 200\ \text{J}.$$

Since specific latent heat of vaporization $= 2257$ kJ/kg

$$= 2{\cdot}257 \times 10^6\ \text{J/kg},$$

heat required to convert 2 kg of water at 100°C into steam at 100°C

$$= 2{\cdot}257 \times 10^6\ [\text{J/kg}] \times 2\ [\text{kg}] = 4{\cdot}514 \times 10^6\ \text{J}.$$

$$\therefore \quad \text{total heat required} = 0{\cdot}3352 \times 10^6 + 4{\cdot}514 \times 10^6$$

$$= 4{\cdot}849 \times 10^6\ \text{J} = 4{\cdot}849\ \text{MJ}.$$

**Example 11.13** *Dry saturated steam (i.e. saturated steam containing no water particles) at 100°C is passed into water contained in a copper calorimeter having a mass of 45 g. The initial mass and temperature of the water are 120 g and 14°C respectively. After the steam has been passed, the mass of the water is 128 g. If the specific latent heat of vaporization of the steam is 2257 kJ/kg, what is the final temperature of the water, assuming the loss of heat to be negligible? The specific heat capacity of copper is 390 J/kg °C.*

From expression (11.4),

water equivalent of calorimeter

$$= 0{\cdot}045\ [\text{kg}] \times 390\ [\text{J/kg °C}]/4190\ [\text{J/kg °C}]$$

$$= 0{\cdot}0042\ \text{kg}.$$

If $t$ be the final temperature of the water, then from expression (11.2),

heat absorbed by water and calorimeter

$$= 4190\ [\text{J/kg °C}] \times (0{\cdot}12 + 0{\cdot}0042)\ [\text{kg}] \times (t - 14)\ [\text{°C}]$$

$$= (520\ t - 7280)\ \text{J}.$$

$$\text{Mass of steam condensed} = 128 - 120 = 8\ \text{g}$$

$$= 0{\cdot}008\ \text{kg}.$$

Since specific latent heat of vaporization = 2257 kJ/kg

$$= 2{\cdot}257 \times 10^6\ \text{J/kg},$$

∴ heat given out by steam in condensing at 100°C

$$= 0{\cdot}008\ [\text{kg}] \times 2{\cdot}257 \times 10^6\ [\text{J/kg}] = 18\,056\ \text{J},$$

and heat given out by condensed steam in cooling from 100°C to $t$

$$= 4190\ [\text{J/kg °C}] \times 0{\cdot}008\ [\text{kg}] \times (100 - t)\ [\text{°C}]$$

$$= (3350 - 33{\cdot}5\ t)\ \text{J}.$$

Equating the heat given out by the steam to that absorbed by the water and calorimeter, we have:

$$18\,056 + 3350 - 33{\cdot}5\ t = 520\ t - 7280$$

$$\therefore \quad t = 51{\cdot}8\text{°C}.$$

## *Summary of Chapter* 11

The SI unit of temperature is the kelvin (K).

Temperature interval of 1°C = temperature interval of 1 K.

$$t \text{ (degrees Celsius)} = T \text{ (kelvins)} - 273{\cdot}15 \qquad (11.1)$$

Temperature of melting ice is 273·15 K or 0°C, and temperature of boiling water at standard atmospheric pressure is 373·15 K or 100°C.

Specific heat capacity, $c$, of a substance is the heat required to raise the temperature of 1 kg of that substance by 1°C and is expressed in joules per kilogram degree Celsius (J/kg °C).

Heat required to raise the temperature of $m$ kilograms of a substance by $t$ degrees = $mct$ joules (11.3)

= 4190 $mt$ joules for water (11.2)

Water equivalent of a vessel = $mc/4190$ (11.4)

Specific latent heat of fusion of a substance is the heat required to change 1 kg of the substance from the solid state to the liquid state at the same temperature.

Specific latent heat of fusion of ice = 335 kJ/kg.

Specific latent heat of vaporization of a substance is the heat required to change 1 kg of the substance from the liquid state to vapour at the same temperature.

Specific latent heat of vaporization of water at standard atmospheric pressure = 2257 kJ/kg.

## EXAMPLES 11

1. Express 40°C in the thermodynamic scale and 290 K in the Celsius scale.
2. Calculate the heat, in megajoules, required to raise the temperature of 200 kg of water from 15°C to 90°C.
3. Calculate the heat required to raise the temperature of 30 kg of copper from 12°C to 70°C. Assume the specific heat capacity of copper to be 390 J/kg °C.
4. 90 MJ of heat are absorbed by a body having a mass of 1 Mg to raise its temperature from 20°C to 200°C. Assuming no loss of heat, calculate the specific heat capacity of the body.
5. Calculate the final temperature when 60 kg of water at 80°C are mixed with 200 kg of water at 20°C.
6. When 0·5 kg of hot water is mixed with 2·4 kg of cold water at 8°C, the final temperature is 16°C. Calculate the initial temperature of the hot water. Neglect any loss of heat.
7. A piece of metal having a mass of 0·4 kg is heated to 100°C and then

immersed in 0·6 kg of water, at 11·2°C, contained in a calorimeter having a water equivalent of 0·05 kg. The temperature of the water rises to 16·7°C. Calculate the specific heat capacity of the metal, assuming no loss of heat.

8. A brass cylinder, having a mass of 180 g, is heated to 156°C and then immersed in 0·3 litre of water at 14°C. The container has a water equivalent of 16 g. Calculate the temperature attained by the water, assuming no loss of heat. The specific heat capacity of brass is 370 J/kg °C.
9. How much heat is required to raise the temperature of 0·2 kg of aluminium from 18°C to 630°C, assuming the specific heat capacity of aluminium to be 950 J/kg °C?

   If this heated aluminium is immersed in 1·4 kg of water contained in a copper calorimeter at 12°C, calculate the temperature attained by the water, assuming no loss of heat. The mass of the calorimeter is 0·25 kg and the specific heat capacity of copper is 390 J/kg °C.
10. During an experiment in which an iron block was immersed in water, the following readings were obtained: mass of iron, 0·9 kg; initial temperature of iron, 120°C; initial temperature of water, 14°C; final temperature of water and iron, 22°C. Assuming the specific heat capacity of the iron to be 500 J/kg °C and neglecting heat losses and the water equivalent of the container, calculate (*a*) the mass of water in the container and (*b*) the heat, in kilojoules, originally in the iron, measured from 0°C.
11. A piece of steel, having a mass of 0·15 kg, is placed in a hot flue until it attains the same temperature as that of the flue gases. It is then quickly transferred into a copper calorimeter having a mass of 0·4 kg and containing 1 kg of water at a temperature of 15°C. The resultant steady temperature of the steel and water is 22°C. Calculate the temperature of the flue gases, neglecting any losses. Assume the specific heat capacity of steel to be 500 J/kg °C and that of copper to be 390 J/kg °C.
12. In a cooling system, 50 litres of oil per hour are cooled from 120°C to 25°C by means of circulating water. If the water is not to rise in temperature by more than 6°C, calculate the mass of water, in kilograms, required per hour. Assume the oil to have a specific heat capacity of 2100 J/kg °C and its density to be 880 kg/m$^3$.
13. During a brake test on an engine, it was found that 50 per cent of the output power of the engine was absorbed by the water supplied to the brake wheel, the remaining 50 per cent being dissipated as heat to the surrounding atmosphere. The output of the engine was 20 kW, the water flow was 300 litres/hour and the inlet temperature of the water was 15°C. Calculate the final temperature of the cooling water.
14. A bearing absorbs 6 kW in overcoming frictional resistance. Calculate the amount of heat, in kilojoules, produced per minute. If 24 kg of oil are passed through the bearing per minute and absorb 80 per cent of this heat, calculate the rise of temperature of the oil. Assume the specific heat capacity of the oil to be 2000 J/kg °C.
15. When 1·54 g of coal were burnt in a special calorimeter containing 1·63 kg of water, the temperature of the water rose from 14·3°C to 20·4°C. The water equivalent of the metal parts of the calorimeter was 0·23 kg. Calculate the calorific value of the coal in megajoules per kilogram.

16. The temperature of the water flowing through a gas water-heater has to be raised from 15°C to 90°C. The heater uses 90 litres of gas per minute and the heat given out by the gas is 18·6 MJ/m³ Calculate the volume of water, in litres, heated per hour, if 80 per cent of the heat released from the gas passes into the water.
17. If a petrol engine converts 25 per cent of the energy available in the petrol into output energy at the shaft, calculate the output power per kilogram of petrol consumed per hour. Assume the calorific value of the petrol to be 42 MJ/kg.
18. If petrol has a calorific value of 28 MJ/l and if the overall efficiency of a certain petrol engine is 22 per cent when the engine is developing 7·5 kW, calculate the petrol consumption in litres per hour.
19. A lump of dry ice, having a mass of 40 g and an initial temperature of 0°C, is immersed in 600 g of water at 30°C. The container has a water equivalent of 30 g. Calculate the final temperature when all the ice has melted. Assume the latent heat of fusion of ice to be 335 kJ/kg and neglect any loss of heat.
20. A lump of dry ice at 0°C is placed in a calorimeter having a water equivalent of 18 g and containing 300 g of water at 37°C. After the ice has melted, the temperature of the mixture is 14·7°C, and the calorimeter and water are heavier by 76 g. Calculate the specific latent heat of fusion of the ice. Neglect any loss of heat.
21. Calculate the heat, in kilojoules, required to convert 60 g of water at 30°C into saturated steam at 100°C. The container has a water equivalent of 5 g. Assume the specific latent heat of vaporization at 100°C to be 2257 kJ/kg and neglect any loss of heat.
22. Two kilograms of dry saturated steam at 100°C are blown into a vessel containing 120 kg of water at 20°C. Calculate the final temperature of the mixture. Neglect the heat absorbed by the vessel and assume the specific latent heat of vaporization at 100°C to be 2257 kJ/kg.

## ANSWERS TO EXAMPLES 11

1. 313·15 K, 16·85°C.
2. 62·85 MJ.
3. 679 kJ.
4. 500 J/kg °C.
5. 33·8°C.
6. 54·4°C.
7. 450 J/kg °C.
8. 20·8°C.
9. 116·3 kJ, 31·1°C.
10. 1·32 kg, 54 kJ.
11. 428°C.
12. 349 kg.
13. 43·6°C.
14. 360 kJ, 6°C.
15. 30·8 MJ/kg.
16. 256 l/h.
17. 2·92 kW.
18. 4·39 l/h.
19. 23·4°C.
20. 329 kJ/kg.
21. 154·4 kJ.
22. 30·2°C.

CHAPTER 12

# Expansion of solids, liquids and gases

## 12.1 Expansion of solids

Most substances increase in volume when they are heated. It has already been mentioned (section 11.3) that the mercury thermometer depends for its action upon the fact that mercury expands more than glass when its temperature is raised. In the case of solid materials, we are usually concerned only with their expansion or contraction in one direction, i.e. with *linear* expansion or contraction. An example of linear expansion and contraction is the shrinking of a steel ring on to a steel shaft. The ring is bored so that its internal diameter, when cold, is slightly less than the diameter of the shaft. The ring is then heated and the increased diameter enables it to be slipped over the shaft; but when the ring cools down, it grips the shaft firmly.

Another example of expansion that has many applications is shown in fig. 12.1, where strips of brass and steel are riveted together.

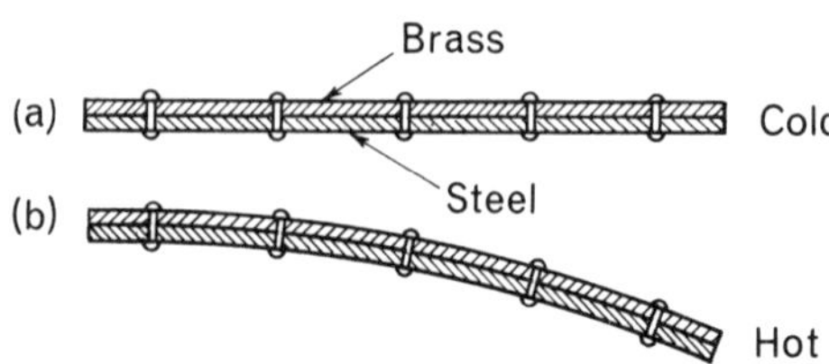


Fig. 12.1 Bimetallic strip.

The expansion of brass is nearly twice that of steel for a given increase of temperature; consequently, when the bimetallic strip is heated, it bends into an arc, as shown in fig. 12.1(*b*). The deflection of the bimetallic strip, for a given increase of temperature, can be practically doubled by using the nickel-steel alloy, *invar*, which contains about 36 per cent nickel, and its coefficient of linear expansion (section 12.2) is about one-sixth of that of ordinary steel. Bimetallic strips are often used in thermostats for controlling the temperature of ovens, etc.

## 12.2 Coefficient of linear expansion

The amount by which unit length of a material expands when the temperature is raised one degree is termed the *coefficient of linear expansion* of that material and is represented by the Greek letter $\alpha$ (alpha). Hence, if a rod has an initial length $l_0$ and its temperature is increased by $t$ degrees,

$$\text{increase in length} = \alpha t l_0$$

$$\therefore \quad \text{new length} = l = l_0 + \alpha t l_0$$

$$= l_0(1 + \alpha t) \qquad (12.1)$$

The variation in the length of such a rod is represented by the straight line AB of fig. 12.2, where OA represents the length $l_0$ at the initial temperature and BC represents the length $l$ after the temperature has increased by $t$ degrees.

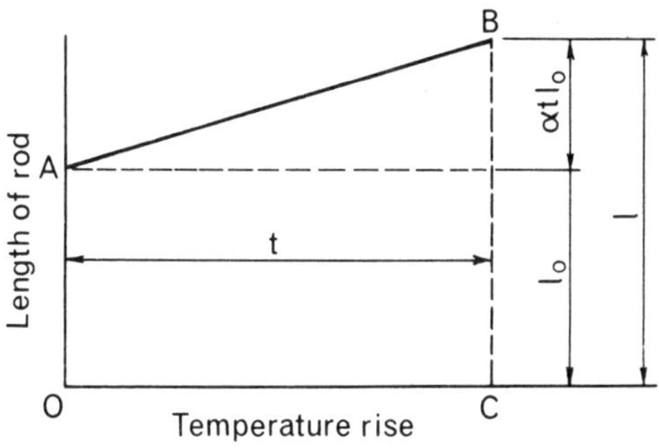


Fig. 12.2 Linear expansion.

The following table gives approximate values of the coefficient of linear expansion for materials that are commonly used in engineering:

| *Material* | *Coefficient of linear expansion* |
|---|---|
| Mild steel and wrought iron | $11 \times 10^{-6}/°C$ |
| Copper | $17 \times 10^{-6}/°C$ |
| Aluminium | $23 \times 10^{-6}/°C$ |
| Brass | $20 \times 10^{-6}/°C$ |

**Example 12.1** *A rod is found to be* 1·5342 *m long at* 12°*C and* 1·5358 *m at* 100°*C. Calculate the value of the coefficient of linear expansion.*

$$\text{Increase in temperature} = 100 - 12 = 88°\text{C}.$$

Substituting in expression (12.1), we have:

$$1{\cdot}5358 = 1{\cdot}5342\ (1 + 88\alpha)$$
$$\alpha = 11{\cdot}85 \times 10^{-6}/°\text{C}.$$

**Example 12.2** *The length of a copper wire forming one span of an electric transmission line is* 40 *m at* 10°*C. If the coefficient of linear expansion of the copper is* $17 \times 10^{-6}/°C$, *what is the increase in the length of wire when the temperature rises to* 45°*C?*

$$\text{Increase in temperature} = 45 - 10 = 35°\text{C}.$$
$$\text{Initial length of wire} = 40\ \text{m}$$
$$\therefore \quad \text{increase in length of wire} = 17 \times 10^{-6}[/°\text{C}] \times 35[°\text{C}] \times 40[\text{m}]$$
$$= 23{\cdot}8 \times 10^{-3}\ \text{m}$$
$$= 23{\cdot}8\ \text{mm}.$$

## 12.3 Superficial expansion of a solid

The amount by which unit *area* of a material increases when the temperature is raised by one degree is termed the *coefficient of superficial expansion* and is represented by the Greek letter $\beta$ (beta). If the initial surface area of the material is $a_0$ and the temperature is raised by $t$ degrees, the increase in area is $\beta t a_0$, and the new area $a$ is given by:

$$a = a_0 + \beta t a_0$$
$$= a_0(1 + \beta t) \qquad (12.2)$$

If we consider a square plate of a material having a coefficient

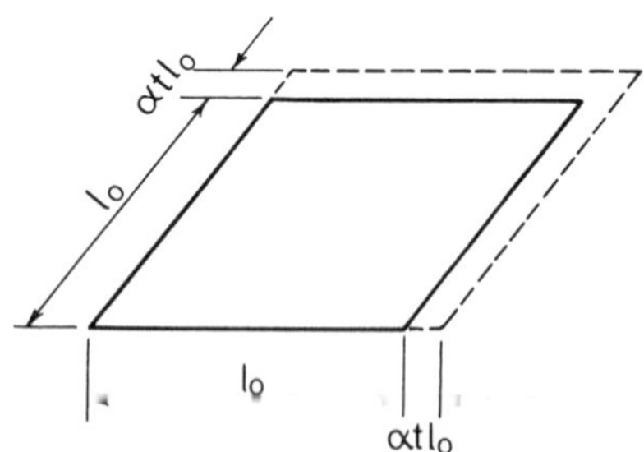


Fig. 12.3 Superficial expansion.

of *linear* expansion $\alpha$, and if the length of each side of the plate is initially $l_0$, as shown in fig. 12.3, then $a_0 = l_0^2$. If the temperature is increased $t$ degrees, the increase in length of each side is $\alpha t l_0$.

If $a$ is the new area, then:

$$
\begin{aligned}
a &= (l_0 + \alpha t l_0)^2 = l_0^2(1 + \alpha t)^2 \\
&= a_0(1 + 2\alpha t + \alpha^2 t^2) \\
&\simeq a_0(1 + 2\alpha t)
\end{aligned} \tag{12.3}
$$

since $\alpha t$ is usually very small compared with unity and $(\alpha t)^2$ is therefore negligible. From a comparison of expressions (12.2) and (12.3), it is seen that

$$\beta \simeq 2\alpha \tag{12.4}$$

i.e. the coefficient of superficial expansion is approximately twice the coefficient of linear expansion.

## 12.4 Cubic expansion of solids and liquids

The amount by which unit *volume* of a material increases for one degree rise of temperature is termed the *coefficient of cubic expansion,* and is represented by the Greek letter $\gamma$ (gamma). Hence, if the initial volume of a material is $V_0$ and the temperature is increased by $t$ degrees, the increase in volume is $\gamma t V_0$, and the new volume $V$ is given by:

$$
\begin{aligned}
V &= V_0 + \gamma t V_0 \\
&= V_0(1 + \gamma t)
\end{aligned} \tag{12.5}
$$

If we consider a cube of solid material having a coefficient of linear expansion $\alpha$, and if the length of each side is initially $l_0$, as shown in fig. 12.4, then $V_0 = l_0^3$. For an increase of temperature of $t$ degrees, the increase in the length of each side is $\alpha t l_0$, so that the new volume $V$ is given by:

$$
\begin{aligned}
V &= (l_0 + \alpha t l_0)^3 = l_0^3(1 + \alpha t)^3 \\
&= V_0(1 + 3\alpha t + 3\alpha^2 t^2 + \alpha^3 t^3) \\
&\simeq V_0(1 + 3\alpha t)
\end{aligned} \tag{12.6}
$$

since $\alpha t$ is usually very small compared with unity for solids. From a comparison of expressions (12.5) and (12.6), it is seen that

$$\gamma \simeq 3\alpha \tag{12.7}$$

i.e. the coefficient of cubic expansion of solids is approximately three times the coefficient of linear expansion.

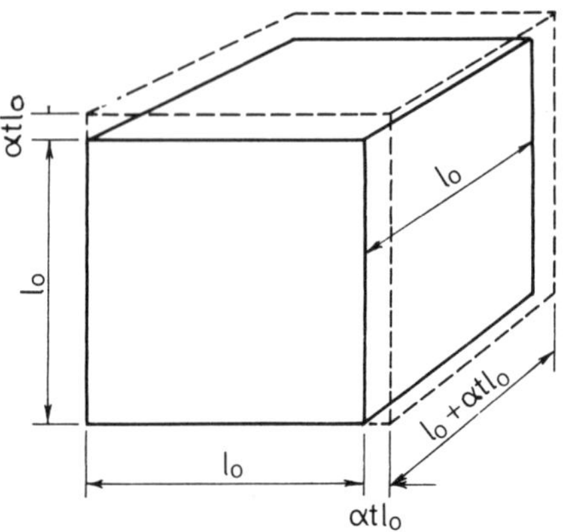


Fig. 12.4 Cubic expansion.

When water cools, contraction occurs until the volume is a minimum at 4°C. Further cooling causes the water to expand until the freezing point is reached. When water is converted into ice, considerable expansion occurs and this often results in the bursting of pipes in very cold weather.

**Example 12.3** *A sphere of copper has a diameter of 40 mm. Calculate the increase in volume when the temperature is raised by 180°C. Assume tne linear coefficient of expansion of copper to be 17 × $10^{-6}$/°C.*

From expression (12.7),

coefficient of cubic expansion of copper $\simeq 3 \times 17 \times 10^{-6}$/°C
$= 51 \times 10^{-6}$/°C.

Initial volume of sphere $= \frac{4}{3}\pi \times (\text{radius})^3$
$= 1{\cdot}333 \times 3{\cdot}14 \times 8000 = 33\,500$ mm$^3$

increase in volume $= \gamma t V_0$
$= 51 \times 10^{-6} \times 180 \times 33\,500$
$= 308$ mm$^3$.

## 12.5 Cubic expansion of gases

For solids and liquids, the value of the coefficient of cubic expansion is so small that no appreciable error is introduced by assuming the initial volume as being that at, or near, the room temperature instead of at 0°C; but in the case of gases, the coefficient of cubic expansion is much larger and it is therefore necessary to be more precise in our definition. Thus *the coefficient of cubic expansion of a gas is the amount by which unit volume of the gas* **at 0°C** *increases when the temperature is increased* 1 *degree, the pressure remaining constant.*

It is found that the volume of a gas, at constant pressure, varies as shown by the straight line AB in fig. 12.5, and the slope of this line is such that if the line were extended backwards, it would cut the horizontal axis at a point C corresponding to about —273°C.

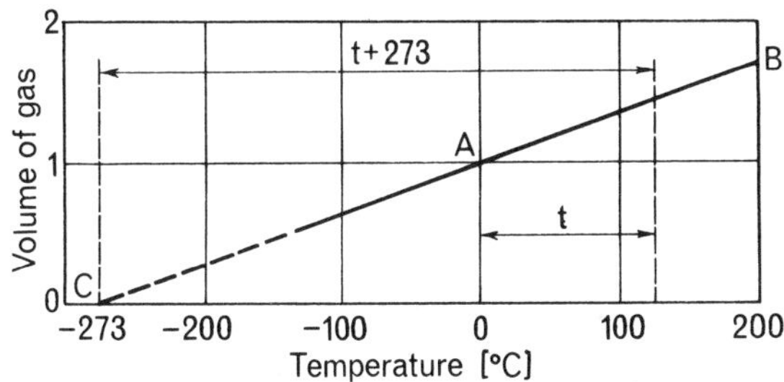


Fig. 12.5 Cubic expansion of a gas at constant pressure.

This suggests that if the gas could be cooled to the thermodynamic zero temperature (—273·15°C), its volume would be zero. Actually, the gas would have been first liquefied and then solidified before this temperature would have been reached. From fig. 12.5, it follows that the volume of a gas, at constant pressure, is directly proportional to the thermodynamic temperature or ($t + 273{\cdot}15$), where $t$ is the temperature in degrees Celsius.

This relationship, first observed by Jacques Charles (1746–1823), a French physicist, is referred to as *Charles' Law* and may be expressed thus: the volume, $V$, of a given mass of gas, at constant pressure, is directly proportional to the thermodynamic temperature $T$,

i.e. $$V = T \times \text{a constant} \tag{12.8}$$

It follows from fig. 12.5 that when a gas remains at constant pressure, the variation of volume for 1°C variation of temperature is $\frac{1}{273}$ of its volume at 0°C,

i.e., coefficient of cubic expansion of a gas at constant pressure

$$= \gamma = (\tfrac{1}{273})/°\text{C}.$$

**Example 12.4** *A litre of gas at 18°C is heated to a temperature of 150°C, the pressure remaining constant. Calculate the new volume.*

Initial thermodynamic temperature $= 18 + 273{\cdot}15 \simeq 291$ K.

Final thermodynamic temperature $= 150 + 273{\cdot}15 \simeq 423$ K.

From Charles' law,

$$\frac{\text{volume at }150°\text{C}}{\text{volume at }18°\text{C}} = \frac{423\ [\text{K}]}{291\ [\text{K}]}$$

$$\therefore \qquad \text{volume at }150°\text{C} = 1\ [\text{l}] \times 423\ [\text{K}]/291\ [\text{K}]$$

$$= 1{\cdot}454 \text{ litres.}$$

## 12.6 Boyle's law

This law, enunciated by Sir Robert Boyle, an Irish scientist, in 1661, states that *the volume of a given mass of gas at a constant temperature is inversely proportional to the pressure*; thus, if the pressure on a given mass of gas is doubled, the volume is halved. Hence, if $V$ be the volume and $p$ be the pressure, then as long as the gas remains at a constant temperature,

$$V \propto 1/p$$

i.e.

$$pV = \text{a constant} \qquad (12.9)$$

## 12.7 Combination of Boyle's and Charles' laws

Let us consider a given mass of gas enclosed in a cylinder fitted with a piston, as in fig. 12.6, such that there is no leakage of gas. With the piston in the position shown in fig. 12.6(*a*), let us assume the absolute pressure, the volume and the thermodynamic temperature to be represented by $p_1$, $V_1$ and $T_1$ respectively.

Suppose that the three quantities—pressure, volume and temperature—vary simultaneously to $p_2$, $V_2$ and $T_2$ respectively, and that we require to know the value of one of the quantities when the new values of the other two quantities are known.

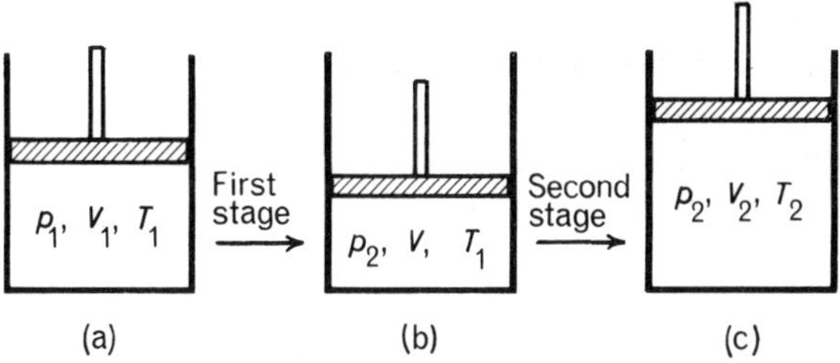


Fig. 12.6 Boyle's and Charles' laws.

The relationship between the three quantities can be derived by imagining the change to take place in two stages in such a way that Boyle's law can be applied to the first stage and Charles' law to the second stage, thus:

(*a*) if the pressure is varied from $p_1$ to $p_2$, with the temperature remaining constant, the volume changes from $V_1$ to a value $V$, fig. 12.6(*b*); then by Boyle's law,

$$p_1 V_1 = p_2 V$$

$$\therefore \qquad V = \frac{p_1 V_1}{p_2} \qquad (12.10)$$

(*b*) if the temperature is then varied from $T_1$ to $T_2$, the pressure remaining constant, the volume changes from $V$ to $V_2$, fig. 12.6(*c*); then by Charles' law,

$$\frac{V}{T_1} = \frac{V_2}{T_2}$$

$$\therefore \qquad V = \frac{V_2 T_1}{T_2} \qquad (12.11)$$

Equating expressions (12.10) and (12.11), we have:

$$\frac{p_1 V_1}{p_2} = \frac{V_2 T_1}{T_2}$$

so that
$$\frac{p_1 V_1}{T_1} = \frac{p_2 V_2}{T_2} = \frac{p_3 V_3}{T_3} \text{ etc.}$$

Hence, in general,
$$\frac{pV}{T} = \text{a constant} \qquad (12.12)$$

This relationship is very useful for converting the volume of a gas determined experimentally at convenient temperature and pressure to the volume at standard temperature of 0°C and standard atmospheric pressure of 101·325 kN/m² (or 760 mm of mercury). Such a condition is referred to as *standard temperature and pressure*, abbreviated to *s.t.p.* The standard atmospheric pressure may also be expressed as 1·013 25 bars, where 1 bar = $10^5$ N/m² or Pa.

In all problems involving gas laws, it is the *absolute pressure* and not the *gauge pressure* that must be used. The gauge pressure—as the name implies—is the reading on a pressure gauge and represents the amount by which the actual pressure differs from the atmospheric pressure. Thus, if a gauge attached to a steam pipe reads 1200 kPa (or 12 bars) and if the actual atmospheric pressure outside the pipe is 100 kPa (or 1 bar), the absolute pressure of the steam is 1300 kPa (or 13 bars). In general,

$$\text{absolute pressure} = \text{gauge pressure} + \text{atmospheric pressure.}$$

**Example 12.5** *Air at* 8°C *and a gauge pressure of* 300 *kPa is passed through a heater. If the temperature is raised to* 50°C *and the gauge pressure falls to* 230 *kPa, what is the percentage change in volume? Assume the atmospheric pressure to be* 101·3 *kPa.*

$$\begin{aligned} \text{Initial absolute pressure} &= 300 + 101{\cdot}3 = 401{\cdot}3 \text{ kPa.} \\ \text{Final absolute pressure} &= 230 + 101{\cdot}3 = 331{\cdot}3 \text{ kPa.} \\ \text{Initial temperature} &= 8 + 273{\cdot}15 \simeq 281 \text{ K} \\ \text{Final temperature} &= 50 + 273{\cdot}15 \simeq 323 \text{ K.} \end{aligned}$$

Suppose the initial volume of the air to be 100 m³, then if $V$ be the final volume in cubic metres, we have from expression (12.12),

$$\frac{401{\cdot}3\,[\text{kPa}] \times 100\,[\text{m}^3]}{281\,[\text{K}]} = \frac{331{\cdot}3\,[\text{kPa}] \times V}{323\,[\text{K}]}$$

$$\therefore \qquad V = 139 \text{ m}^3.$$

Hence, increase in volume = 39 per cent.

**Example 12.6** *A quantity of gas at a pressure of* 1150 *mm of mercury and a temperature of* 75°C *occupies a volume of* 2·4 *l. Calculate the volume at standard temperature and pressure. Standard pressure corresponds to a pressure of* 760 *mm of mercury.*

Initial temperature of gas $\simeq 75 + 273 = 348$ K.

Let $V$ be the volume of the gas, in litres, at standard temperature and pressure. Substituting in expression (12.12), we have:

$$\frac{1150\ [\text{mmHg}] \times 2{\cdot}4\ [\text{l}]}{348\ [\text{K}]} = \frac{760\ [\text{mmHg}] \times V}{273\ [\text{K}]}$$

$$\therefore \qquad V = 2{\cdot}85 \text{ litres.}$$

**Example 12.7** *Calculate the mass of dry air in a room,* 8 *m* × 6 *m* × 4 *m, when the atmospheric pressure is* 96 *kPa and the temperature is* 20°*C. The mass of* 1 *m*³ *of dry air at s.t.p. is* 1·29 *kg.*

$$\text{Volume of room} = 8 \times 6 \times 4 = 192 \text{ m}^3.$$

$$\text{Temperature of room} = 20°\text{C}$$
$$\simeq 20 + 273 = 293 \text{ K.}$$

Standard atmospheric pressure $\simeq$ 101·3 kPa.

If $V$ be the volume of the air, in cubic metres, at s.t.p., then:

$$\frac{96\ [\text{kPa}] \times 192\ [\text{m}^3]}{293\ [\text{K}]} = \frac{101{\cdot}3\ [\text{kPa}] \times V}{273\ [\text{K}]}$$

$$\therefore \qquad V = 169{\cdot}7 \text{ m}^3$$

and $\qquad$ mass of air $= 169{\cdot}7 \times 1{\cdot}29 = 219$ kg.

## *Summary of Chapter* 12

Coefficient of linear expansion of a solid is the amount by which unit length increases for one degree rise of temperature.

$$l = l_0(1 + \alpha t) \qquad (12.1)$$

Coefficient of superficial expansion of a solid is the amount by which unit area increases for one degree rise of temperature.

$$a = a_0(1 + \beta t) \qquad (12.2)$$

Coefficient of cubic expansion of a substance is the amount by which unit volume increases for one degree rise of temperature.

$$V = V_0(1 + \gamma t) \qquad (12.5)$$

In the case of a gas, $V_0$ in expression (12.5) is the volume at 0°C, $t$ is the temperature above 0°C and the pressure is assumed to remain constant.

For solids: $\beta \simeq 2\alpha$ (12.4)

and $\gamma \simeq 3\alpha$ (12.7)

For a gas: $\gamma = (\frac{1}{273})/°\text{C}$.

*Charles' law.* The volume of a given mass of gas, at constant pressure, is directly proportional to the thermodynamic temperature,

i.e. $V/T = \text{a constant}$ (12.8)

*Boyle's law.* The volume of a given mass of gas, at constant temperature, is inversely proportional to the absolute pressure,

i.e. $pV = \text{a constant}$ (12.9)

For a given mass of gas, the product of the absolute pressure and the volume divided by the thermodynamic temperature is a constant,

i.e. $pV/T = \text{a constant}$ (12.12)

EXAMPLES 12

1. A metal rod is 500 mm long at 30°C and 500·5 mm at 130°C. Calculate the coefficient of linear expansion of the metal.
2. If a brass scale is correct length at 5°C, calculate the percentage increase in length at 60°C. The coefficient of linear expansion of brass is 0·000 02/°C.
3. A steel pipe is 20 m long at 12°C. Calculate the increase in length when steam at 130°C is passing through it, assuming the coefficient of linear expansion of the steel to be 0·000 011/°C.
4. In an experiment, it was found that a copper rod, 400 mm long at the initial temperature, increased in length by 0·544 mm when the temperature was raised by 80°C. Calculate the coefficient of linear expansion of the copper.
5. The length of an aluminium conductor forming one span between the towers of an electric transmission line is 120 m. What is the variation in the length of the conductor when the temperature varies between —20°C and 55°C? Assume $\alpha = 0{\cdot}000\ 023/°\text{C}$ for the aluminium.
6. A steel wire, 2 mm in diameter, is stretched without appreciable tension between two fixed points at a temperature of 25°C. Calculate the tension when the temperature falls to 0°C. Assume $\alpha = 0{\cdot}000\ 011/°\text{C}$ and $E =$ 220 GPa.
7. A brass plate is 400 mm square at 12°C. If the temperature is raised to 100°C, what is the increase in the superficial area of the plate? Assume the coefficient of linear expansion of brass to be 0·000 02/°C.
8. A block of cast iron measures 24 mm × 15 mm × 10 mm when placed in melting ice. What is the increase in volume when the block is placed in steam at standard atmospheric pressure, if the coefficient of linear expansion of cast iron is 0·000 01/°C?

9. The mercury contained in a certain thermometer has a volume of 540 mm³ at 5°C. What is the increase in volume when the temperature is raised to 85°C? Assume the coefficient of cubic expansion of mercury to be 0·000 18/°C.
10. Ethyl alcohol has a coefficient of cubic expansion of 0·0011/°C. Calculate the reduction in volume when the temperature of 0·05 litre of ethyl alcohol is reduced from 40°C to −15°C.
11. A volume of 0·4 litre of dry air at 10°C is heated to a temperature of 120°C, the pressure remaining constant. Calculate the new volume.
12. A quantity of gas at constant pressure and a temperature of 27°C is heated to 60°C and becomes 0·5 m³. What was the original volume?
13. If the pressure on a gas remains constant and its initial volume and temperature are 1 m³ and 12°C respectively, what will be (*a*) its volume if the temperature is raised to 70°C and (*b*) the temperature when its volume increases to 1·3 m³?
14. A certain gas has a volume of 50 m³ at 200°C. If the pressure is maintained constant while the temperature is reduced to −50°C, what is the final volume? What is the thermodynamic temperature of the gas when the volume is reduced to 40 m³?
15. The density of air at s.t.p. is 1·29 g/l. Calculate the mass of air contained in a flask of volume 2·5 litres at standard pressure when the temperature is (*a*) 80°C and (*b*) 230 K.
16. Four cubic metres of gas are compressed from a pressure of 100 kPa to a pressure of 500 kPa, the temperature remaining constant. What is the final volume?
17. Air is compressed in a cylinder from a pressure of 105 kPa and a temperature of 20°C to one-quarter of its original volume at constant temperature. What is the resulting pressure? The air is then heated at constant pressure until it occupies its original volume. Calculate the final temperature of the air in degrees Celsius.
18. A certain mass of air has a volume of 0·56 litre at a pressure equivalent to 780 mm of mercury and a temperature of 5°C. What is the volume when the pressure is equivalent to 710 mm of mercury and the temperature is 25°C?
19. An oxygen cylinder contains 110 litres of gas at an absolute pressure of 1000 kPa and a temperature of 12°C. What would be the volume of the gas at standard atmospheric pressure of 101·3 kPa and a temperature of 28°C?
20. Express a gauge pressure of 2·5 bars as an absolute pressure and an absolute pressure of 1·6 bars as a gauge pressure. Assume the atmospheric pressure to be 1·013 bars.
21. Compressed air at a pressure of 400 kPa and a temperature of 15°C is passed through a heater. If the temperature is raised to 70°C and the pressure falls to 300 kPa, what is the percentage change in volume?
22. Gas at s.t.p. is compressed to one-third its initial volume and its temperature is raised to 150°C. What is the final pressure of the gas in bars? Assume the standard atmospheric pressure to be 1·013 bars.

23. A gas has a volume of 0·8 litre at 30°C and a pressure of 90 kPa. Calculate the volume at s.t.p.
24. A volume of gas occupying 0·8 litre at s.t.p. is heated until its pressure is 115 kPa. If the final volume is 0·92 litre, what is the temperature of the gas?
25. A compressed-air cylinder is 2 m long internally and has an internal diameter of 0·6 m. It contains air at a pressure of 50 bars and a temperature of 15°C. If 1 $m^3$ of air at s.t.p. has a mass of 1·29 kg, calculate the mass of air in the cylinder.
26. A motor tyre was pumped up to a gauge pressure of 1·75 bars when the temperature was 10°C. During a run, the temperature of the air rose to 38°C. Assuming that the tyre did not change in capacity, calculate the tyre gauge pressure at the end of the run. Assume the atmospheric pressure to be 1·013 bars.

    If 1 $m^3$ of air has a mass of 1·29 kg at s.t.p., calculate the mass of air in the tyre if the capacity of the tyre is 40 litres.

## ANSWERS TO EXAMPLES 12

1. $10 \times 10^{-6}$/°C.
2. 0·11 per cent.
3. 25·96 mm.
4. $17 \times 10^{-6}$/°C.
5. 207 mm.
6. 190 N.
7. 563 $mm^2$.
8. 10·8 $mm^3$.
9. 7·78 $mm^3$.
10. 3025 $mm^3$.
11. 0·555 litre.
12. 0·45 $m^3$.
13. 1·203 $m^3$, 97·5°C.
14. 23·6 $m^3$, 378·6 K.
15. 2·49 g, 3·825 g.
16. 0·8 $m^3$.
17. 420 kPa, 900°C.
18. 0·66 litre.
19. 1145 litres.
20. 3·513 bars, 0·587 bar.
21. 59 per cent increase.
22. 4·71 bars.
23. 0·64 litre.
24. 83°C.
25. 34·2 kg.
26. 2·03 bars, 0·136 kg.

CHAPTER 13

# Electric charges

## 13.1 Electrification by friction

The fact that amber, when rubbed, acquires the property of attracting light bodies was referred to over 2500 years ago by a Greek philosopher, named Thales, as a phenomenon that was quite familiar at that time. It was not until about A.D. 1600 that Dr Gilbert of Colchester discovered that other bodies, such as glass, could also be electrified by rubbing.

If a glass rod is rubbed with silk and placed on a stirrup hung from a wooden stand (fig. 13.1), and if the end of another similarly rubbed glass rod is brought near one end of the suspended rod, they are found to repel each other. If an ebonite rod rubbed with fur is brought near the suspended glass rod, attraction takes place. Similarly, if the charged ebonite rod is supported on the stirrup, the approach of another charged ebonite rod causes repulsion, whereas a charged glass rod causes attraction.*

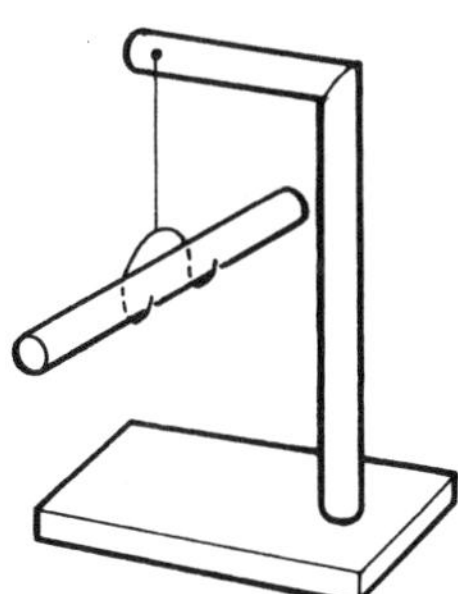

Fig. 13.1 Attraction and repulsion of electric charges.

The glass and the ebonite appear to be charged with different kinds of electricity; and the above experiments show that bodies

* These experiments are difficult to perform satisfactorily if the atmosphere is humid, but they do form a very simple and useful introduction to the idea of positive and negative charges of electricity and to the existence of forces of repulsion between like charges and of attraction between unlike charges.

charged with the same kind of electricity repel, while bodies charged with opposite kinds of electricity attract one another.

About 1750, Benjamin Franklin suggested that electricity was some form of fluid which passed from one body to the other when they were rubbed together, and that in the case of glass rubbed with silk, the electric fluid passed from the silk into the glass so that the glass contained a 'plus' or 'positive' amount of electricity. On the other hand, when ebonite was rubbed with fur, the electric fluid passed from the ebonite to the fur, leaving the ebonite with a 'minus' or 'negative' amount of electricity. Franklin's one-fluid theory has long been discarded, but his convention still remains: thus, the glass is said to be positively charged and the ebonite negatively charged, and it is this convention which governs the signs used for the terminals of batteries and direct-current generators.

We may now summarise the results of the above experiments by stating that like charges repel and unlike charges attract each other, as shown in fig. 13.2, where the arrows represent the directions of the forces on the charges.

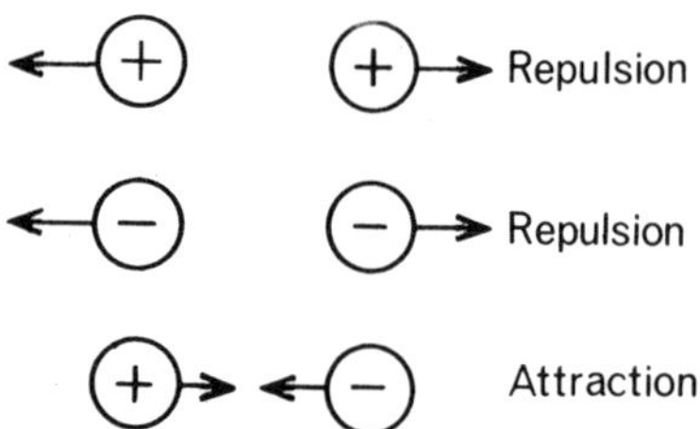


Fig. 13.2 Repulsion and attraction between electric charges.

## 13.2 Structure of the atom

It was mentioned in section 11.1 that all substances consist of molecules, the molecule being the smallest particle of a substance that can have a separate existence and still retain the characteristics of that substance. A molecule consists of one or more particles called *atoms*, an atom being the smallest particle of an element which can take part in a chemical reaction. Molecules of helium, neon and argon contain one atom each and are therefore termed *monatomic* molecules. Molecules of hydrogen, oxygen and nitrogen consist of two atoms each and are therefore said to be *diatomic*. A molecule of sulphur consists of eight atoms.

Every atom consists of a relatively massive core or nucleus carrying a positive charge, around which *electrons* move in orbits at distances that are great compared with the size of the nucleus. Each *electron* has a mass of $9{\cdot}11 \times 10^{-31}$ kg and a *negative* charge, $-\mathbf{e}$. The nucleus of every atom except that of hydrogen consists of *protons* and *neutrons*. Each *proton* carries a *positive* charge, **e**, equal in magnitude to that of an electron and its mass is $1{\cdot}673 \times 10^{-27}$ kg, namely 1836 times that of an electron. A *neutron*, on the other hand, carries *no* resultant charge and its mass is approximately the same as that of a proton. Under normal conditions, an atom is neutral, i.e. the total negative charge on its electrons is equal to the total positive charge on the protons.

The atom possessing the simplest structure is that of hydrogen. It consists merely of a nucleus of one proton together with a single electron which may be thought of as revolving in an orbit, of about $10^{-7}$ mm diameter, around the proton, as in fig. 13.3(a).

Fig. 13.3(b) shows the arrangement of a helium atom. In this case, the nucleus consists of two protons and two neutrons, with two electrons orbiting in what is termed the *K shell*. This shell is complete with only two electrons. Consequently the helium atom is stable and does not combine with any other atom, i.e. helium is an *inert* element.

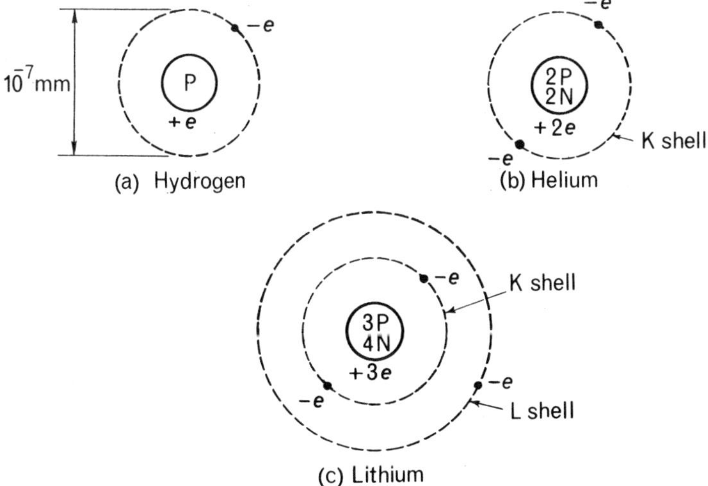


Fig. 13.3 Hydrogen, helium and lithium atoms.

The element with *three* orbital electrons is lithium. Two of these electrons form the K shell, as before, and the third electron starts another shell, known as the L shell, as shown in fig. 13.3(c).

The number of electrons in the L shell of the elements beryllium, boron, carbon, nitrogen, oxygen, fluorine and neon are 2, 3, 4, 5, 6, 7 and 8 respectively. With eight electrons, the L shell is complete; hence the reason why neon is another inert element.

Further increase in the number of orbital electrons results in the formation of a third shell, referred to as the M shell. For instance, the element sodium has eleven orbital electrons, two of which form the K shell, eight the L shell and the remaining electron forms the M shell.

The electrons orbiting in the outermost shell are termed *valence electrons*. The number of valence electrons determines the properties of an element; for instance, the elements copper, silver and gold, which are good conductors of electricity, have a single electron orbiting in the N, O and P shells respectively.

When atoms are packed tightly together, as in a metal, each outer electron experiences a small force of attraction towards neighbouring nuclei, with the result that such an electron is no longer bound to any individual atom, but can move at random within the metal. These electrons are termed *free* or *conduction* electrons, and only a slight external influence is required to cause them to drift in a desired direction.

The full lines AB, BC, CD, etc. in fig. 13.4 represent paths of the random movement of one of the free electrons in a *metal* rod when there is no cell connected across the terminals EF; i.e. the electron is accelerated in direction AB until it collides with an atom with the result that it may rebound in direction BC, etc. Different free electrons move in different directions so as to maintain the electron density constant throughout the metal; in other words, there is no resultant drift of electrons towards either E or F.

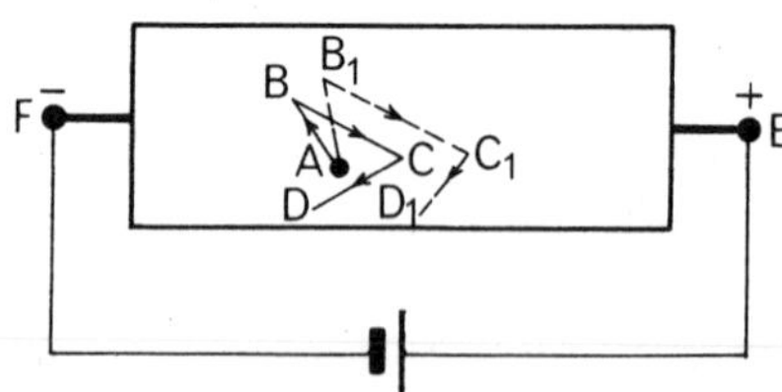


Fig. 13.4 Movement of a free electron.

If a cell, such as that used in an electric torch, is connected across terminals E and F, the cell terminal marked '+' being connected to terminal E, the latter becomes positive relative to F. The effect is to modify the random movement of the electron as shown by the dotted lines $AB_1$, $B_1C_1$ and $C_1D_1$ in fig. 13.4, i.e. there is superimposed on the random movement a drift of the electron towards the positive terminal E. The number of electrons reaching terminal E from the rod is the same as that entering the rod at terminal F. It is this drift of the electrons that constitutes an electric current in the circuit.

An atom which has lost or gained one or more electrons is referred to as an *ion*; thus, for an atom which has lost one or more electrons, its negative charge is less than its positive charge and such an atom is therefore termed a *positive ion*.

In materials such as glass and porcelain at normal temperature, the electrons are tightly bound to their respective nuclei so that there are very few free electrons present. Consequently such materials are very poor conductors of electric charges and are therefore termed *insulators*. The higher the temperature of an insulating material, the higher is the random velocity at which the free electrons move from one atom to another, and therefore the greater is the impact when an electron collides with an atom. Such impacts may be sufficiently violent as to release electrons from the atoms and thus increase the number of free electrons available to carry electric charges. In other words, the material becomes a better conductor of electricity.

## 13.3 Movement of free electrons in a metal

It was explained in section 13.2 how the drift of electrons in a desired direction can be produced by connecting a cell across the circuit. In fig. 13.5, DE represents an enlarged view of a metal rod forming part of a circuit which includes a battery B, shown as two cells connected in series.

The circles with crosses represent positive ions, namely atoms which have lost one or more of their outermost electrons. These ions are locked in the structure of the metal and are therefore unable to move. The small black circles represent the free electrons moving from left to right. This procession or drift of the electrons takes place round the whole circuit, including battery B; i.e. the number of electrons emerging per second from the negative terminal of B is exactly the same as that entering the positive terminal per second.

It was mentioned in section 13.1 that the convention governing the marking of the terminals of a battery as 'positive' and 'negative' is based upon a theory propounded over 200 years ago by Benjamin Franklin. This theory suggested that the positive terminal of a battery is at a *higher electrical potential* than the negative terminal. Consequently, when a metal wire is connected across the terminals, electricity is assumed to flow through the wire *from* the positive terminal *to* the negative terminal. Unfortunately, this theory has proved incorrect. We now know that the electric current in a metal wire connected across a battery consists of the movement of electrons from the negative terminal of the battery to the positive terminal, i.e. from a point at the *lower* potential to a point at the *higher* potential, as indicated in fig. 13.5.

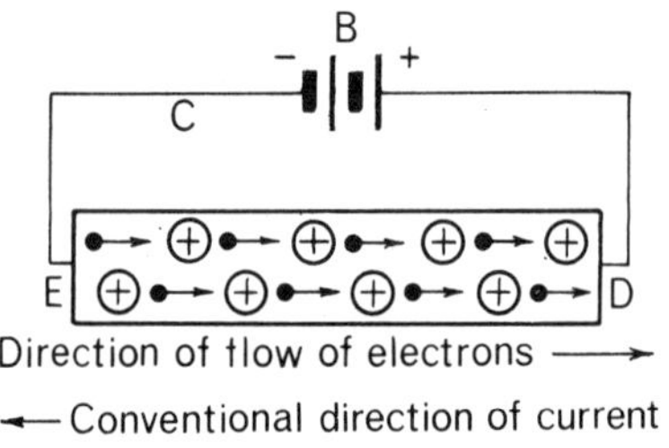


Fig. 13.5 Movement of free electrons in a metal.

The convention for the direction of an electric current, based on Franklin's theory, had been universally adopted long before the discovery of the electron in 1897, and so we continue to say that an electric current flows from a point at the *higher* potential to that at the *lower* potential, as shown in fig. 13.5.

## 13.4 Units of quantity of electricity and of electric current

When a current is flowing in a circuit, suppose $n$ to be the number of electrons passing a given point of the circuit in a given time. Since each electron carries a negative charge, $-\mathbf{e}$, it follows that the total electric charge that has passed the given point is $-n\mathbf{e}$. This total charge is referred to as the *quantity of electricity* that has passed that point.

The unit of quantity of electricity is the *coulomb* (symbol, C), named after a French physicist, C. A. de Coulomb (1736–1806).

The magnitude of the coulomb is enormous compared with that of the charge on an electron; in fact, the coulomb is 6·24 million million million ($6{\cdot}24 \times 10^{18}$) times the charge on an electron.

Expressed in another way, the charge on an electron

$$= \frac{1}{6{\cdot}24 \times 10^{18}} = 1{\cdot}602 \times 10^{-19} \text{ coulomb.}$$

The rate at which electricity flows past any given point of an electric circuit is termed an *electric current*. The unit of current is the *ampere* (symbol, A), named after another French scientist, André-Marie Ampere (1775–1836), and is one of the seven SI base units listed in section 1.1. We shall defer giving the definition of the ampere until a later stage. All we need say at this stage is that if $I$ represents the value of the current, in amperes, flowing in a circuit for time $t$, in seconds, then the quantity of electricity $Q$, in coulombs, that has passed any given point of the circuit in that time is:

$$Q = I \times t$$

or

$$I = Q/t \tag{13.1}$$

**Example 13.1** *The current through a certain lamp is* 0·6 *A. If this current remains constant for* 2 *hours, what is the quantity of electricity, in coulombs, that flows through the lamp during that time?*

$$\text{Time} = 2\,[\text{h}] \times 3600\,[\text{s/h}] = 7200 \text{ s,}$$

$$\therefore \quad \text{quantity of electricity} = 0{\cdot}6\,[\text{A}] \times 7200\,[\text{s}] = 4320 \text{ C.}$$

**Example 13.2** *If the current in a circuit is* 1 *A, calculate the number of electrons passing any given point of the circuit per second.*

$$\text{Quantity of electricity per second} = 1\,[\text{A}] \times 1\,[\text{s}] = 1 \text{ C.}$$

Since each electron carries a negative charge of $1{\cdot}602 \times 10^{-19}$ C,
∴ number of electrons passing a given point of the circuit per second

$$= \frac{1}{1{\cdot}602 \times 10^{-19}} = 6{\cdot}24 \times 10^{18},$$

i.e. when the current in a circuit is 1 ampere, electrons are passing any given point of the circuit at the rate of $6{\cdot}24 \times 10^{18}$ per second.

## *Summary of Chapter* 13

Matter consists of an aggregate of atoms, each of which has a nucleus of protons and neutrons surrounded by electrons moving in orbits around the nucleus. The protons carry positive charges and the electrons negative charges, but the neutrons carry no resultant charge. When the atom is in its normal or neutral state, its positive and negative charges are equal in magnitude. In metals, one or more of the outermost electrons are loosely held to the atom and can be made to drift from one atom to another in a desired direction by connecting an electric cell across the circuit.

The unit of quantity of electricity is the coulomb and that of current is the ampere.

The current is the rate at which electricity flows in a circuit,

i.e. $$I = Q/t \tag{13.1}$$

An electron has a negative charge of $1{\cdot}602 \times 10^{-19}$ C, and when the current in a circuit is 1 A, the number of electrons passing a given point of the circuit is $6{\cdot}24 \times 10^{18}$ per second.

## EXAMPLES 13

1. The current through an electric heater is 4 A. If this current remains constant for 3 hours, calculate the quantity of electricity, in coulombs, that has passed a given point of the circuit.
2. If the quantity of electricity that passes a given point of a circuit in 40 s is 600 C, what is the average value of the current?
3. If the number of electrons flowing per second through a circuit is $6 \times 10^{16}$, what is the value of the current in milliamperes?
4. The current in a circuit is 30 μA. Calculate the rate at which electrons pass any given point of the circuit.
5. If there are $3 \times 10^{14}$ electrons passing per second across the space between two metal surfaces, calculate (*a*) the current, in microamperes and (*b*) the quantity of electricity, in microcoulombs, that passes in 5 s.

## ANSWERS TO EXAMPLES 13

1. 43 200 C.
2. 15 A.
3. 9·6 mA.
4. $187{\cdot}2 \times 10^{12}$ electrons/second.
5. 48 μA, 240 μC.

# CHAPTER 14

# Electric current

## 14.1 Effects of an electric current

All the phenomena which an electric current may produce can be grouped under the following headings:

(*a*) magnetic effect,
(*b*) heating effect,
(*c*) chemical effect.

These three effects can be demonstrated by connecting the following items in series as shown in fig. 14.1:

(i) a coil C wound on an iron core bent into a U shape,
(ii) an electric fire element H,
(iii) a double-pole change-over switch D, to which are also connected a flash-lamp bulb L and a pair of lead plates, P and Q, dipping into a dilute solution of sulphuric acid in water.

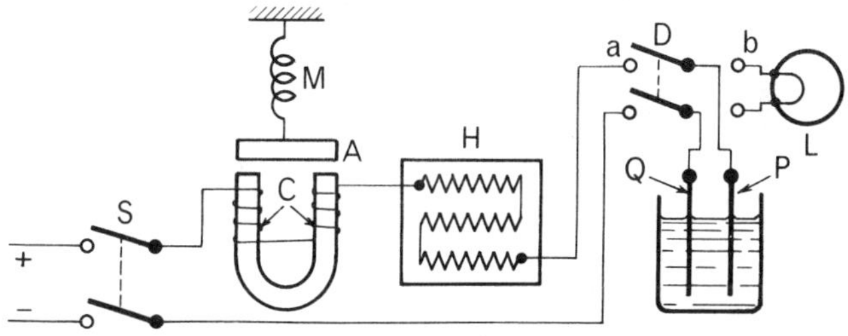


Fig. 14.1 Magnetic, heating and chemical effects of an electric current.

The circuit is connected through a switch S to a direct-current supply, namely a supply that produces an electric current in one direction only. This supply can be obtained either from a direct-current generator or from a battery of secondary cells (section 19.4). When switch S is closed and switch D is put over to side *a*, an electric current flows through coil C, heater H and the liquid between plates P and Q.

If an iron plate or armature A is suspended by a spring M a little above the ends of the iron core, it will be attracted downwards immediately S is closed. In other words, the iron core becomes magnetized and mechanical work is done in extending M.

The element H is found to become warm and its temperature may even rise sufficiently for the wire to glow, thereby giving out light energy as well as thermal energy.

After the current has been flowing for a few minutes, let us open S and switch D over to side *b*. The filament of lamp L becomes incandescent, but its brightness gradually fades away. If the plates P and Q are withdrawn from the solution, P is found to have a faint chocolate-colour coating while Q appears unaltered. In this case, electrical energy has been converted into chemical energy in changing the lead on the surface of plate P into an oxide of lead. This chemical action happens to be of a kind that is reversible. Consequently, when the plates are connected to the lamp, chemical energy is converted back into electrical energy which is then converted into thermal and light energies. This reversible chemical action is the basis upon which secondary batteries operate (Chapter 19).

## 14.2 Measurement of an electric current

It is important that we should have some idea of the magnitude of the current in a circuit and we shall now consider how the magnetic and chemical effects can be utilized for determining the value of the current.

*Magnetic effect.* In fig. 14.2, E and F represent the cross-section of two co-axial coils, E being suspended from a spring balance G about 2 cm above F which rests on a table or bench. The coils are connected so that they carry current in the *same* direction, the current being led into and out of coil E by flexible wires. Another coil, B, also shown in section in fig. 14.2, is wound on a hollow cylindrical former and an iron core C is suspended from a spring balance S.

The three coils, E, F and B, are connected in series with a variable resistor R, an ammeter A (i.e. an instrument calibrated to indicate the magnitude of the current directly in amperes) and a switch N across a direct-current supply.

With switch N closed, an electric current flows in the circuit. The magnetic effect produced by the current in coils E and F causes them to attract each other (the reason for this behaviour is given in section 17.13). Also, the iron core C is attracted towards coil B. The greater the current, the larger are the readings on balances G and S, and marks can be inserted on the two scales to register the various currents indicated by ammeter A.

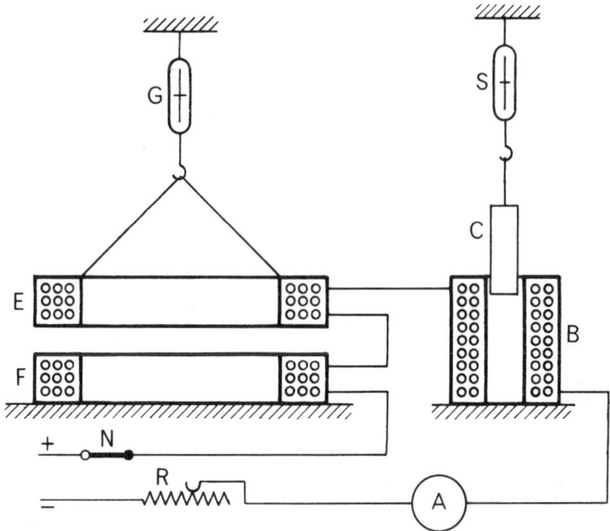


Fig. 14.2 Use of the magnetic effect to measure an electric current.

These two methods are comparatively crude but are given at this stage merely to indicate the principles of methods applying the magnetic effect to the measurement of an electric current. The method using the two co-axial coils, E and F, illustrates the principle of operation of the balance used in National Laboratories for determining the value of the unit of electric current as defined in section 14.4.

The second method, namely the attraction of an iron core towards a coil carrying a current, illustrates the principle of action of one type of moving-iron instrument referred to in section 17.8(c).

*Chemical effect.* The glass vessel B in fig. 14.3 contains a solution of copper sulphate in water. Two copper plates, P and Q, are partially immersed in the *electrolyte* (an electrolyte being any liquid that can be decomposed electrically) and are connected in series with an ammeter A, a variable resistor R and a switch N to a direct-current supply.

Plate Q, namely the plate connected to the negative terminal of the supply, is washed, dried and weighed on a beam balance (section 1.7) before it is placed in the copper sulphate solution. Switch N is then closed and the current is maintained constant at a known value for, say 10 minutes. Switch N is then opened and plate Q is removed, washed, dried and again weighed. It is found that the mass of the plate has increased and the exact amount is determined from the difference between the balance readings.

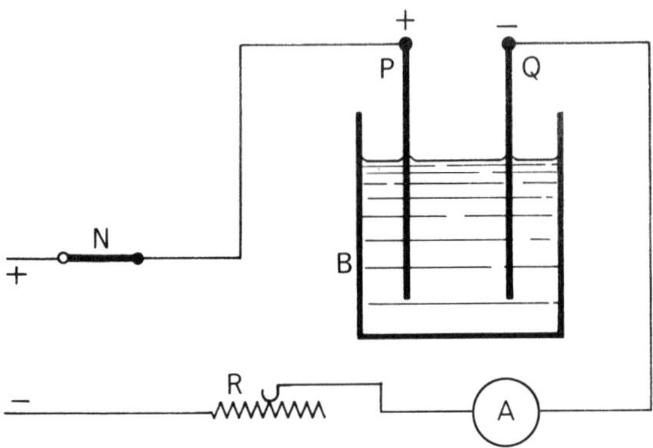


Fig. 14.3 Use of the chemical effect to measure an electric current.

In this way we can measure accurately the amount of copper deposited by a *given current* in, say, 10, 20 and 30 minutes, and also the amount deposited in a *given time* by different values of the current. It is found that the mass of copper deposited is directly proportional to the current and to the time; in other words, it is proportional to the *quantity* of electricity.

Owing to the accuracy and the ease with which the increase in the mass of a plate can be determined, this chemical effect of an electric current may be employed for measuring the quantity of electricity with a high degree of precision. At one time, this was the principle of the method used for defining the magnitude of the unit of current.

### 14.3 Effect of reversing the current

Let us change over the supply connections to the circuits of figs. 14.2 and 14.3, so that the supply wire marked '—' will now go to coil F in fig. 14.2 and to plate P in fig. 14.3.

It is found that the effect of the reversed current is still to attract coil E and core C downwards in fig. 14.2. In fact, a given current, as indicated by ammeter A, produces exactly the same readings as before on the two scales.

In the case of fig. 14.3, it is found that the mass of plate Q decreases, so that the chemical action at this plate must now be taking place in the reverse direction. Further, it is found that the decrease in the mass of plate Q due to a given current for a given time is the same as the increase that took place before the connections were reversed.

Let us go a step further and connect the circuits of figs. 14.2 and 14.3 across an alternating-current supply. The electric current now reverses its direction many times every second. Coil E and core C are still found to be attracted downwards, but no change takes place in the mass of plate Q. The chemical method is therefore useless for measuring an alternating current, whereas it is possible to construct instruments of the magnetic type that will measure direct and alternating currents equally accurately.

If the heater H of fig. 14.1 is connected across an alternating-current supply, it is found that for a given current, as indicated by a suitable ammeter connected in series with the heater, the heating effect is the same as that produced by direct current.

## 14.4 The ampere and the coulomb

It was stated in section 13.4 that the SI unit of current is the *ampere*, that the unit of quantity of electricity is the *coulomb* and that if a current $I$, in amperes, is maintained for time $t$, in seconds, the corresponding quantity of electricity, in coulombs, is represented by $Q$, where

$$Q = I \times t$$

or

$$I = Q/t \qquad (13.1)$$

In 1948 it was decided internationally that the *ampere* should be defined as *the constant current which, if maintained in two straight parallel conductors of infinite length, of negligible circular cross-section and placed at a distance of* 1 *metre apart in a vacuum, would produce between them a force equal to* $2 \times 10^{-7}$ *newton per metre length*. The reason for the existence of a force between two parallel current-carrying conductors is given in section 17.12.

The apparatus required to measure a current accurately in terms of the above definition is based upon the principle of the attraction between two current-carrying coils, E and F, of fig. 14.2; but its construction is so elaborate and expensive that it is only available at the principal national laboratories, such as the National Physical Laboratory in this country.

From carefully-conducted experiments based upon the chemical effect of an electric current described in section 14.2, it has been found that for every coulomb of electricity passing between two copper plates immersed in a copper sulphate solution, 0·3294 mg of copper is deposited on the negative plate (plate Q in fig. 14.3). An arrangement such as that shown in fig. 14.3, when used to measure the quantity of electricity by the amount of a substance liberated electro-chemically, is known as a *coulometer* or *voltameter*.

Previous to 1948, the ampere was defined in terms of the chemical effect produced when current was passed through a silver coulometer consisting of two silver plates immersed in a silver nitrate solution. The silver coulometer can still be used as a sub-standard for measuring the value of the current on the assumption that 1 coulomb of electricity deposits 1·1182 mg of silver on the negative plate.

Since 1 ampere is 1 coulomb/second, the unit quantity of electricity may also be termed an *ampere second.* For many purposes, such as for stating the quantity of electricity a battery is capable of supplying, the coulomb is inconveniently small, and a larger unit, known as the *ampere hour* (A h), is preferable. Thus, if a battery supplies 4 A for 10 h, the quantity of electricity

$$= 4\ [\text{A}] \times 10\ [\text{h}] = 40\ \text{A h}.$$

**Example 14.1** *The current through a silver coulometer is maintained constant for* 15 *min. The initial and final masses of the negative plate are* 16·347 *g and* 17·518 *g respectively. Calculate the value of the current, assuming that* 1 *C of electricity deposits* 1·1182 *mg of silver on the negative plate.*

$$\text{Mass of silver deposited} = 17{\cdot}518 - 16{\cdot}347 = 1{\cdot}171\ \text{g}$$
$$= 1171\ \text{mg}.$$

$$\therefore \quad \left.\begin{matrix}\text{quantity of electricity}\\ \text{through coulometer}\end{matrix}\right\} = \frac{1171\ [\text{mg}]}{1{\cdot}1182\ [\text{mg/C}]}$$
$$= 1047{\cdot}2\ \text{C}.$$

$$\text{Duration of test} = 15 \text{ min}$$
$$= 15 \times 60 = 900 \text{ s},$$
$$\therefore \qquad \text{current} = \frac{1047{\cdot}2\ [\text{C}]}{900\ [\text{s}]} = 1{\cdot}163 \text{ A}.$$

**Example 14.2** *If a current of* 15 *A is maintained constant for* 20 *min, calculate the quantity of electricity in* (a) *coulombs and* (b) *ampere hours.*

(*a*) Quantity of electricity, in coulombs

$$= 15\ [\text{A}] \times (20 \times 60)\ [\text{s}] = 18\,000 \text{ C}.$$

(*b*) Quantity of electricity, in ampere hours

$$= 15\ [\text{A}] \times (20/60)\ [\text{h}] = 5 \text{ A h}.$$

## 14.5 Electrochemical equivalent

In section 14.2 it was found that the mass of copper deposited on the negative plate (or *cathode*) was proportional to the quantity of electricity passing through the electrolyte. This relationship was discovered by Michael Faraday in 1832 when he enunciated two laws:

(*a*) the amount of chemical change produced by an electric current is proportional to the quantity of electricity,

(*b*) the amounts of different substances liberated by a given quantity of electricity are proportional to their chemical equivalent mass, where

$$\text{chemical equivalent mass} = \frac{\text{relative atomic mass}}{\text{valency}}$$

The relative atomic masses and valencies of some of the most common elements are given in the table on p. 226.

The mass of a substance liberated from an electrolyte by 1 coulomb is termed the *electrochemical equivalent* of that substance; thus, the electrochemical equivalents of copper and silver are respectively 0·3294 and 1·1182 milligrams/coulomb.

If $z$ = electrochemical equivalent of a substance in milligrams per coulomb

and $I$ = current, in amperes, for time $t$ seconds,
mass of substance liberated = $zIt$ milligrams. (14.1)

| ELEMENT | RELATIVE ATOMIC MASS | VALENCY | ELECTRO-CHEMICAL EQUIVALENT |
|---|---|---|---|
| Aluminium | 27·0 | 3 | |
| Chlorine | 35·5 | 1 | |
| Chromium | 52·0 | 3 or 6 | |
| Copper (cuprous) | 63·6 | 1 | |
| Copper (cupric) | 63·6 | 2 | 0·3294 mg/C |
| Gold | 197·2 | 3 | |
| Hydrogen | 1·008 | 1 | |
| Iron | 55·8 | 2 or 3 | |
| Lead | 207·2 | 2 | |
| Nickel | 58·7 | 2 | 0·304 mg/C |
| Oxygen | 16·0 | 2 | |
| Potassium | 39·1 | 1 | |
| Silver | 107·9 | 1 | 1·1182 mg/C |
| Sodium | 23·0 | 1 | |
| Tin | 118·7 | 2 or 4 | |
| Zinc | 65·4 | 2 | 0·338 mg/C |

**Example 14.3** *A steady current of* 6·3 *A is passed for* 45 *min through a solution of copper sulphate. Calculate the mass of copper deposited.*

Quantity of electricity, in coulombs

$$= 6{\cdot}3\ [\text{A}] \times (45 \times 60)\ [\text{s}]$$
$$= 17\,010\ \text{C}.$$

Since 1 coulomb deposits 0·3294 mg of copper,

$$\therefore \quad \text{mass of copper deposited} = 0{\cdot}3294\ [\text{mg/C}] \times 17\,010\ [\text{C}]$$
$$= 5600\ \text{mg}$$
$$= 5{\cdot}6\ \text{g}.$$

**Example 14.4** *A current of* 2 *A is passed for* 30 *min between two platinum plates (or electrodes) immersed in a dilute solution of* *sulphuric acid in water. Calculate the mass of hydrogen and oxygen released.*

The effect of electrolysis in this case is to decompose water into its constituents, hydrogen and oxygen, the former being liberated at the negative plate (or cathode) and the latter at the positive plate (or anode).

$$\text{Quantity of electricity} = 2\,[\text{A}] \times (30 \times 60)\,[\text{s}] = 3600\ \text{C}.$$

From the data given in the table on page 226, the relative atomic mass and the valency of hydrogen are 1·008 and 1 respectively,

$$\therefore \quad \text{chemical equivalent mass of hydrogen} = 1{\cdot}008/1 = 1{\cdot}008.$$

Also, for silver, the electrochemical equivalent is 1·1182 mg/C and the chemical equivalent mass is 107·9; hence, by Faraday's second law,

$$\begin{aligned}&\text{electrochemical equivalent of hydrogen}\\ &\qquad = 1{\cdot}1182\ [\text{mg/C}] \times 1{\cdot}008/107{\cdot}9\\ &\qquad = 0{\cdot}010\,45\ \text{mg/C},\end{aligned}$$

$$\begin{aligned}\therefore\quad &\text{mass of hydrogen released}\\ &\qquad = 0{\cdot}010\,45\ [\text{mg/C}] \times 3600\ [\text{C}]\\ &\qquad = 37{\cdot}6\ \text{mg} = 0{\cdot}0376\ \text{g}.\end{aligned}$$

Similarly, chemical equivalent mass of oxygen

$$= 16/2 = 8$$

$$\begin{aligned}\therefore\quad &\text{electrochemical equivalent of oxygen}\\ &\qquad = 1{\cdot}1182\ [\text{mg/C}] \times 8/107{\cdot}9\\ &\qquad = 0{\cdot}0829\ \text{mg/C}\end{aligned}$$

$$\begin{aligned}\text{and}\quad &\text{mass of oxygen released}\\ &\qquad = 0{\cdot}0829\ [\text{mg/C}] \times 3600\ [\text{C}]\\ &\qquad = 298\ \text{mg} = 0{\cdot}298\ \text{g}.\end{aligned}$$

## *Summary of Chapter* 14

A simple circuit enables the magnetic, heating and chemical effects of a direct current to be demonstrated. The chemical effect is shown to be reversible. The principles of current measurement are also demonstrated. It is shown that an alternating current produces the magnetic and heating effects but not the chemical effect of a direct current. The chemical effect can be used to determine the magnitude of a direct current.

Mass of substance liberated from electrolyte, in milligrams

$$= z\ [\text{milligrams/coulomb}] \times Q\ [\text{coulombs}]$$
$$= z\ [\text{mg/C}] \times I\ [\text{amperes}] \times t\ [\text{seconds}] \qquad (14.1)$$

## EXAMPLES 14

1. If 4 A h of electricity flow in an electrical circuit in 10 min, calculate (*a*) the quantity of electricity in coulombs and (*b*) the average value of the current.
2. If a deposit of 0·18 g of metal forms on a cathode in 10 min and a current of 1 A is flowing, what is the electrochemical equivalent of this metal? (U.E.I., G1)
3. A constant current of 3 A passes for 16 min through a silver coulometer. If 1 A deposits 1·118 mg of silver per second, what is the increase in the mass of the cathode? Calculate also the quantity of electricity (*a*) in coulombs, (*b*) in ampere hours.
4. A copper voltameter connected in series with an ammeter was used as a means of checking the calibration of the ammeter. During the test the ammeter reading was 2 A, and in 15 min the mass of copper deposited was 0·625 g. Take the electrochemical equivalent of copper as 0·33 mg/C and calculate (*a*) the true current, (*b*) the error in the ammeter reading and (*c*) the quantity of electricity which flowed during the test. (U.E.I., G1)
5. Define the *coulomb*. In an experiment with a copper voltameter, the initial mass of the cathode was 40 g. A current of 5 A was then passed through the voltameter for one hour. If the electrochemical equivalent of copper is 0·33 mg/C, determine: (*a*) the number of coulombs of electricity which passed through the apparatus, (*b*) the new mass of the cathode. (U.L.C.I., G1)
6. Calculate the quantity of zinc used when a Leclanché cell supplies a current of 0·05 A for one hour. Assume the electrochemical equivalent of zinc to be 0·339 mg/C. (N.C.T.E.C., G2)
7. A metal plate having a surface of 12 000 mm$^2$ is to be copper-plated. If a current of 2 A is passed for one hour, what thickness of copper is deposited? Density of copper is 8900 kg/m$^3$ and the electrochemical equivalent of copper is 0·3294 mg/C.
8. It is required to deposit a layer of nickel, 0·2 mm thick, on a surface area of 15 000 mm$^2$. Calculate the minimum time required if the maximum permissible current is 8 A. Assume the electrochemical equivalent of nickel to be 0·304 mg/C and the density of nickel to be 8800 kg/m$^3$. Also determine the quantity of electricity required (*a*) in coulombs and (*b*) in ampere hours.
9. A metal plate, having a total surface of 20 000 mm$^2$, is to be chromium-plated. If a current of 5 A is maintained for one hour, what thickness of chromium is deposited on the plate? Assume the electrochemical equivalent of chromium to be 0·09 mg/C and the relative density of chromium to be 6·6.

10. A constant current flowing through acidulated water liberates 1·248 litres of hydrogen, measured at standard temperature and pressure, per hour. Assuming the electrochemical equivalent of hydrogen to be 0·0104 mg/C and its density at standard temperature and pressure to be 90 $g/m^3$, calculate the value of the current.
11. From the data given in the table on page 226, determine the electrochemical equivalents of aluminium, gold and chlorine.

## ANSWERS TO EXAMPLES 14

1. 14 400 C, 24 A.
2. 0·3 mg/C.
3. 3·22 g, 2880 C, 0·8 A h.
4. 2·104 A; 0·104 A, ammeter reading low; 1894 C.
5. 18 000 C, 45·94 g.
6. 61 mg.
7. 0·0222 mm.
8. 3·015 h; 86 800 C, 24·12 A h.
9. 0·012 28 mm.
10. 3 A.
11. 0·0932 mg/C, 0·681 mg/C, 0·3675 mg/C.

CHAPTER 15

# Electric circuit

## 15.1 Conductors and insulators

An Englishman, Stephen Gray, was the first person to transmit electricity a distance of about 300 metres. He did this in 1729, using a brass wire suspended by silk threads. His 'transmission line' possessed the two essential requisites for such purpose, namely (*a*) a good conductor to allow electrons to flow easily along it and (*b*) a good insulator to ensure that the movement of the electrons was confined to the conducting path along which they were to be transmitted.

It was explained in section 13.2 that materials such as copper, which are good conductors, possess a plentiful supply of free electrons, and that it is the drift of these electrons that constitutes an electric current. Materials, such as paper and rubber, possess very few free electrons and are therefore very good insulators. Hence, if a copper wire has a covering of, say, rubber or paper, it is said to be insulated, and practically none of the electrons travelling along the wire can 'leak' from it.

It is usual to divide materials into two categories, conductors and insulators; but it should be realized that these are only relative terms. No material is a perfect conductor and no material is a perfect insulator. In general, metals such as copper and aluminium are very good conductors, whereas non-metallic materials such as rubber and oil are good insulators. Air, in its normal atmospheric condition, is a very good insulator.

The insulating property of oil and of fibrous materials, such as paper and cotton, is greatly affected by the amount of moisture they contain. The greater the moisture content, the poorer is the insulating property.

Certain materials such as germanium and silicon are neither good conductors nor good insulators and are referred to as *semi-conductors*. Germanium and silicon crystals, doped with certain impurities, have attained considerable importance during recent years as constituents of rectifiers and transistors.

## 15.2 Heating effect of an electric current

It was demonstrated in section 14.1 that an electric current flowing in a circuit produces heat. This is due to the fact that the material of which the circuit is made tends to resist the drift of electrons through it, and the electrical energy required to overcome this resistance is converted into thermal energy. It follows that good conductors have a low electrical resistance, whereas good insulators have a high electrical resistance.

The laws relating to the heating effect of an electric current were discovered by James Prescott Joule whose name is perpetuated by the unit of energy, the *joule* (section 1.9). The general principle of the apparatus used by Joule is shown in fig. 15.1. A copper calorimeter B contains a known mass of water, and any loss of heat is minimized by a layer of cotton wool C between B and an outer

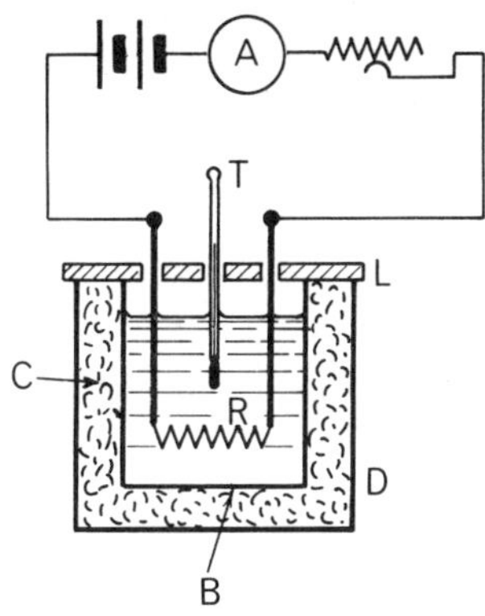


Fig. 15.1 Verification of Joule's laws.

container D. Stout copper wires pass through a wooden lid L and are attached at the lower ends to a helix R of known resistance. The temperature of the water is measured by a thermometer T. A constant current is passed through R for a known time and the rise of temperature of the water is noted. It is necessary to stir the water very thoroughly during the test to ensure uniform distribution of the heat generated in R.

From his researches, Joule deduced that the heat generated in a wire is proportional to:

(*a*) the square of the current, e.g. if the current is doubled, the rate of heat generation is increased fourfold,

(*b*) the resistance of the wire; e.g. if the length of the wire is doubled, the rate of heat generation by a given current is also doubled,
(*c*) the time during which the current is flowing.

Consequently, if a current $I$ flows through a circuit having resistance $R$ for time $t$,

$$\text{heat generated} \propto I^2 Rt.$$

The *unit of resistance* is that resistance in which a current of 1 ampere flowing for 1 second generates 1 joule of thermal energy. This unit of resistance is termed the *ohm* in commemoration of Georg Simon Ohm (1787–1854), the German physicist who enunciated Ohm's law (section 15.7). Hence, if a current $I$, in amperes, flows through a circuit having a resistance $R$, in ohms, for a time $t$, in seconds,

$$\text{heat generated, in joules} = I^2 Rt \qquad (15.1)$$

The symbol for ohm is the Greek letter Ω (capital omega). Sometimes it is more convenient to express the resistance in millionths of an ohm, in which case the resistance is said to be so many *microhms* (μΩ). On the other hand, when we are dealing with the resistance of insulating materials, the ohm is inconveniently small; consequently, another unit called the *megohm* (MΩ) is used, one megohm being a million ohms.

**Example 15.1** *A current of* 5 *A was maintained for* 6 *min through a* 1·3-Ω *resistor immersed in* 0·44 *l of water. The water equivalent of the vessel and heater was* 17·2 *g. Assuming no loss of heat, calculate* (a) *the heat generated and* (b) *the temperature rise of the water.*

(*a*) Heat generated in resistor $= I^2 Rt$ joules

$$= 5^2 \times 1{\cdot}3 \times 6 \times 60 = 11\,700 \text{ J}.$$

(*b*) Assuming the mass of 1 litre of water to be 1 kg,

mass of water in vessel = 0·44 kg

and water equivalent of vessel and heater

= 0·0172 kg,

∴ water equivalent of water, vessel and resistor

= 0·44 + 0·0172 = 0·4572 kg.

From expression (11.2), heat required to raise the temperature of 0·4572 kg of water by $t$ degrees Celsius

$$= 4190\ [\text{J/kg °C}] \times 0{\cdot}4572\ [\text{kg}] \times t$$

But heat generated in resistor = heat absorbed by water, vessel and resistor

i.e. $$11\,700\ [\text{J}] = 4190\ [\text{J/kg °C}] \times 0{\cdot}4572\ [\text{kg}] \times t$$

$$\therefore \quad \text{temperature rise} = t = 6{\cdot}11\text{°C}.$$

### 15.3 Electrical power

Since power is the rate of doing work, it follows that power is expressed in joules/second or *watts* (section 1.10); and since the electrical energy converted into thermal energy when a current $I$ amperes flows for $t$ seconds through a circuit having a resistance $R$ ohms is $I^2Rt$ joules,

$$\therefore \quad \text{electrical power} = \frac{I^2Rt}{t}\ \text{joules/second}$$

$$= I^2R\ \text{watts} \qquad (15.2)$$

### 15.4 Unit of electrical energy

In mechanics and heat we have used the SI unit of energy, namely the joule and its multiples the kilojoule and the megajoule. In electrical work, however, it has been the practice to express the energy in kilowatt hours and this practice will undoubtedly continue.

$$1\ \text{kilowatt hour} = 1000\ \text{watt hours}$$
$$= 1000 \times 3600\ \text{watt seconds or joules}$$
$$= 3\,600\,000\ \text{J} = 3600\ \text{kJ} = 3{\cdot}6\ \text{MJ}.$$

**Example 15.2** *The wire used in a heater element has a resistance of* 57 Ω. *Calculate* (a) *the power, in kilowatts, when the heater is taking a current of* 3·8 *A,* (b) *the energy absorbed in* 4 *hours, in kilowatt hours,* (c) *the cost of energy consumed if the charge is* 1·5 *pence per kilowatt hour.*

(*a*) $$\text{Power} = (3{\cdot}8)^2\ [\text{A}]^2 \times 57\ [\Omega] = 823\ \text{W}$$
$$= 0{\cdot}823\ \text{kW}.$$

(*b*) Energy absorbed $= 0{\cdot}823\ [\text{kW}] \times 4\ [\text{h}]$
$= 3{\cdot}292$ kW h.

(*c*) Cost of energy $= 3{\cdot}292\ [\text{kW h}] \times 1{\cdot}5\ [\text{p/kW h}]$
$= 4{\cdot}94$ p.

## 15.5 Fall of electrical potential along a circuit

Suppose CD in fig. 15.2 to represent a long thin wire of uniform diameter made of an alloy such as Eureka (60 per cent copper and 40 per cent nickel) having a much higher resistance than the same length and diameter of copper. The wire is connected across the terminals of an accumulator. A milliammeter A in series with a resistor R of, say, 1000 Ω is connected between terminal D and a contact K that can be moved along CD. The function of R is to limit the current through ammeter A, thereby protecting it from an excessive current and preventing an appreciable fraction of the current being diverted from length KD of the wire.

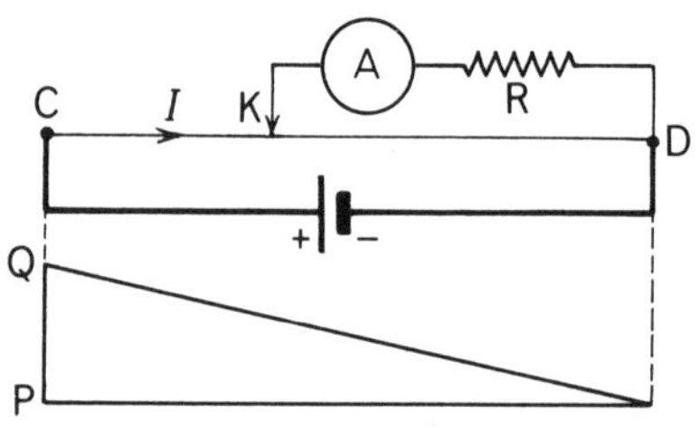


15.2 Fall of electrical potential.

It is found that as the distance between K and D is increased, the current through A also increases, as indicated by the height of the graph shown in fig. 15.2, where PQ represents the ammeter reading when K is at C. Since the direction of the current $I$ in CD is assumed to be from C to D, C is said to be at a higher electrical *potential* than D. In other words, when two points at different electrical potentials are connected together by a conductor, electricity is assumed to flow from the one at the higher potential to that at the lower potential (section 13.3). In fig. 15.2, the difference of potential between K and D is directly proportional to the deflection on A and therefore to the distance between D and the movable contact, the latter being at a higher potential than D.

## 15.6 Hydraulic analogy of fall of potential

A brass tube T (fig. 15.3) has a number of glass tubes attached to it. The tube is connected to a large jar A filled with a solution, such as methylene blue, which stands out clearly against a white background. At the other end of T, there is a tap C by which the flow of the liquid can be controlled. When C is shut, the level of the liquid in the glass tubes is the same as that in the jar; i.e. the whole pressure

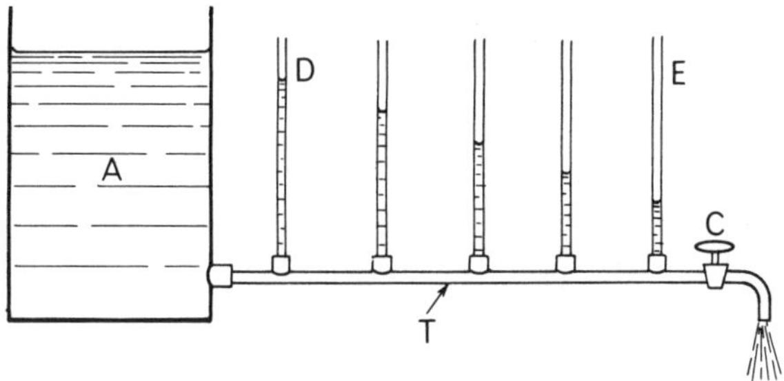


Fig. 15.3 Hydraulic analogy.

head of the liquid is available at C and there is no pressure drop in the pipe. Such a condition corresponds to a cell on open circuit, namely when there is no conducting path between the terminals. The whole of the electromotive force of the cell (section 15.8) then appears as a difference of electrical potential between the terminals.

If tap C is opened gradually, the liquid flows out at an increasing rate, and it is found that the height of the liquid in the glass tubes varies from a minimum in E to a maximum in D; in fact, if a straight rod be placed opposite the heights of the liquid columns in D and E, it will also coincide with the heights in the intermediate tubes, showing that the difference of pressure between any two points along the tube is proportional to the distance between them. Also, it is found that the more rapidly the liquid is allowed to run out at C, the greater is the difference of pressure between two adjacent tubes; in other words, the greater is the fall of pressure in a given length of pipe. These effects are somewhat similar to the electrical relationships discussed in the next section, where it is shown that the difference of potential across a resistor is proportional to the current and to the resistance.

## 15.7 Ohm's law

As long ago as 1827 Dr G. S. Ohm discovered that the current through a conductor, under constant conditions, was proportional to the difference of potential across the conductor. This fact can be demonstrated by connecting, as shown in fig. 15.4, a fixed resistor X, made of Eureka or other material whose resistance is not affected by temperature, in series with a variable resistor B and an ammeter A across the terminals of an accumulator. A voltmeter* V is connected across X.

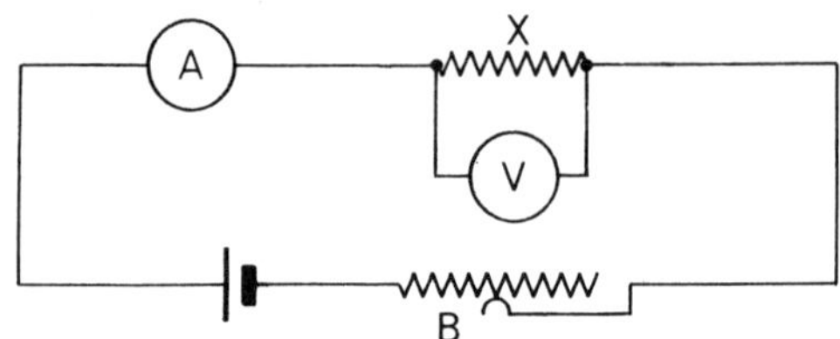


Fig. 15.4 Variation of p.d. with current.

Different currents are obtained by varying B, and for each current the reading on V is noted. It is found that the ratio

$$\frac{\text{potential difference across X}}{\text{current through X}}$$

remains constant within the limits of experimental error, i.e. the current through a circuit having a constant resistance is proportional to the difference of potential across the circuit.

Suppose X in fig. 15.5 to be a resistance box so constructed that the resistance of the wire between each pair of adjacent studs is 1 ohm and that, by means of an arm C, the total resistance of X can be varied in steps of 1 Ω up to, say, 5 Ω. With C on stud 1, the resistance of B is adjusted to give a reading of, say, 0·3 A on ammeter A and the reading on voltmeter V is noted. Arm C is then moved to stud 2, B is readjusted to bring the current back to 0·3 A and the reading on V is again noted. The test is repeated with C on each of the other studs. It is found that the ratio

* All voltmeters, except the electrostatic type, consist of a milliammeter connected in series with a resistor having a high resistance, as shown in fig. 15.2. At this stage, however, a voltmeter may be regarded merely as an instrument that indicates the difference of electric potential between the two points across which it is connected.

$$\frac{\text{potential difference across X}}{\text{resistance of X}}$$

remains constant, i.e. for a given current, the difference of potential between two points is directly proportional to the resistance of the circuit between those points.

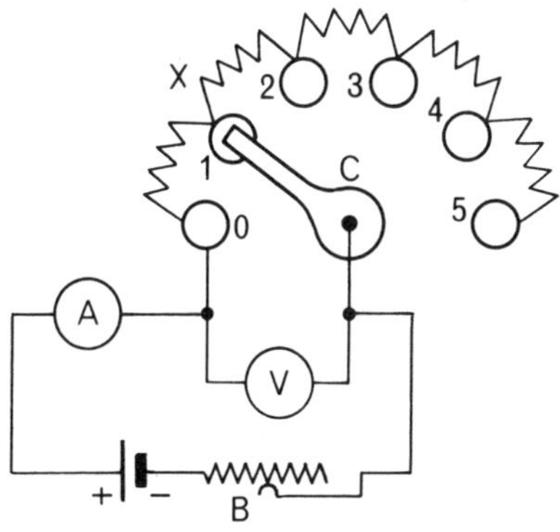


Fig. 15.5 Variation of p.d. with resistance.

The *unit of potential difference* (or p.d.) is the voltage* across a 1-ohm resistor carrying a current of 1 ampere and is termed the *volt* (V) after Count Alessandro Volta (1745–1827), an Italian physicist who was the first to discover how to make an electric battery (section 19.1). It follows from the above experiments that if the current through a resistor of 1 ohm is increased to, say, 3 A, the p.d. is 3 volts, and that if the resistance is increased to, say, 4 Ω with the current maintained at 3 A, the p.d. across the resistor becomes 3 × 4, namely 12 V. Hence if a circuit having a resistance of $R$ ohms is carrying a current of $I$ amperes, the p.d., $V$ volts, across the circuit is given by:

* The term *voltage* originally meant a difference of potential expressed in volts; but it is now used as a synonym for potential difference irrespective of the unit in which it is expressed. For instance, the voltage between the lines of a transmission system may be 400 kV, while in communication and electronic circuits, the voltage between two points may be 5 μV.

$$V = IR, \quad \text{or} \quad I = V/R, \quad \text{or} \quad R = V/I\dagger \qquad (15.3)$$

This relationship, known as *Ohm's Law*, is more complete and useful than that originally enunciated by Ohm. At this stage it is best to memorize Ohm's Law in one form only; and for this purpose, the form $I = V/R$ is probably the most convenient.

From (15.2), electrical power

$$= I^2 R \text{ watts}$$

$$= I \times IR = IV \text{ watts}\ddagger \qquad (15.4)$$

or

$$= \left(\frac{V}{R}\right)^2 \times R = \frac{V^2}{R} \text{ watts} \qquad (15.5)$$

**Example 15.3** *An electric kettle takes* 3 *kW from a* 240-*V supply. Calculate the time required to raise the temperature of* 1·7 *l of water from* 8°*C to the boiling point, if the efficiency of the kettle is* 85 *per cent. Also calculate the value of the current and the cost of the electrical energy consumed if the tariff is* 1·5 *pence per kilowatt hour.*

Assuming the mass of 1 litre of water to be 1 kg,

mass of water in kettle = 1·7 kg.

Rise of temperature = 100 − 8 = 92°C.

From expression (11.2),

useful heat = 4190 [J/kg °C] × 1·7 [kg] × 92 [°C]
= 655 000 J.

Since efficiency of kettle is 85 per cent,

input energy to kettle = 655 000/0·85 = 771 000 J

$$= \frac{771\,000\ [\text{J}]}{3\,600\,000\ [\text{J/kW h}]} = 0{\cdot}214 \text{ kW h.}$$

† It follows from this relationship that the *ohm* can be defined as *the resistance between two points of a circuit when a constant p.d. of* 1 *V, applied between these points, produces in the circuit a current of* 1 *A, assuming that there is no source of e.m.f. in the circuit between the two points.*

‡ This relationship gives us an alternative definition of the *volt*, namely *the difference of potential between two points of a circuit carrying a constant current of* 1 *A when the power dissipated between these points is* 1 *W.*

$$\text{Time required, in hours} = \frac{\text{energy, in kilowatt hours}}{\text{power, in kilowatts}}$$
$$= 0{\cdot}214\ [\text{kW h}]/3\ [\text{kW}] = 0{\cdot}0713\ \text{h}$$
$$= 0{\cdot}0713 \times 60 = 4{\cdot}28\ \text{min.}$$

$$\text{Alternatively, } 771\ 000\ [\text{J}] = 3000\ [\text{W}] \times \text{time in seconds}$$
$$\therefore \quad \text{time required} = 257\ \text{s} = 4{\cdot}28\ \text{min.}$$
$$\text{Current} = \frac{\text{power, in watts}}{\text{p.d., in volts}}$$
$$= 3000\ [\text{W}]/240\ [\text{V}] = 12{\cdot}5\ \text{A.}$$

$$\text{Cost of energy} = 0{\cdot}214\ [\text{kW h}] \times 1{\cdot}5\ [\text{p/kW h}]$$
$$= 0{\cdot}32\ \text{p.}$$

**Example 15.4** *An electric furnace is required to raise the temperature of* 2 *kg of brass from* 12°*C to* 500°*C in* 15 *min. The supply voltage is* 230 *V and the efficiency of the furnace is* 80 *per cent. Assuming the specific heat capacity of brass to be* 370 *J/kg °C, calculate* (a) *the energy absorbed, in kilowatt hours,* (b) *the power supplied to the furnace,* (c) *the current and* (d) *the resistance of the heating element.*

(*a*) Rise of temperature = 500 − 12 = 488°C.

From expression (11.3),

$$\text{useful energy} = 370\ [\text{J/kg °C}] \times 2\ [\text{kg}] \times 488\ [°\text{C}]$$
$$= 361\ 000\ \text{J.}$$

Since $1\ \text{kW h} = 3\ 600\ 000\ \text{J}$,

$$\text{useful energy} = \frac{361\ 000\ [\text{J}]}{3\ 600\ 000\ [\text{J/kW h}]} = 0{\cdot}1002\ \text{kW h.}$$

Allowing 80 per cent for the furnace efficiency,

$$\text{energy absorbed} = 0{\cdot}1002/0{\cdot}8 = 0{\cdot}125\ \text{kW h.}$$

$$(b)\ \text{Input power, in kilowatts} = \frac{\text{energy, in kilowatt hours}}{\text{time, in hours}}$$
$$= \frac{0{\cdot}125\ [\text{kW h}]}{(15/60)\ [\text{h}]} = 0{\cdot}5\ \text{kW.}$$

(*c*) From expression (15.4),

$$(0{\cdot}5 \times 1000)\ [\text{W}] = I \times 230\ [\text{V}]$$
$$\therefore \quad I = 2{\cdot}175\ \text{A.}$$

(*d*) From expression (15.3),

$$R = 230\ [\text{V}]/2{\cdot}175\ [\text{A}]$$
$$= 105{\cdot}8\ \Omega.$$

## 15.8 Electromotive force

It was shown in section 14.1 that when a current is passed for several minutes between two lead plates immersed in a dilute solution of sulphuric acid in water, a chocolate-colour coating is formed on the positive plate and that an electric current is obtained when the plates are then connected to a separate circuit; i.e. the combination of plates and acid became a voltaic cell* capable of converting chemical energy into electrical energy. Such an arrangement is a source† of an *electromotive force* (e.m.f.); in other words, an electromotive force represents something in the cell which impels electricity through a conductor connected across the terminals of that cell. Electromotive force is measured in *volts* and is represented by the symbol $E$. Difference of potential between two points is represented by the symbol $V$.

Consideration of theories accounting for the presence of an e.m.f. between the plates of a voltaic cell is outside the scope of this book. As far as we are concerned, the fact has to be accepted that when plates of different materials, such as lead and lead dioxide (as used in a lead-acid accumulator) or zinc and carbon (as used in the Leclanché primary cell), are placed in suitable solutions, an e.m.f. exists between the plates; and if a resistor is connected across them, an electric current flows through it.

Suppose $E$, in volts, to be the e.m.f. of cell B in fig. 15.6 and $I$ to be the current, in amperes, when a circuit having a resistance $R$, in ohms, is connected across the terminals. If the *internal resistance of the cell is negligible*,‡ the terminal voltage of the cell is the same as its e.m.f., namely $E$; hence

$$E = IR \quad \text{or} \quad I = E/R \tag{15.6}$$

* See section 19.1.

† There are sources of e.m.f. other than voltaic cells, e.g., magnetic flux cutting a conductor and junctions of dissimilar metals at different temperatures.

‡ The effect of the internal resistance of a cell is considered in section 16.3.

The e.m.f. of a cell can be measured by connecting a voltmeter across the terminals of the cell when the latter is on open circuit, i.e. when there is no other circuit connected across the terminals of the cell.

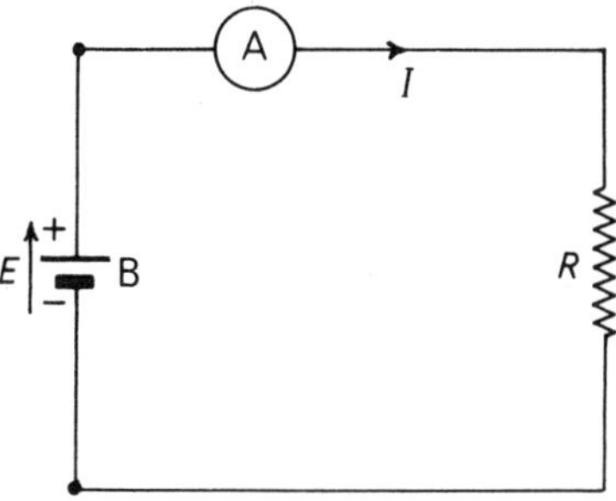


Fig. 15.6 E.M.F. of a cell.

## 15.9 Series connection of cells

Three accumulators were connected in series as in fig. 15.7, i.e. the negative terminal of the first cell was connected to the positive

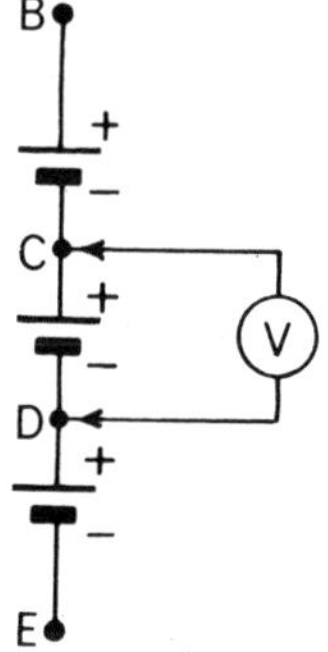


Fig. 15.7 Cells in series.

terminal of the second, and the negative of the second to the positive of the third. A voltmeter V was connected across various pairs of terminals in turn and the following results were obtained:

| *Terminals* | *E.M.F., volts* |
|---|---|
| ED | 2·16 |
| DC | 2·08 |
| CB | 2·1 |
| EB | 6·32 |

The sum of the e.m.f.s across ED, DC and CB = 6·34 volts, which is the same—within experimental error—as the total e.m.f. across EB. It is therefore evident that the e.m.f. of a number of cells connected in series is the sum of their individual e.m.f.s.

## 15.10 Parallel connection of cells

The positive ends of two similar accumulators were joined together to one end of a resistor R as shown in fig. 15.8. The negative ends were connected through milliammeters to the other end of R. When R was adjusted to 100 ohms, the readings on A, B and C were found to be 11·3, 8·5 and 19·8 milliamperes respectively. The milliam-

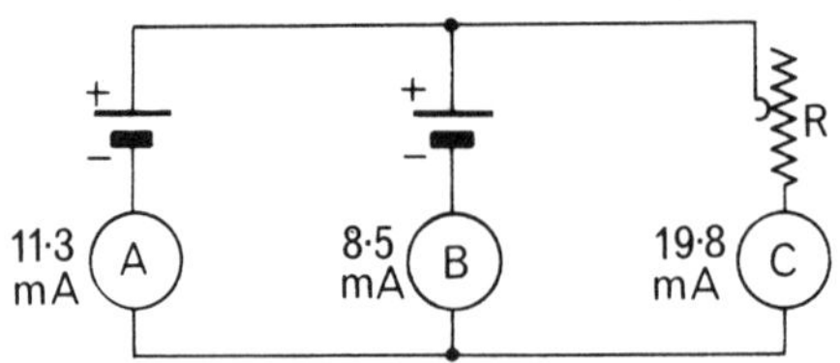


Fig. 15.8 Cells in parallel.

meters had relatively low resistance, so that the e.m.f. of the parallel cells is approximately (19·8/1000) [A] × 100 [Ω] = 1·98 V. These results indicate that when similar cells are in parallel, the e.m.f. is the same as that of one cell. On the other hand, the current is divided between the cells and the total current is the sum of the currents through the individual cells.

### *Summary of Chapter* 15

Conductors are materials which allow electricity to pass freely, whereas insulators are materials which almost entirely obstruct the passage of electricity.

The heating effect of a current is proportional to the square of the current, to the resistance and to the time; i.e.

$$\text{heat generated} = I^2Rt \text{ joules} \quad (15.1)$$

$$= IVt \text{ joules.}$$

$$\text{Power} = I^2R \text{ watts} \quad (15.2)$$

$$= IV \text{ watts} \quad (15.4)$$

$$= V^2/R \text{ watts.} \quad (15.5)$$

Ohm's Law:

$$I = V/R, \quad V = IR \quad \text{or} \quad R = V/I \quad (15.3)$$

For a battery having e.m.f. $E$ and negligible internal resistance,

$$I = E/R \quad (15.6)$$

When two or more cells are connected in series,

total e.m.f. = sum of their e.m.f.s.

When two or more cells are connected in parallel,

total current = sum of their currents.

## EXAMPLES 15

1. A certain circuit has a resistance of 50 Ω. Calculate the voltage across the circuit when the current is 3 A.
2. A certain circuit has a resistance of 800 Ω. Calculate the current when the voltage across the circuit is 2 kV.
3. When the voltage across a certain circuit is 5 mV, the current is 2 A. Calculate the resistance of the circuit in microhms.
4. Using Ohm's Law, complete the following table:

| Voltage, $V$ volts | Current, $I$ amperes | Resistance, $R$ ohms |
|---|---|---|
| 25 | 0·25 | – |
| – | 0·75 | 4 |
| 240 | – | 120 |
| – | 0·5 | 1000 |

(U.E.I., G1)

5. A coil of insulated copper wire has a resistance of 150 Ω. If the current through the coil is 0·5 A, calculate (*a*) the p.d. across the coil and (*b*) the power absorbed by the coil.
6. Calculate the current through a lamp which is taking 100 W from a 230-V supply. Also, find the corresponding resistance of the filament.

I

7. How many coulombs of electricity will flow in 10 hours when a p.d. of 2 mV is applied across a resistor having a resistance of 4 μΩ?
8. If a voltmeter has a resistance of 30 kΩ, calculate the current and the power absorbed when the p.d. across the voltmeter is 460 V.
9. How many joules of energy are taken from a 12-V battery supplying 2 A for 50 min? (N.C.T.E.C., G.2)
10. An ammeter of resistance 20 Ω is connected in series with a coil whose resistance is to be measured. When a current of 3 A passes through the coil, the potential difference across both coil and ammeter is 240 V. What is the true resistance of the resistor? What is the percentage error in calculating the resistance directly from the meter readings? (U.L.C.I., G1)
11. How many 150-W electric lamps could safely be used on a 250-V supply which will fuse at 5 A? (U.E.I., G2)
12. How many appliances rated at 250 V and 0·5 kW could safely be run from a 13-A supply? (U.E.I., G2)
13. An electric motor takes a current of 38 A at 240 V. Neglecting all losses, calculate the output power of the motor and the number of kilojoules of work done by the motor in 10 min.
14. An electric motor connected across a 460-V supply is developing 40 kW with an efficiency of 90 per cent. Calculate (*a*) the current, (*b*) the input power and (*c*) the cost of running the motor at that load for 6 h, if the charge for electrical energy is 1 p/kW h.
15. A generator is supplying 80 lamps, each taking 60 W at 240 V. Calculate (*a*) the total current supplied by the generator, (*b*) the number of kilowatt hours consumed in 4 h, (*c*) the output power of the engine driving the generator, if the efficiency of the latter is 85 per cent.
16. The energy absorbed in 10 min by an electric heater is 1·5 MJ. The supply voltage is 240 V. Calculate (*a*) the current, (*b*) the quantity of electricity, in coulombs, taken in 5 min and (*c*) the energy, in kilowatt hours, absorbed in 100 h.
17. The heating element of an electric kettle has a resistance of 80 Ω. Calculate the time required by a current of 3·1 A to raise the temperature of 1·2 litres of water from 14°C to the boiling point, if the efficiency of the kettle is 78 per cent. Also, calculate (*a*) the power and (*b*) the cost of the energy consumed at 1·2 p/kW h.
18. A current of 2·6 A was passed through a coil of wire immersed in 0·726 kg of water for 10 min. The initial and final temperatures were 13·2°C and 17·8°C respectively. Assuming the water equivalent of the containing vessel and heater to be 24 g and neglecting any loss of heat, calculate (*a*) the resistance of the coil, (*b*) the electrical power.
19. An electric kettle takes 3 kW when the terminal voltage is 240 V. Calculate (*a*) the resistance of the heating element, (*b*) the quantity of water, in litres, which could be heated from 5°C to the boiling point in 5 min, assuming the efficiency of the kettle to be 80 per cent.
20. An electric furnace is required to raise the temperature of 3 kg of iron from 16°C to 750 °C in 20 min. The supply voltage is 240 V. The efficiency of the furnace is 76 per cent and the specific heat capacity of the iron is

500 J/kg °C. Calculate (*a*) the current, (*b*) the resistance of the heating element, (*c*) the power and (*d*) the energy absorbed in kilowatt hours.

## ANSWERS TO EXAMPLES 15

1. 150 V.
2. 2·5 A.
3. 2500 μΩ.
4. 100 Ω, 3 V, 2 A, 500 V.
5. 75 V, 37·5 W.
6. 0·435 A, 529 Ω.
7. 18 000 000 C.
8. 15·33 mA, 7·05 W.
9. 72 000 J.
10. 60 Ω, 33·3 per cent.
11. 8 lamps.
12. 6 appliances.
13. 9·12 kW, 5472 kJ.
14. 96·6 A, 44·4 kW, 267 p.
15. 20 A, 19·2 kW h, 5·65 kW.
16. 10·42 A, 3126 C, 250 kW h.
17. 723 s, 769 W, 0·185 p.
18. 3·56 Ω, 24·1 W.
19. 19·2 Ω, 1·81 litres.
20. 5·03 A, 47·7 Ω, 1·207 kW, 0·402 kW h.

CHAPTER 16

# Electrical resistance

## 16.1 Resistors in series

Two resistors, $R_1$ and $R_2$, were connected in series across a battery, as shown in fig. 16.1, the current being indicated by an ammeter A. Also, voltmeters $V_1$, $V_2$ and $V_3$ were connected to measure the voltages across $R_1$, $R_2$ and the whole circuit respectively. The following readings were obtained:

| A | $V_1$ | $V_2$ | $V_3$ |
|---|---|---|---|
| 0·92 ampere | 2·2 volts | 3·6 volts | 5·8 volts |

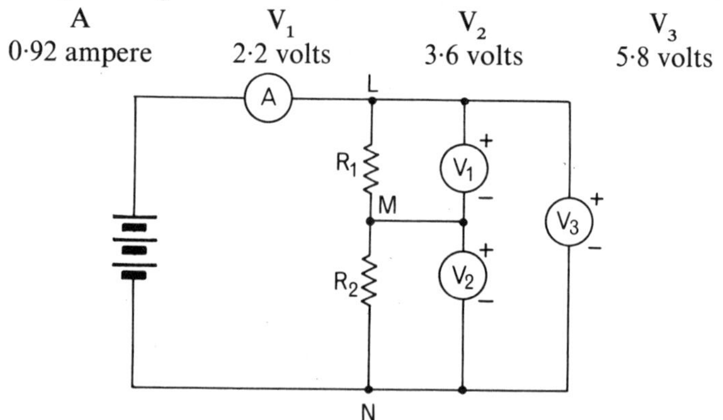


Fig. 16.1 Resistors in series.

By applying Ohm's law, we find that

$$R_1 = 2{\cdot}2/0{\cdot}92 = 2{\cdot}39\ \Omega,$$
$$R_2 = 3{\cdot}6/0{\cdot}92 = 3{\cdot}91\ \Omega$$

and the resistance of the whole circuit

$$= 5{\cdot}8/0{\cdot}92 = 6{\cdot}30\ \Omega.$$

The sum of $R_1$ and $R_2$

$$= 2{\cdot}39 + 3{\cdot}91 = 6{\cdot}30\ \Omega.$$

Hence it is seen that the total resistance of a circuit is the sum of the resistances in series; in other words, if resistors $R_1$, $R_2$ and $R_3$ be in series, the total resistance $R$ is given by

$$R = R_1 + R_2 + R_3 \tag{16.1}$$

Also, if the voltmeters are of the moving-coil type (section 17.10), $V_1$ and $V_2$ indicate that L is at a higher potential than M and that M is at a higher potential than N; and $V_3$ indicates that L is at a higher potential than N by an amount equal to the sum of the potential differences across LM and MN.

## 16.2 Resistors in parallel

Two resistors $R_1$ and $R_2$ were connected in parallel as in fig. 16.2, the currents being measured by ammeters $A_1$ and $A_2$. The total current was read on $A_3$ and the voltage across the resistors was read on voltmeter V. It was found that the instrument readings were:

| Voltmeter | $A_1$ | $A_2$ | $A_3$ |
|---|---|---|---|
| 5·9 volts | 1·5 amperes | 0·9 ampere | 2·4 amperes |

It will be seen that the total current is equal to the sum of the currents in the parallel circuits.

From Ohm's law, it follows that

$$R_1 = 5{\cdot}9/1{\cdot}5 = 3{\cdot}93\ \Omega$$

and

$$R_2 = 5{\cdot}9/0{\cdot}9 = 6{\cdot}55\ \Omega.$$

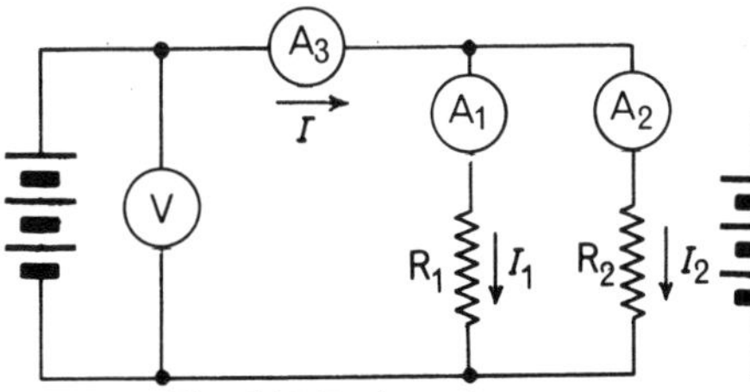

Fig. 16.2 Resistors in parallel

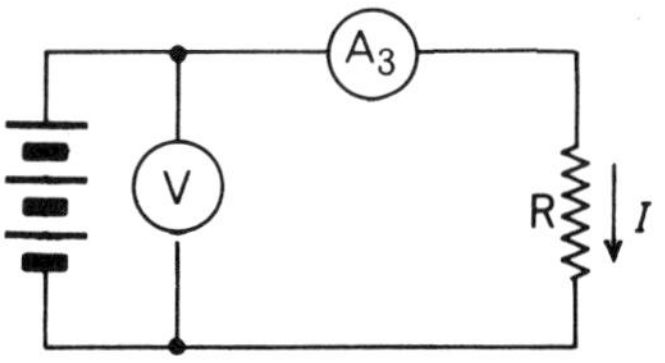

Fig. 16.3 Equivalent circuit of Fig. 16.2

The two resistors $R_1$ and $R_2$ in fig. 16.2 can be replaced by a single resistor R, as in fig. 16.3, the only condition being that the resistance of R must be such that the total current remains unaltered; i.e. in the above case:

$$2{\cdot}4\ [\text{A}] = \frac{5{\cdot}9\ [\text{V}]}{\text{resistance of R in ohms}}$$

$$\therefore \qquad \text{resistance of R} = 5{\cdot}9/2{\cdot}4 = 2{\cdot}46\ \Omega.$$

Hence, 2·46 Ω may be said to be *equivalent* to 3·93 Ω and 6·55 Ω in parallel. It is evident that the equivalent resistance is less than either of the parallel resistances, but there does not seem to be any obvious connection between the values. For this problem it is more satisfactory to derive the relationship by considering the general case than by taking particular values.

Suppose $I_1$ and $I_2$ amperes to be the currents in parallel resistors $R_1$ and $R_2$ respectively when the p.d. is $V$ volts (fig. 16.2). Then, by Ohm's law, $I_1 = V/R_1$ and $I_2 = V/R_2$. If $I$ is the total current indicated by $A_3$,

$$I = I_1 + I_2$$

$$= \frac{V}{R_1} + \frac{V}{R_2} = V\left(\frac{1}{R_1} + \frac{1}{R_2}\right)$$

If $R$ in fig. 16.3 represents the resistance of a single resistor through which a p.d. of $V$ volts produces the same current $I$ amperes, then $I = V/R$.

We have now derived two expressions for $I$; and by equating these expressions, we have

$$\frac{V}{R} = V\left(\frac{1}{R_1} + \frac{1}{R_2}\right)$$

$$\therefore \quad \frac{1}{R} = \frac{1}{R_1} + \frac{1}{R_2} \qquad (16.2)$$

Let us apply this expression to the experimental results considered above:

$$\frac{1}{R_1} = \frac{1}{3{\cdot}93} = 0{\cdot}254 \quad \text{and} \quad \frac{1}{R_2} = \frac{1}{6{\cdot}55} = 0{\cdot}1527$$

$$\therefore \quad \frac{1}{R} = \frac{1}{R_1} + \frac{1}{R_2} = 0{\cdot}254 + 0{\cdot}1527 = 0{\cdot}4067$$

$$\text{and} \quad R = \frac{1}{0{\cdot}4067} = 2{\cdot}46\ \Omega,$$

which is the same as the value previously derived.

The reciprocal of the resistance, that is, 1/resistance, is termed the *conductance*, the unit of conductance being 1 *siemens* (abbreviation, S). From expression (16.2) it follows that for resistors connected in parallel, the conductance of the equivalent resistor is the sum of the conductances of the parallel resistors.

The current in each of the parallel resistors $R_1$ and $R_2$ in fig. 16.2 can be expressed in terms of the total current thus:

$$V = I_1 R_1 = I_2 R_2 = (I - I_1)R_2$$

$$\therefore \quad (I - I_1) = \frac{I_1 R_1}{R_2}$$

so that $$I = I_1\left(1 + \frac{R_1}{R_2}\right)$$

$\therefore$ $$I_1 = I.\frac{R_2}{R_1 + R_2} \tag{16.3}$$

Similarly, $$I_2 = I.\frac{R_1}{R_1 + R_2}$$

**Example 16.1** *Three coils,* A, B *and C have resistances of* 8 Ω, 12 Ω *and* 15 Ω *respectively. Find the equivalent resistance when they are connected* (a) *in series,* (b) *in parallel.*

(*a*) With the three coils in series,

total resistance = 8 + 12 + 15 = 35 Ω.

(*b*) If $R$ be the equivalent resistance of the three coils in parallel, then

$$\frac{1}{R} = \frac{1}{8} + \frac{1}{12} + \frac{1}{15} = 0{\cdot}125 + 0{\cdot}0833 + 0{\cdot}0667$$
$$= 0{\cdot}275 \text{ siemens,}$$

$\therefore$ $$R = 3{\cdot}64\ \Omega.$$

**Example 16.2** *If coils* B *and* C *of Example* 16.1 *are connected in parallel and coil* A *is connected in series, as in fig.* 16.4(a), *across a* 20-*V supply, find* (a) *the resistance of the combined circuit,* (b) *the current in each coil.*

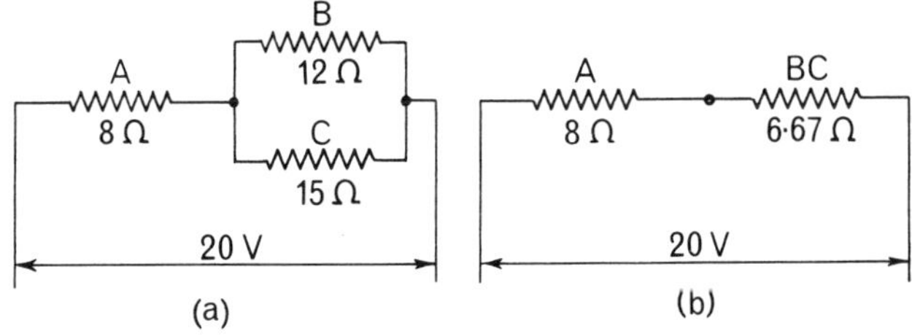


Fig. 16.4 Circuit of Example 16.2.

(*a*) Let $R$ be the equivalent resistance of B and C, then

$$\frac{1}{R} = \frac{1}{12} + \frac{1}{15} = 0{\cdot}0833 + 0{\cdot}0667 = 0{\cdot}15 \text{ S,}$$

$\therefore \qquad R = 6{\cdot}67\ \Omega,$

and $\qquad$ total resistance $= 8 + 6{\cdot}67 = 14{\cdot}67\ \Omega.$

(*b*) Total current $= \dfrac{20\ [\text{V}]}{14{\cdot}67\ [\Omega]} = 1{\cdot}364$ A, which is the current in coil A.

The p.d. across B and C in fig. 16.4(a) is the same as that across the equivalent 6·67-Ω resistor in fig. 16.4(b),

$$\therefore \qquad \text{total current} = \frac{\text{p.d. across B and C}}{\text{equivalent resistance of B and C}}$$

$$\text{i.e.} \qquad 1{\cdot}364\ [\text{A}] = \frac{\text{p.d. across B and C}}{6{\cdot}67\ [\Omega]}$$

$$\therefore \qquad \text{p.d. across B and C} = 1{\cdot}364 \times 6{\cdot}67 = 9{\cdot}098\ \text{V}.$$

Hence,
$$\text{current in B} = \frac{\text{p.d. across B}}{\text{resistance of B}}$$
$$= 9{\cdot}098\ [\text{V}]/12\ [\Omega] = 0{\cdot}758\ \text{A}$$

and
$$\text{current in C} = 1{\cdot}364 - 0{\cdot}758 = 0{\cdot}606\ \text{A}.$$

Alternatively, using expression (16·3), we have:

$$\text{current in B} = 1{\cdot}364 \times 15/(12 + 15)$$
$$= 0{\cdot}758\ \text{A}.$$

**Example 16.3** *The resistance of the heating element of an electric iron is* 180 Ω. *It is connected across a* 240-*V supply by two conductors, each having a resistance of* 1·3 Ω. *Calculate:* (a) *the voltage across the heating element,* (b) *the voltage drop in the cable,* (c) *the power absorbed by the electric iron and* (d) *the power wasted in the cable.*

(*a*) The circuit is shown in fig. 16.5.

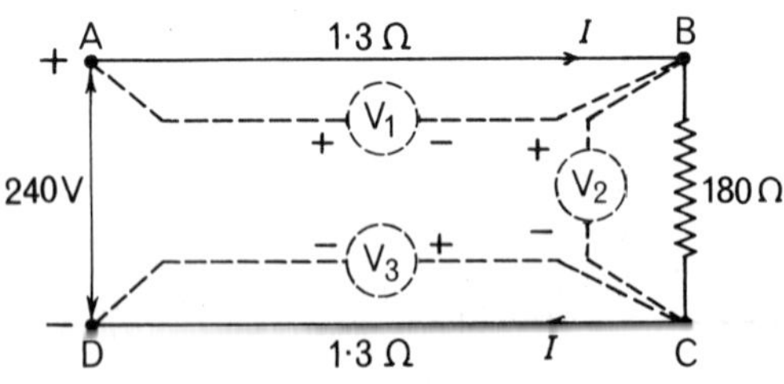


Fig. 16.5 Circuit of Example 16.3.

$$\text{Total resistance of circuit} = 1{\cdot}3 + 180 + 1{\cdot}3 = 182{\cdot}6\ \Omega.$$

$$\therefore \quad \text{current} = \frac{\text{p.d. (volts) between A and D}}{\text{resistance (ohms) between A and D}}$$

$$= 240\ [\text{V}]/182{\cdot}6\ [\Omega] = 1{\cdot}314\ \text{A}.$$

$$\text{Current also} = \frac{\text{p.d. between B and C}}{\text{resistance between B and C}}$$

$$\therefore \quad 1{\cdot}314\ [\text{A}] = \frac{\text{p.d. between B and C}}{180\ \ [\Omega]}$$

$$\therefore \quad \text{p.d. between B and C} = 1{\cdot}314 \times 180 = 236{\cdot}6\ \text{V}.$$

(*b*)
$$\text{Current also} = \frac{\text{p.d. between A and B}}{\text{resistance between A and B}}$$

$$\therefore \quad 1{\cdot}314\ [\text{A}] = \frac{\text{p.d. between A and B}}{1{\cdot}3\ [\Omega]}$$

$$\therefore \quad \text{p.d. between A and B} = 1{\cdot}314 \times 1{\cdot}3 = 1{\cdot}7\ \text{V}.$$

Similarly,
$$\text{p.d. between C and D} = 1{\cdot}314 \times 1{\cdot}3 = 1{\cdot}7\ \text{V}.$$

Hence, $\quad \text{total voltage drop in cable} = 1{\cdot}7 + 1{\cdot}7 = 3{\cdot}4\ \text{V}.$

The existence of these potential differences can be demonstrated by connecting moving-coil voltmeters $V_1$, $V_2$ and $V_3$, as shown dotted in fig. 16.5. The readings on the instruments indicate that the potential of A is 1·7 V above that of B, the potential of B is 236·6 V above that of C and the potential of C is 1·7 V above that of D. Further, it is seen that the sum of the voltmeter readings is equal to the total potential difference between A and D.

(*c*) $\text{Power absorbed by electric iron} = 1{\cdot}314\ [\text{A}] \times 236{\cdot}6\ [\text{V}] = 311\ \text{W}.$

(*d*) $\text{Power wasted in cable} = \text{current} \times \text{voltage drop in cable} = 1{\cdot}314\ [\text{A}] \times 3{\cdot}4\ [\text{V}] = 4{\cdot}47\ \text{W}.$

### 16.3 Effect of the internal resistance of a cell

In section 15.8 it was stated that when a cell is supplying a current and the internal resistance is negligibly small, the terminal voltage is equal to the e.m.f. of the cell. In practice, however, the resistance of the electrolyte of the cell is seldom negligible and can easily be taken into account. Thus, in fig. 16.6, TT represent the terminals

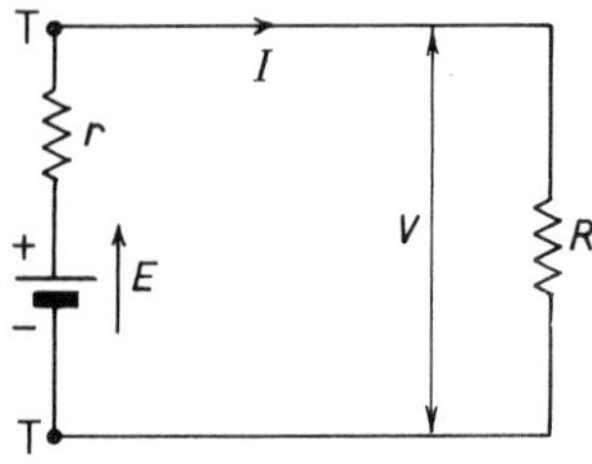


Fig. 16.6 E.M.F. of a cell

of a cell having an e.m.f. $E$ and an internal resistance $r$. If a circuit of resistance $R$ is connected across TT, then from expression (16.1), the total resistance of the circuit is $R + r$, and the current $I$ is given by:

$$I = \frac{E}{R + r}$$

and the terminal voltage $= V = IR$

$$= E - Ir.$$

The internal resistance of a cell can be determined by connecting a voltmeter across the terminals of the cell and noting the terminal voltage (*a*) with the cell on open circuit and (*b*) when the cell is supplying a known current $I$. The open-circuit reading gives the e.m.f., $E$, of the cell. If $V$ be the terminal voltage when the current is $I$, then:

$$V = E - Ir$$

so that $r = (E - V)/I.$

**Example 16.4** *Two resistors,* A *and* B, *having resistances of* 10 Ω *and* 15 Ω *respectively, are connected in parallel across a battery of four cells in series, as in fig.* 16.7. *Each cell has an e.m.f. of* 2 V *and an internal resistance of* 0·2 Ω. *Calculate:* (a) *the terminal voltage of the battery,* (b) *the current through each resistor and* (c) *the total power dissipated in the resistors.*

(*a*) Total e.m.f. of battery = 2 [V] × 4 = 8 V

and total internal resistance of battery = 0·2 [Ω] × 4 = 0·8 Ω.

If $R$ is the equivalent resistance of A and B:

$$\frac{1}{R} = \frac{1}{10} + \frac{1}{15} = 0{\cdot}1 + 0{\cdot}0667 = 0{\cdot}1667 \text{ S}$$

$\therefore \quad R = 6\ \Omega.$

Hence, total resistance of circuit $= 6 + 0{\cdot}8 = 6{\cdot}8\ \Omega$

and $\quad \text{current} = \dfrac{\text{total e.m.f.}}{\text{total resistance}}$

$= 8\ [\text{V}]/6{\cdot}8\ [\Omega] = 1{\cdot}177$ A.

Voltage across A and B

= total current × equivalent resistance of A and B
$= 1{\cdot}177\ [\text{A}] \times 6\ [\Omega] = 7{\cdot}06$ V
= potential difference across battery terminals PQ.

Alternatively, voltage across the 0·8-Ω resistor in fig. 16.7

$= 1{\cdot}177\ [\text{A}] \times 0{\cdot}8\ [\Omega] = 0{\cdot}94$ V,

∴ terminal voltage of battery
= battery e.m.f. — voltage drop due to internal resistance
$= 8 - 0{\cdot}94 = 7{\cdot}06$ V.

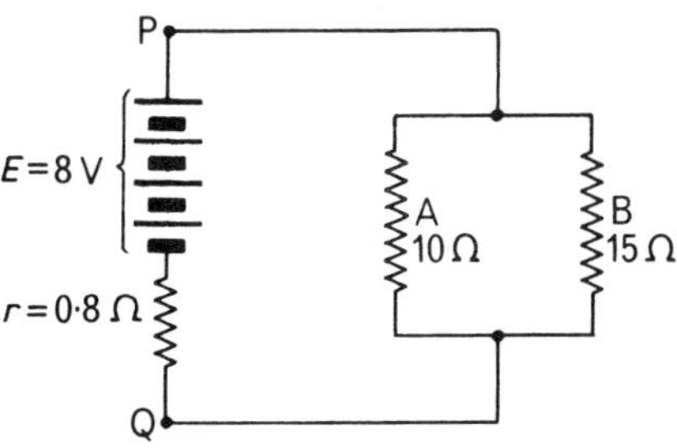


Fig. 16.7 Circuit diagram for Example 16.4.

(*b*) $\quad \text{Current through A} = \dfrac{\text{p.d. across A}}{\text{resistance of A}}$

$= 7{\cdot}06\ [\text{V}]/10\ [\Omega] = 0{\cdot}706$ A

and $\quad$ current through B $= 7{\cdot}06\ [\text{V}]/15\ [\Omega] = 0{\cdot}471$ A

or alternatively,

current through B $= 1{\cdot}177 - 0{\cdot}706 = 0{\cdot}471$ A.

(*c*) Power dissipated in resistors
= total current × p.d. across resistors
$= 1{\cdot}177\ [\text{A}] \times 7{\cdot}06\ [\text{V}] = 8{\cdot}3$ W.

## 16.4 Comparison of the resistance of different materials

In fig. 16.8, CD represents, say, one metre of Eureka wire having a diameter of about 0·5 mm, and DE and EF represent the same length and diameter of iron and copper wires respectively. The current is adjusted to about 0·5 A by means of a variable resistor R, and the differences of potential are measured by means of voltmeter V.

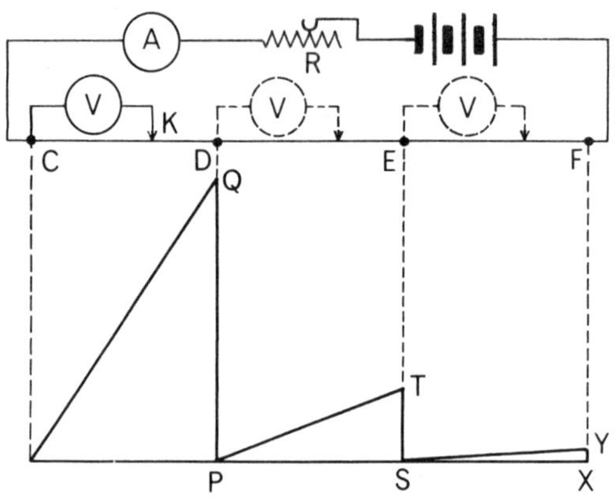


Fig. 16.8 Fall of potential in wires of different materials.

One end of V is connected to terminal C and the other end to a sliding contact K. As the latter is moved from C to D, the p.d. increases uniformly from zero to PQ, as shown by the graph.

The test is repeated with one end of V connected to terminal D, and K is moved from D to E. The p.d. again increases uniformly from zero to ST. A repetition of the test on wire EF gives a p.d. increasing from zero to XY.

Since both the currents and the dimensions are the same for the three wires, it follows that different materials having the same dimensions offer different resistances to the passage of an electric current. Thus, in the above experiment, ST is found to be about 7·5 times XY, while PQ is about 30 times XY: in other words, an iron wire of given length and diameter has 7·5 times the resistance of a similar copper wire, while a Eureka wire of the same dimensions has 30 times the resistance of the copper wire.

## 16.5 Relationship between the resistance and the dimensions of a conductor

From the experiment described in the preceding article, it follows that for a uniform wire of a given material the resistance between any two points of the wire (i.e. the p.d. between the two points divided by the current) is proportional to the distance between the two points.

Also, in section 16.2 it is explained that if two resistors, each of resistance $R$, are connected in parallel, the equivalent resistance $R_e$ is given by:

$$\frac{1}{R_e} = \frac{1}{R} + \frac{1}{R} = \frac{2}{R}$$

$$\therefore \qquad R_e = \tfrac{1}{2}R.$$

Hence, if two wires of the same material, having the same length and diameter, are connected in parallel, the resistance of the parallel wires is half that of one wire alone. But the effect of connecting two wires in parallel is exactly similar to doubling the area of the conductor. In just the same way the effect of connecting, say, five wires in parallel is the same as increasing the cross-sectional area of a wire five times, and the result is to reduce the resistance to a fifth of that of the original wire. In general, we may therefore say that the resistance of a conductor is inversely proportional to its cross-sectional area.

Apart from the effect of temperature, referred to in section 16.6, the only other factor that influences the value of the resistance is the nature of the material, as shown experimentally in the last article; hence we may now say that:

$$\left\{\begin{matrix}\text{resistance} \\ \text{of a wire}\end{matrix}\right\} = \frac{\text{length of wire}}{\text{cross-sectional area}} \times \left\{\begin{matrix}\text{a constant for a} \\ \text{given material}\end{matrix}\right\}$$

i.e.
$$R\ [\text{ohms}] = \frac{l\ [\text{metres}]}{a\ [\text{metres}^2]} \times \rho$$

so that
$$\rho = Ra/l \text{ ohm metres} \qquad (16.4)$$

where $\rho$ (Greek letter, 'rho') represents the constant.

If $l$ is 1 m and $a$ is 1 $\text{m}^2$ (e.g., if the resistance is being measured between the opposite faces of a metre cube of the material), the value of the resistance $= \dfrac{1\ [\text{m}]}{1\ [\text{m}^2]} \times \rho\ [\Omega\ \text{m}] = \rho$ ohm. Consequently the constant may be regarded as the resistance of a specimen 1 m long, and 1 $m^2$ in cross-sectional area, and is termed the *resistivity* of the material. For example, the resistivity of annealed copper at 20°C is 0·000 000 017 25 ohm metre. It is generally more convenient to use microhms rather than ohms, so that the above value then becomes 0·017 25 μΩ m (or 17·25 μΩ mm).

**Example 16.5** *Calculate the length of copper wire,* 1·5 *mm diameter, to have a resistance of* 0·3 Ω, *given that the resistivity of copper is* 0·017 μΩ m.

$$\text{Cross-sectional area of wire} = (\pi/4) \times (1{\cdot}5)^2 = 1{\cdot}766\ \text{mm}^2$$
$$= 1{\cdot}766 \times 10^{-6}\ \text{m}^2.$$

From expression (16.4), we have:

$$0{\cdot}3\ [\Omega] = 0{\cdot}017 \times 10^{-6}\ [\Omega\ \text{m}] \times \frac{\text{length}}{1{\cdot}766 \times 10^{-6}\ [\text{m}^2]}$$

$$\therefore \quad \text{length} = 31{\cdot}2\ \text{m}.$$

## 16.6 Effect of temperature on resistance

Let us connect a tungsten-filament lamp L in series with an ammeter A and a variable resistor R across a 240-volt supply, as in fig. 16.9. A voltmeter V is connected across the lamp. Assuming the current through the voltmeter to be very small compared with that through the lamp, we can determine the resistance of the lamp by dividing

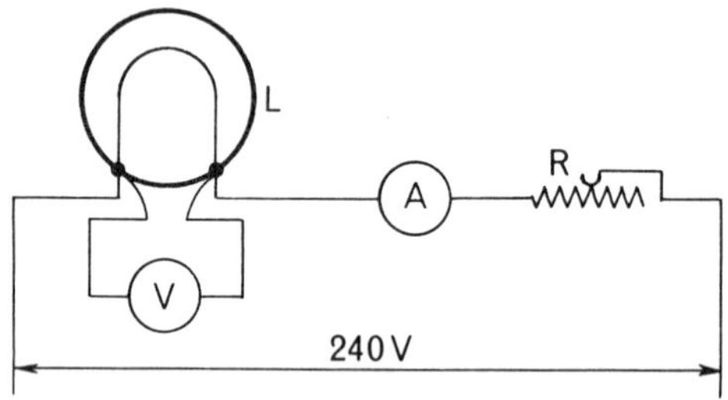


Fig. 16.9 Measurement of filament resistance at different voltages.

the voltmeter reading by the corresponding ammeter reading. By varying the value of R, we can vary the current through the lamp and thus vary the filament temperature. In this way we can determine the resistance of the tungsten filament over a wide range of temperature.

The graph in fig. 16.10 shows the results obtained with a 100-W, 240-V gas-filled lamp having a tungsten filament. It will be seen that as the temperature of the filament increases, so also does its resistance, and that the resistance at normal working temperature, i.e. with a terminal voltage of 240 V, is about ten times that of the filament when cold.

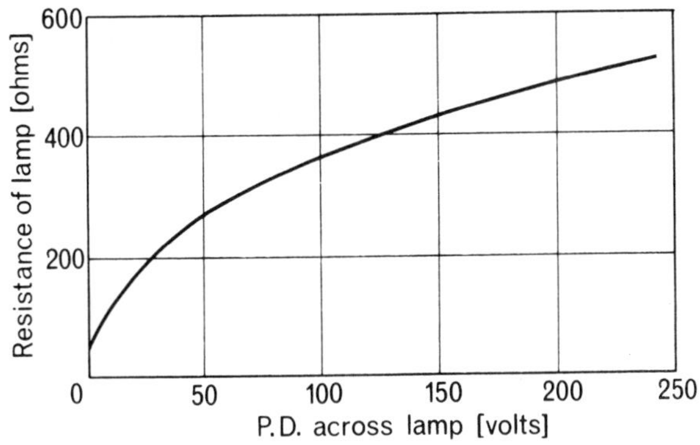


Fig. 16.10 Variation of filament resistance with voltage.

The resistance of all pure metals, such as copper, iron, tungsten, etc., increases with increase of temperature. On the other hand, the resistance of carbon, electrolytes and insulating materials, such as rubber, paper, etc., decreases with increase of temperature. The resistance of certain alloys, such as manganin (copper, manganese and nickel), remains practically constant for a considerable variation of temperature; consequently these alloys are employed whenever the resistance is required to remain practically independent of temperature, for instance in the construction of resistance boxes used for electrical measurements.

## 16.7 Temperature coefficient of resistance

If the resistance of a coil of insulated copper wire is measured at various temperatures up to, say, 200°C, it is found to vary as shown in fig. 16.11, the resistance at 0°C being, for convenience, taken as 1 ohm. The resistance increases uniformly with increase of temperature until it reaches 1·426 Ω at 100°C; i.e. the increase of resistance is 0·426 Ω for an increase of 100°C in the temperature, or 0·004 26 Ω/°C rise of temperature.

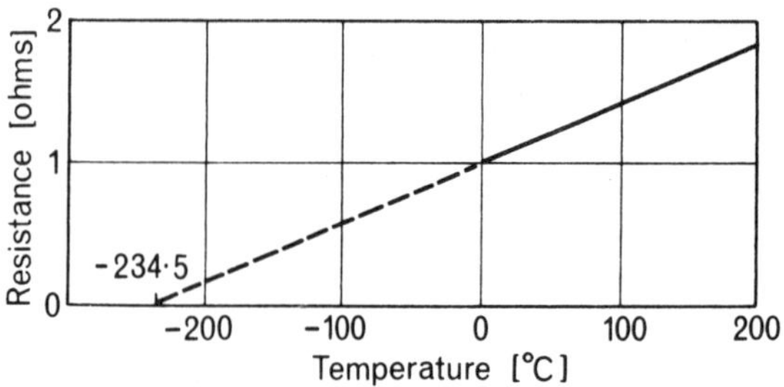


Fig. 16.11 Variation of resistance of copper with temperature.

The ratio of the increase of resistance per degree Celsius rise of temperature to the resistance at some definite temperature, adopted as standard, is termed *the temperature coefficient of resistance* and is represented by the Greek letter $\alpha$. From the above figures, it follows that *if the standard temperature is assumed to be* 0°C, the temperature coefficient of resistance of annealed copper

$$= \frac{0{\cdot}004\ 26\ \Omega/°\text{C}}{1\ \Omega} = 0{\cdot}004\ 26/°\text{C}.$$

If the straight line of fig. 16.11 is extended backwards, the point of intersection with the horizontal axis is found to be —234·5°C. This means that for the range of temperature over which copper conductors are usually operated, the resistance varies as if it would be zero at —234·5°C. Actually, the resistance/temperature relationship is not linear below about —50°C.

In general, if a material has a resistance $R_0$ at 0°C and a temperature coefficient of resistance $\alpha_0$ at 0°C, the increase of resistance for 1°C rise of temperature is $R_0\alpha_0$. If the temperature rises to $t$, the increase of resistance is $R_0\alpha_0 t$. Hence, if $R_t$ be the resistance at $t$,

$$\begin{aligned} R_t &= \text{resistance at } 0°\text{C} + \text{increase of resistance} \\ &= R_0 + R_0\alpha_0 t = R_0(1 + \alpha_0 t) \qquad (16.5) \\ &= R_0(1 + 0{\cdot}004\ 26\ t) \text{ for annealed copper.} \end{aligned}$$

It is usually inconvenient and unnecessary to measure the resistance at 0°C; for instance, in the case of the windings of electrical machines, it is often the practice to calculate the temperature rise after, say, three hours' operation at full load by measuring the resistance of the field coils before the commencement of the test and again immediately it is concluded. If $t_1$ be the initial temperature—usually taken as the temperature of the surrounding atmosphere—

and $t_2$ be the average temperature throughout the coils at the conclusion of the test, and if $R_1$ and $R_2$ be the corresponding resistances (fig. 16.12), then:

$$R_1 = R_0(1 + \alpha_0 t_1)$$

and
$$R_2 = R_0(1 + \alpha_0 t_2)$$

$$\therefore \qquad \frac{R_2}{R_1} = \frac{1 + \alpha_0 t_2}{1 + \alpha_0 t_1} \qquad (16.6)$$

Hence, if $R_1$, $R_2$ and $t_1$ are known, the value of $t_2$ can be calculated.

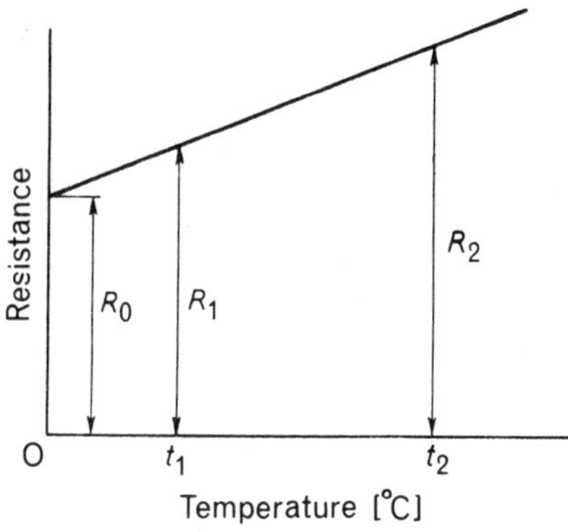


Fig. 16.12 Variation of resistance with temperature.

For rough estimate of the increase of resistance for a given increase of temperature or of the increase of temperature for a given increase of resistance, it is often the practice to assume the standard temperature to be 20°C, i.e. to assume the temperature of the atmosphere surrounding the windings to be 20°C. This means using a different value for the temperature coefficient of resistance; for instance, the temperature coefficient of resistance of annealed copper at 20°C is 0·003 92/°C. Hence for a coil of copper wire having a resistance $R_{20}$ at 20°C, the resistance $R_t$ at temperature $t$ is given by:

$$R_t = R_{20}\{1 + 0{\cdot}003\,92(t - 20)\}$$

In general,
$$R_t = R_{20}\{1 + \alpha_{20}(t - 20)\} \qquad (16.7)$$

where $\alpha_{20}$ = temperature coefficient of resistance at 20°C.

**Example 16.6** *The resistance of a coil of copper wire at the beginning of a heat test is 173 Ω, the temperature being 16°C. At the end of the test, the resistance is 212 Ω. Calculate the temperature rise of the coil. Assume the temperature coefficient of resistance of copper to be 0·004 26/°C at 0°C.*

Substituting in expression (16.6), we have:

$$\frac{212}{173} = \frac{1 + (0{\cdot}004\,26 \times t_2)}{1 + (0{\cdot}004\,26 \times 16)}$$

$$\therefore \qquad t_2 = 72{\cdot}5^\circ\text{C},$$

so that temperature rise of coil $= 72{\cdot}5 - 16 = 56{\cdot}5^\circ$C.

**Example 16.7** *A certain length of aluminium wire has a resistance of 28·3 Ω at 20°C. What is its resistance at 60°C? The temperature coefficient of resistance of aluminium is 0·004 03/°C at 20°C.*

Substituting in expression (16.7), we have:

$$R_{60} = 28{\cdot}3\{1 + 0{\cdot}004\,03(60 - 20)\}$$
$$= 32{\cdot}86\ \Omega.$$

*Summary of Chapter* 16

For resistors in series,

$$R = R_1 + R_2 + R_3 + \text{etc.} \qquad (16.1)$$

For resistors in parallel,

$$\frac{1}{R} = \frac{1}{R_1} + \frac{1}{R_2} + \frac{1}{R_3} + \text{etc.} \qquad (16.2)$$

For two resistors in parallel,

$$I_1 = I . \frac{R_2}{R_1 + R_2} \qquad (16.3)$$

$$R = \frac{\rho l}{a} \qquad (16.4)$$

The resistance of pure metals increases and that of carbon and insulating materials decreases with increase of temperature. The resistance of certain alloys, such as Eureka, is practically independent of temperature variation.

$$R_t = R_0(1 + \alpha_0 t) \quad (16.5)$$

or

$$R_t = R_{20}\{1 + \alpha_{20}(t - 20)\} \quad (16.7)$$

## EXAMPLES 16

1. Two resistors, A and B, have resistances of 40 Ω and 70 Ω respectively. Calculate the resistance of the combination when they are connected (*a*) in series, (*b*) in parallel.
2. A resistor A has a resistance of 12 Ω. Calculate the resistance of another resistor B so that when A and B are in parallel, their combined resistance is 8 Ω.
3. State the THREE main effects of an electrical current, giving one practical example of each.

   Three coils have resistances of 5 Ω, 4 Ω and 2 Ω respectively.

   (*a*) Calculate the resulting resistance when they are connected in parallel.
   (*b*) If a potential difference of 12 V is applied to the circuit, what is the current taken from the supply? (N.C.T.E.C., G1)
4. Three resistors of 10 Ω, 15 Ω and 30 Ω can be connected either (*a*) in series, or (*b*) in parallel across a 140-V supply. Determine the current taken from the supply in each case and the current taken by the 30-Ω resistor in the case of the parallel connection. (U.L.C.I., G1)
5. State Ohm's law.

   Three cells each have an e.m.f. of 2 V and an internal resistance of 0·6 Ω and are connected in series, supplying current to a resistor of 8·2 Ω. Find (*a*) the current flowing, (*b*) the volt drop across the resistor.

   If the cells are now connected in parallel, find the reduction in the current flow. (E.M.E.U., G2)
6. Three resistors are to be connected in parallel to provide an equivalent network resistance of 1·5 Ω. One of the resistors is 3 Ω and another is 5 Ω. Calculate the value of the third resistor. (N.C.T.E.C., G2)
7. Two circuits, A and B, are connected in parallel across a 50-V supply. Circuit A is found to absorb 120 W, and the total current is 4·2 A. Calculate the resistances of A and B and the power absorbed by B.
8. Two resistors, 80 Ω and 20 Ω, are connected in parallel. This parallel group is connected in series with a 5-Ω resistor to a 100-V supply. Calculate the current in each resistor and the power loss in the 5-Ω resistor. (E.M.E.U., G2)
9. Three resistors are in series across a supply of 235 V and a current of 5A is flowing. If two of the resistors are 10 Ω and 25 Ω, calculate (*a*) the resistance of the third resistor and (*b*) the voltage drop across the 25-Ω resistor.

   If the 10-Ω and the 25-Ω resistors were connected in parallel, what fourth resistor in series with the third would be required to maintain the same total resistance? (U.L.C.I., G1)
10. Two resistors of 30 Ω and 20 Ω are connected in parallel. Two further resistors of 75 Ω and 50 Ω are also connected in parallel. The two groups of parallel resistors are then connected in series with an electric supply.

Make a sketch of the circuit, and if the total current in the circuit is 2 A, determine (*a*) the total potential difference across the circuit and (*b*) the current through the 50-Ω resistor. (U.L.C.I., G1)

11. A voltmeter connected across an accumulator reads 2·06 V when the cell is on open circuit. The voltmeter reading immediately falls to 1·92 V when a 0·2-Ω resistor is connected across the terminals of the cell. Calculate (*a*) the current, (*b*) the internal resistance of the cell, (*c*) the total electrical power generated in the cell, (*d*) the power dissipated in the external resistor and (*e*) the power wasted due to the internal resistance of the cell.
12. A battery of 10 primary cells has an e.m.f. of 15 V. When a 30-Ω resistor is connected across the battery, the terminal voltage is 12 V. Calculate (*a*) the current, (*b*) the internal resistance per cell and (*c*) the power dissipated in the external circuit.
13. A battery consists of 6 similar cells in series. Each cell has an e.m.f. of 2 V and an internal resistance of 0·1 Ω. The battery is connected to a lamp of resistance 3·4 Ω. Calculate the current supplied by the battery and the p.d. across the battery terminals. (U.E.I., G1)
14. Four cells, each having an e.m.f. of 1·5 V, are connected in series. Each cell has an internal resistance of 0·5 Ω. A resistor is connected across the four cells and the current in the circuit is 0·75 A. What is the terminal voltage across the four cells and what is the resistance of the resistor? (U.L.C.I., G1)
15. A circuit consists of two resistors of 6 Ω and 4 Ω respectively in parallel, connected in series with a 10-Ω resistor. The circuit is connected across a battery having an e.m.f. of 12 V and an internal resistance of 1 Ω. Determine (*a*) the current in the 4-Ω resistor and (*b*) the p.d. across the 10-Ω resistor.

    Sketch a diagram of the circuit and show on it the positions in which an ammeter and a voltmeter would be placed to check the above calculated values. (U.E.I., G2)
16. The resistance of 20 m of wire of 0·1 mm diameter is 1000 Ω. Calculate the resistivity of the wire.
17. Calculate the length of wire required to make a resistor which takes 5 A from a 230-V supply. The wire is 0·1 mm in diameter and has a resistivity of 0·4 μΩ m.
18. Calculate the resistance of 1 km of aluminium wire having a diameter of 4 mm, given that the resistivity of aluminium is 0·028 μΩ m.
19. Calculate the cross-sectional area of a copper conductor, 300 m long, such that it may carry 500 A with a voltage drop of 8 V. Assume the resistivity of copper to be 0·019 μΩ m.
20. A coil consists of 400 turns of insulated copper wire having a cross-sectional area of 0·5 $mm^2$. The mean length per turn is 450 mm and the resistivity of copper at working temperature is 0·02 μΩ m. Calculate (*a*) the resistance of the coil and (*b*) the power absorbed when the p.d. across the coil is 20 V.
21. A copper rod, 6 mm in diameter and 1 m long, has a resistance of 620 μΩ. If this rod were drawn out to a wire having a cross-sectional area of 0·08 $mm^2$, what would be its resistance?

22. The field coils of a direct-current motor are wound with insulated copper wire and have a resistance of 85 Ω at 10°C. What is their resistance at 80°C? Assume the temperature coefficient of resistance of copper to be 0·0043/°C at 0°C.
23. What is meant by the *temperature coefficient of resistance* of a conductor? During a test on a conductor the following readings were obtained:

| Temperature, °C | 10 | 20 | 30 | 45 | 60 |
|---|---|---|---|---|---|
| Resistance, Ω | 3·65 | 3·8 | 3·94 | 4·16 | 4·39 |

Plot a graph showing the relationship between the resistance of the conductor and the temperature. From the graph, determine the temperature coefficient of resistance at 0°C. (U.E.I., G2)
24. A constant voltage of 240 V is applied to the field windings of a motor. At 15°C the current through the windings is 2 0 A. After the motor has been running for some time, the field current was found to be 1·9 A. Calculate the temperature of the field windings given that the temperature coefficient of resistance for the coil windings is 0·0042/°C at 0°C. (E M E.U , G2)
25. A coil of insulated copper wire has a resistance of 210 Ω at 20°C Calculate the resistance of the coil at 70°C, assuming the temperature coefficient of resistance of copper to be 0·0039/°C at 20°C.
26. An aluminium conductor is 2 km long and has a diameter of 5 mm. Calculate its resistance at 60°C, given that the resistivity of aluminium at 20°C is 0·0283 μΩ m and its temperature coefficient of resistance at 20°C is 0·004 03/°C.

## ANSWERS TO EXAMPLES 16

1. 110 Ω, 25·45 Ω.
2. 24 Ω.
3. 1·053 Ω, 11·4 A.
4. 2·545 A; 28 A, 4·667 A.
5. 0·6 A, 4·92 V; 0·362 A
6. 7·5 Ω.
7. 20·83 Ω, 27·78 Ω, 90 W.
8. 0·952 A, 3 81 A, 4·762 A; 113·5 W
9. 12 Ω, 125 V, 27·86 Ω
10. 84 V 1 2 A.
11. 9 6 A, 0·0146 Ω, 19·78 W, 18·43 W, 1 35 W.
12. 0 4 A, 0·75 Ω, 4·8 W
13. 3 A, 10·2 V.
14. 4·5 V, 6 Ω
15. 0·537 A, 8·95 V.
16. 0·3925 μΩ m.
17. 0·903 m.
18. 2·23 Ω
19. 356 mm$^2$.
20. 7·2 Ω, 55·5 W.
21. 77·3 Ω
22. 109 5 Ω.
23. 0 004 22/°C (approx.).
24. 28·3 C.
25. 251 Ω.
26. 3·35 Ω.

CHAPTER 17

# Electromagnetism

## 17.1 Magnetic field

Before dealing with the magnetic effect of an electric current, it is necessary to explain what is meant by a magnetic field.

If a permanent magnet is suspended so that it is free to swing in a horizontal plane, as in fig. 17.1, it is found that it always takes up a position such that a particular end points towards the earth's North Pole. That end is, therefore, said to be the *north-seeking* end

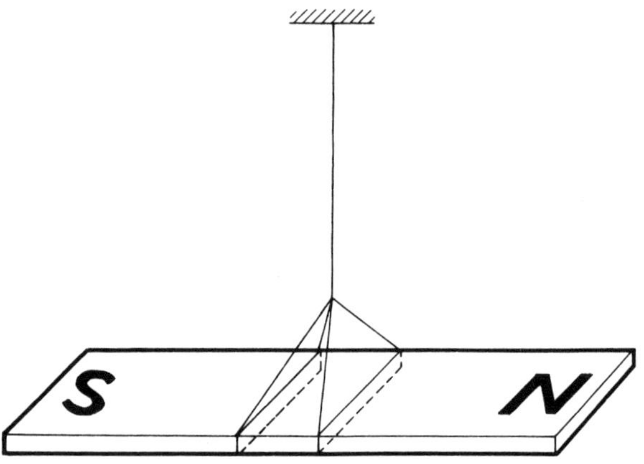


Fig. 17.1 A suspended permanent magnet.

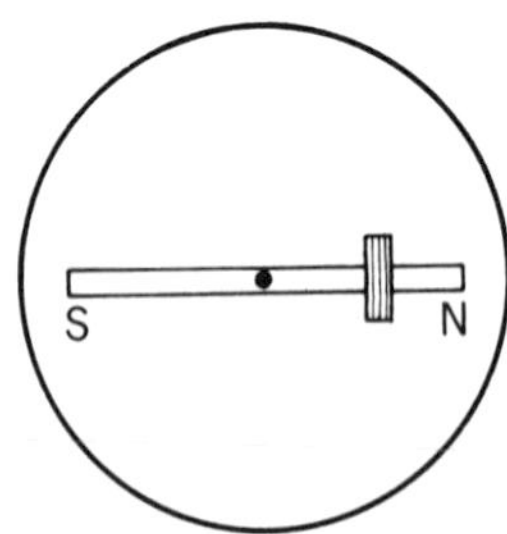


Fig. 17.2 Compass needle.

of the magnet; similarly, the other end is the *south-seeking* end. For short, these are referred to as the *north* (or N) and *south* (or S) *poles* respectively of the magnet. In the case of small compass-needles, the north pole is usually indicated by a small crosspiece, as shown in fig. 17.2, or by an arrowhead.

If the N pole of another magnet is brought near the N pole of the suspended magnet, the latter is repelled; whereas attraction occurs if the S pole of the second magnet is brought near the N pole of the suspended magnet. In general we may therefore say that like poles repel each other whereas unlike poles attract each other.

Let us next place a permanent magnet on a table, cover it over with a sheet of smooth cardboard and sprinkle some iron filings uniformly over the sheet. Slight tapping of the sheet causes the filings to set themselves in curved chains between the poles, as shown in fig. 17.3. The shape and density of these chains enable one to form a mental picture of the magnetic condition of the space or 'field' around a bar magnet and lead to the idea of *lines of magnetic flux*.

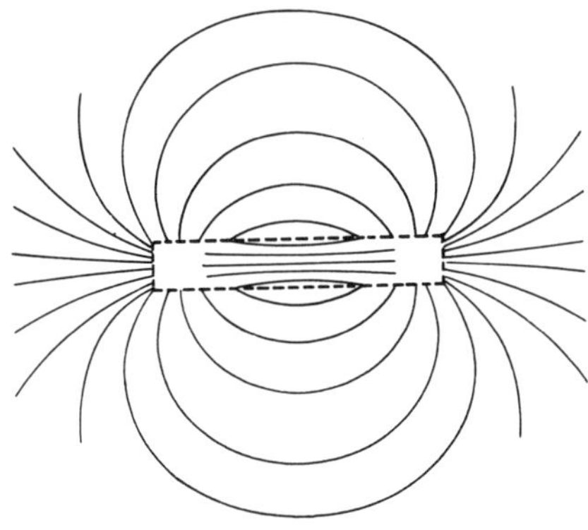

Fig. 17.3 Use of iron filings for determining distribution of magnetic field.

It is necessary to emphasize at this stage that these lines of magnetic flux have no physical existence; they are purely imaginary and were introduced by Michael Faraday as a means of visualizing the distribution and density of a magnetic field. It is important, however, to realize that the magnetic flux permeates the *whole* of the space occupied by that flux.

As to the nature of the magnetic field, all we can say is that it appears to be some form of strain both in the space occupied by the magnet and in that around the magnet—somewhat analogous to the skin of a rubber balloon in that it tends to collapse, and in consequence to keep to the shape that involves the minimum of stress. Thus, if a balloon is dented inwards, say by pressing a finger against it, the skin reacts against the finger, tending to push the latter out of the way. In sections 17.9 and 18.6 it is shown that when a magnetic field is distorted by an electric current, the magnetic field reacts in such a way that it tends to push away the conductor carrying the current.

## 17.2 Direction of magnetic field

The direction of a magnetic field is taken as that in which the north-seeking pole of a magnet points when the latter is suspended in the field. Thus, if a bar magnet NS rests on a table and four compass-needles are placed in positions indicated in fig. 17.4, it is found that the needles take up positions such that their axes coincide with the corresponding chain of filings (fig. 17.3) with their north poles all

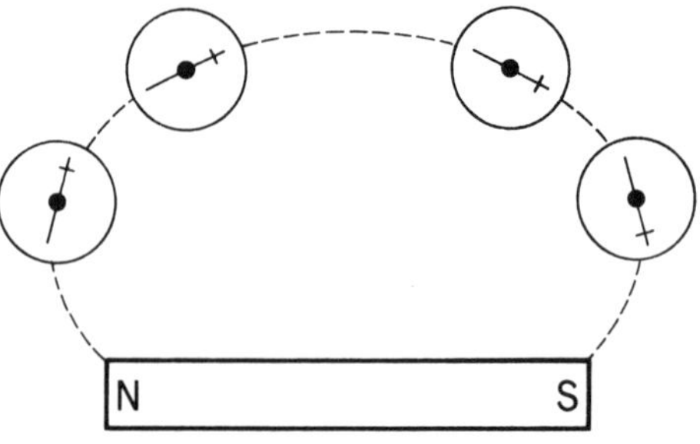


Fig. 17.4 Use of compass needles for determining direction of magnetic field.

pointing along the dotted line, from the N pole of the bar magnet to its S pole. Hence the magnetic flux is assumed to pass through the magnet, emerge from the N pole and return to the S pole. This may be expressed in another way, thus: if a compass-needle is placed near one end of a magnetized iron rod and if its N pole is repelled from the rod, the direction of the magnetic flux in that region is outwards from the rod, and the adjacent surface of the rod has a north polarity.

## 17.3 Characteristics of lines of magnetic flux

In spite of the fact that lines of magnetic flux have no physical existence, they do form a very convenient and useful basis for explaining various magnetic effects. For this purpose, lines of magnetic flux are assumed to have the following properties:

1. *The direction of a line of magnetic flux at any point in a non-magnetic medium, such as air, is that of the north-seeking pole of a compass-needle placed at that point,* as already described in section 17.2.

2. *Each line of magnetic flux forms a closed path,* as shown by the dotted lines in fig. 17.5 and 17.6. This means that a line of flux emerging from any point at the N-pole end of a magnet passes through the surrounding space back to the S-pole end and is then assumed to continue through the magnet to the point at which it emerged at the N-pole end.

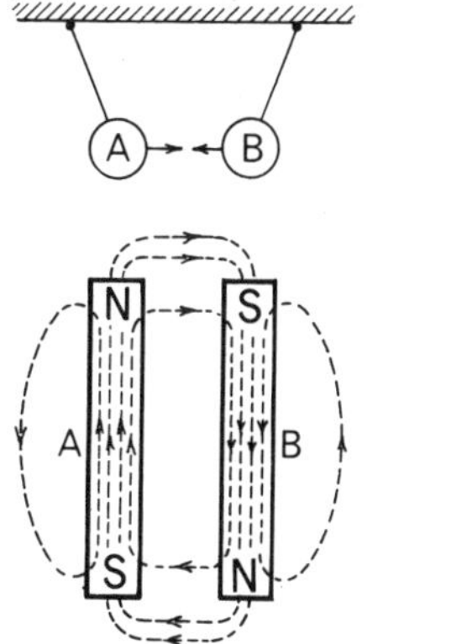


Fig. 17.5 Attraction between magnets.

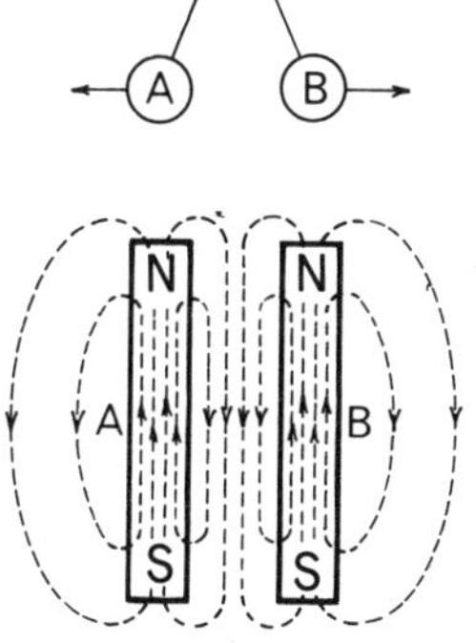


Fig. 17.6 Repulsion between magnets.

3. *Lines of magnetic flux never intersect.* This follows from the fact that if a compass needle is placed in a magnetic field, its north-seeking pole will point in one direction only, namely in the direction of the magnetic flux at that point.

4. *Lines of magnetic flux are like stretched elastic cords, always trying to shorten themselves.* This effect can be demonstrated by suspending two permanent magnets, A and B, parallel to each

other, with their poles arranged as in fig. 17.5. The distribution of the resultant magnetic field is indicated by the dotted lines. The lines of magnetic flux passing between A and B behave as if they were in tension, trying to shorten themselves and thereby causing the magnets to be attracted towards each other. In other words, unlike poles attract each other.

5. *Lines of magnetic flux which are parallel and in the same direction repel one another*. This effect can be demonstrated by suspending the two permanent magnets, A and B, with their N poles pointing in the same direction, as in fig. 17.6. It will be seen that in the space between A and B the lines of flux are practically parallel and are in the same direction. These flux lines behave as if they exerted a lateral pressure on one another, thereby causing magnets A and B to repel each other. Hence like poles repel each other.

## 17.4 Magnetic induction and magnetic screening

In fig. 17.7, N and S are the poles of a U-shaped permanent magnet M, A and B are soft-iron rectangular blocks attached to the magnet and C is a hollow cylinder of soft-iron placed midway between A and B. The dotted lines in fig. 17.7 represent the paths of the magnetic flux due to the permanent magnet. It will be seen that this flux passes through A, B and C, making them into temporary magnets

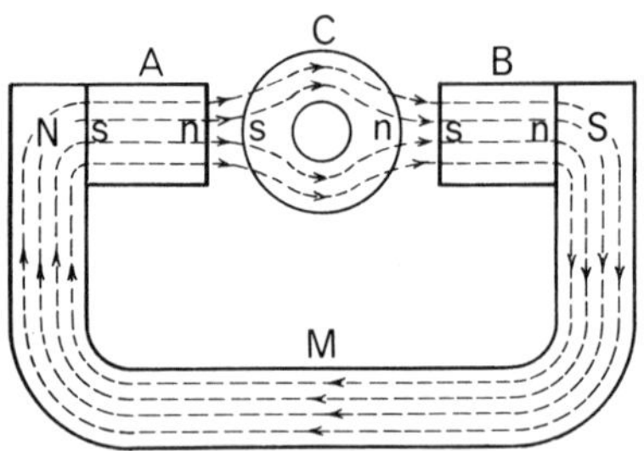


Fig. 17.7 Magnetic induction and screening.

with the polarities indicated by *n* and *s*, i.e. A, B and C are magnetized by *magnetic induction*. Being of soft-iron, A, B and C will lose almost the whole of their magnetism when they are removed from the influence of the permanent magnet M.

Fig. 17.7 also shows that no* flux passes through the air space inside cyclinder C. Consequently, a body placed in this space would be screened from the magnetic field around it. Magnetic screens are used to protect cathode-ray tubes (as used in television) and instruments such as moving-iron ammeters and voltmeters (section 17.8) from external magnetic fields.

The effect of magnetic induction can also be shown by suspending a soft-iron bar A by a spring B above the poles of a U-shaped permanent magnet M, as in fig. 17.8. Bar A is magnetized by induction as shown by the dotted lines and is attracted towards M, thereby increasing the tension in B. This force of attraction may be

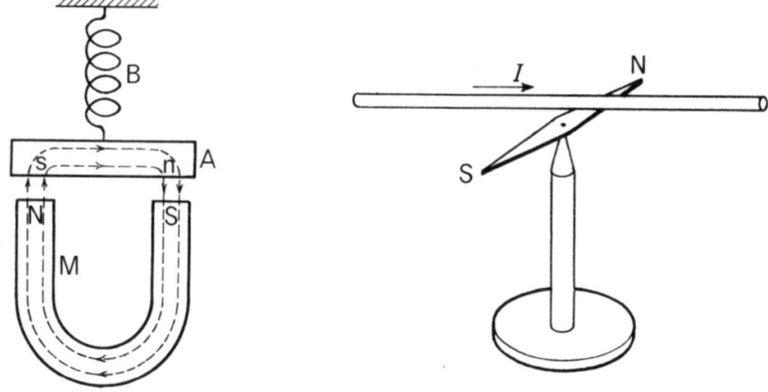


Fig. 17.8 Magnetic induction.

Fig. 17.9 Oersted's experiment.

explained as being due to the tension in the flux between A and the poles of M pulling A towards N and S, or merely as the attraction between N and *s* and between S and *n*; but these are simply two ways of stating the same thing.

## 17.5 Magnetic field due to an electric current

In section 14.1 it was demonstrated that one of the characteristics of an electric current is its ability to produce a magnetic effect. The discovery of this phenomenon by Oersted at Copenhagen, in 1820, was the first definite demonstration of a relationship between electricity and magnetism. Oersted found that if he placed a wire carrying an electric current above a magnetic needle (fig. 17.9) and

* Actually, there must be some magnetic flux across the air space inside the soft-iron cylinder C, but the density of this flux is so low that, for most purposes, it can be assumed to be zero.

in line with the normal direction of the latter, the needle was deflected clockwise or counterclockwise, depending upon the direction of the current. This phenomenon may be better understood from experiments made with the apparatus shown in fig. 17.10.* A stout copper wire W passes vertically through a hole in a sheet of cardboard or glass G placed horizontally, with a number of compass-needles arranged around W.

With no current through the wire, the north-seeking poles of the needles all point towards the earth's north pole. Immediately a current of, say, 10 amperes is switched on, the needles are deflected, and after a number of oscillations they come to rest with their axes lying roughly on a circle having the conductor as its centre—as shown in fig. 17.10.

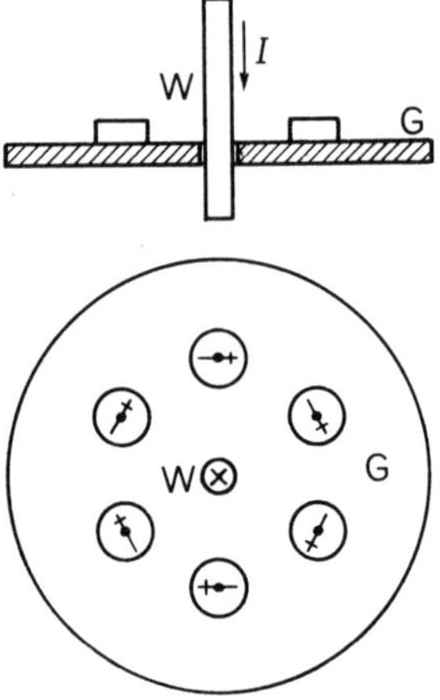


Fig. 17.10 Magnetic field of an electric current.

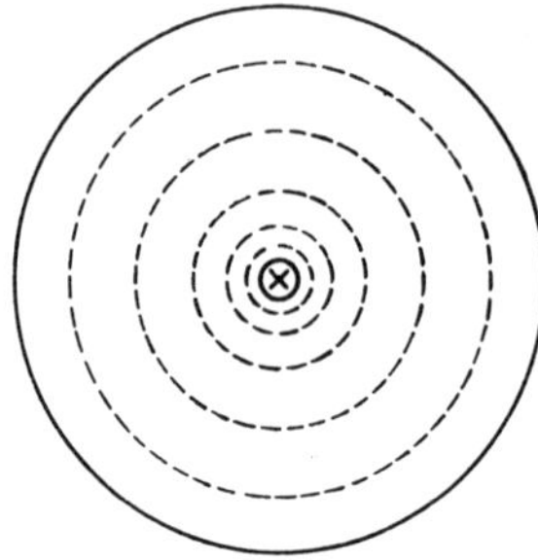

Fig. 17.11 Distribution of magnetic field in Fig. 17.10

If the current is reversed, all the compass-needles reverse their direction.

Let us now remove the compass-needles and sprinkle plate G as uniformly as possible with iron filings. A current of, say, 20 or 30 amperes is then passed through wire W and the plate tapped gently.

* It is usual to represent a current receding from the reader by a cross, as in fig. 17.10, and an approaching current by a dot, as on the left-hand conductor C in fig. 17.21. These conventions are based upon the cross being the back end elevation of the feathers of an arrow and the dot being the front view of the arrow point.

It is found that the filings tend to arrange themselves in concentric circles around the wire (fig. 17.11), this tendency being most pronounced in the vicinity of the conductor.

These experiments indicate that a magnetic state or field is produced around a wire carrying an electric current and that the intensity of this field decreases as the distance from the conductor increases.

## 17.6 Direction of the magnetic field due to an electric current

It was pointed out in section 17.2 that the direction of the magnetic field in any particular space is always taken as the direction of the N pole of a compass-needle placed in that field. Consequently, from an experiment such as that described in connection with fig. 17.10, we can determine the relationship between the direction of the magnetic field and that of the current. Thus, it is seen from fig. 17.10 that if we look along the conductor and if the current is flowing away from us, the magnetic field has a clockwise direction.

A convenient method of representing this relationship is to grip the conductor with the *right* hand, with the thumb outstretched parallel to the conductor and pointing in the direction of the current; the fingers then point in the direction of the magnetic field around the conductor.

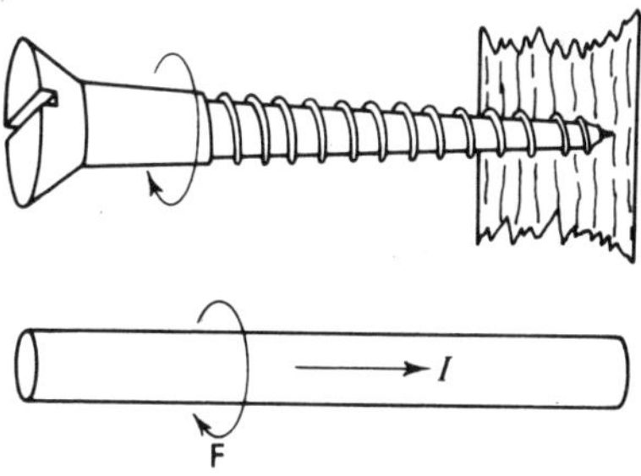


Fig. 17.12 Right-hand screw rule.

Another way of representing the relationship between the direction of a current and that of its magnetic field is to place a cork-screw or a wood-screw (fig. 17.12) alongside the conductor carrying the current. In order that the screw may travel in the same direction as the current, namely towards the right in fig. 17.12, it has to be turned clockwise when viewed from the left-hand side. Similarly, the direction of the magnetic field, viewed from the same side, is clockwise around the conductor, as indicated by curved arrow F.

## 17.7 Magnetic field of a solenoid

A solenoid consists of a number of turns of wire wound in the same direction, so that when the coil is carrying a current, all the turns are assisting one another in producing a magnetic field. Fig. 17.13 shows a few turns of wire wound helically through holes in a horizontal board and connected to a battery capable of supplying 15 to 20 amperes. Iron filings are evenly sprinkled over the board and the latter is gently tapped. The filings arrange themselves in concentric

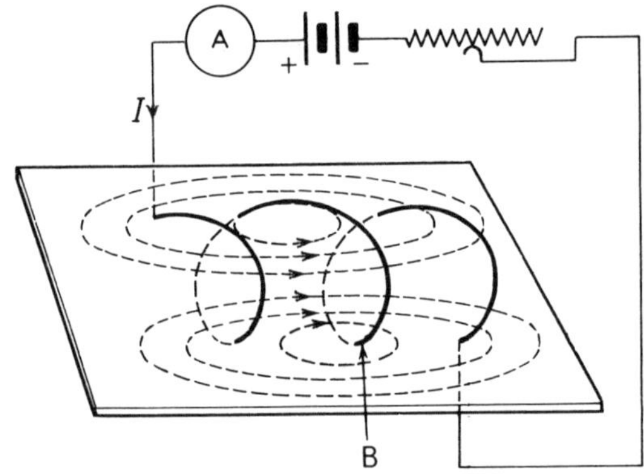


Fig. 17.13 Magnetic field of a solenoid.

paths of the shape shown by the thin dotted lines. The direction of the magnetic field can be determined by placing a compass-needle on the board and noting the direction in which its north-seeking pole is pointing. It is found that this direction is that indicated by the arrow-heads on the dotted lines in fig. 17.13.

The magnetic field may be intensified by inserting a rod of iron inside the solenoid, as shown in fig. 17.14. The magnetic flux is again represented by the dotted lines and the arrowheads indicate the direction of the flux. The iron core thus becomes magnetized with the polarities shown, and behaves like a permanent magnet so long as the current is maintained in the coil.

The direction of the magnetic flux produced by a current in a solenoid may be deduced by applying either the grip or the screw rule. Thus, if the solenoid is gripped with the *right* hand, with the fingers pointing in the direction of the current, then the thumb outstretched parallel to the axis of the solenoid points in the direction of the magnetic flux *inside* the solenoid.

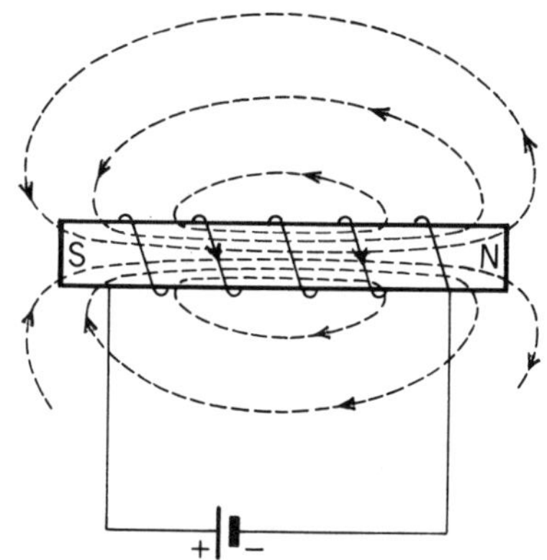


Fig. 17.14 Solenoid with an iron core.

The screw rule can be expressed thus: If the axis of the screw is placed along that of the solenoid and if the screw is turned in the direction of the current, it travels in the direction of the magnetic flux, namely towards the right in figs. 17.13 and 17.14.

## 17.8 Applications of the magnetic effect of a current

The following examples are given to indicate ways in which the magnetic effect of a current may be utilized, and are not intended to be exhaustive.

(*a*) *Lifting Magnet.* A coil C, usually of insulated copper strip, is wound round a central core A forming part of an iron casting shaped as shown in fig. 17.15, where the upper half is a sectional elevation at YY and the lower half is a sectional plan at XX. Over the face of the electromagnet is a disc D of non-magnetic manganese steel which is capable of withstanding considerable impacts when the load L is picked up. The load must be of magnetic material and the dotted lines FF represent the paths of the magnetic flux.

(*b*) *Magnetic circuits of generators and motors.* The magnetic circuits of direct-current generators and motors are similar. Fig. 17.16 shows the magnetic circuit of a four-pole machine. The armature core A consists of iron laminations, assembled on the

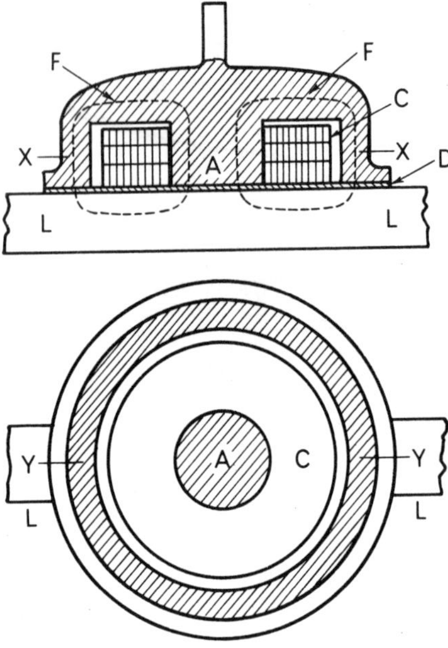


Fig. 17.15 Sectional elevation and plan of a lifting magnet.

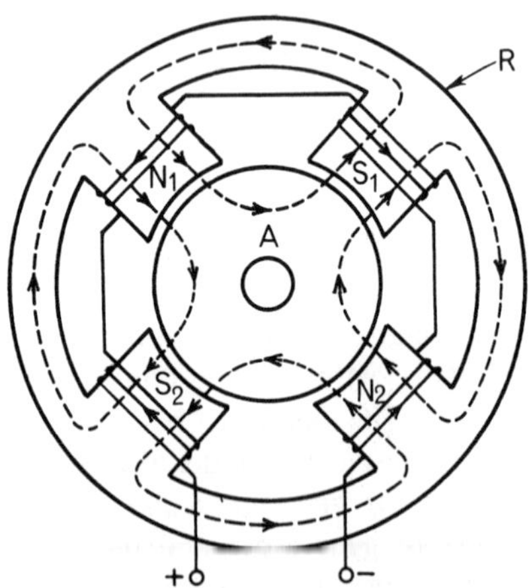


Fig. 17.16 Magnetic circuit of a 4-pole generator or motor.

shaft. The fixed part of the machine is made of four iron cores attached to an iron ring R, called the *yoke.* It is usual to place a coil on each pole core and to connect them in series as shown. Since the poles must be alternately N and S, it is essential that the coils should be so connected that the current flows clockwise—when viewed from the armature end of the poles—round the cores which are to be south, and anticlockwise round those which are to be north.

It will be seen from fig. 17.16 that the magnetic flux which emerges from $N_1$ divides, half going towards $S_1$ and half towards $S_2$. Similarly, the flux which emerges from $N_2$ divides equally between $S_1$ and $S_2$.

(*c*) *Moving-iron ammeters and voltmeters.* In section 14.2 it was shown that the magnetic pull on a piece of iron could be utilized to measure the magnitude of a current. The majority of modern moving-iron instruments, however, are of the *repulsion* type, in which two parallel rods or strips of soft iron, magnetized inside a coil, are regarded as repelling each other. The principal features of this type of instrument are shown in fig. 17.17, where C represents a

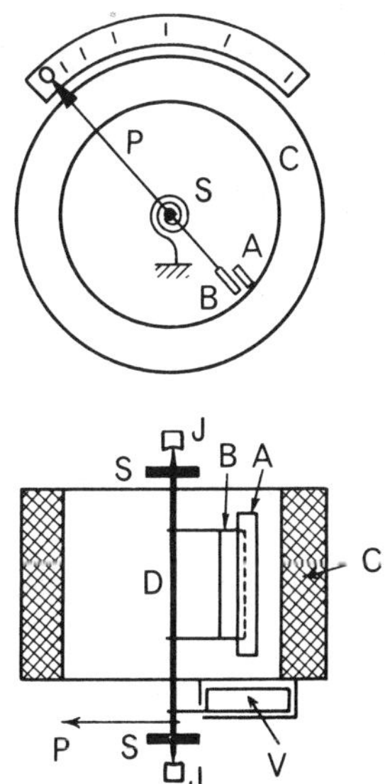


Fig. 17.17 Moving-iron instrument.

coil wound with insulated copper wire. The number of turns on C depends upon the current required to produce full-scale deflection (see Example 17.1).

A soft-iron rod A is fixed to the bobbin on which the coil is wound and another soft-iron rod is attached to spindle D supported in jewelled bearings J. A current through coil C magnetizes rods A and B in the same direction, and it is usual to say that B tries to move away from A because poles of the same polarity repel each other. Such a statement, however, gives no indication as to how the deflecting force is actually produced.

In section 17.3, it was shown that when two *permanent* magnets are placed side by side, with their N poles pointing in the same direction, as in fig. 17.6, the force of repulsion between the magnets is due to the lateral pressure in the magnetic field occupying the space *between* the magnets. In the case of two parallel rods, A and B, situated inside a coil C carrying a current, as in fig. 17.18,* the distribution of the magnetic flux is roughly as shown dotted. There is practically no magnetic flux in the space between A and B, and

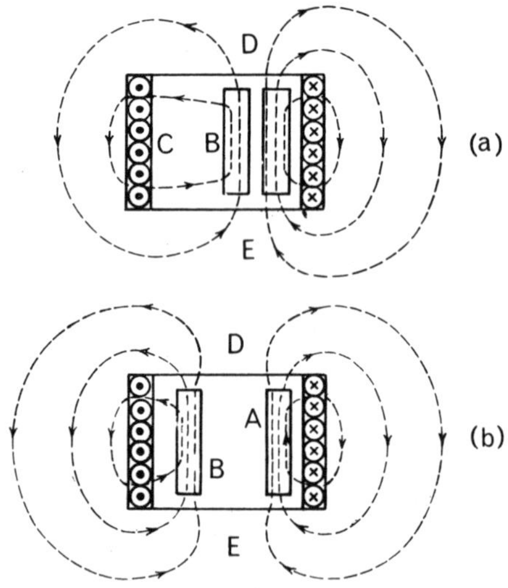


Fig. 17.18 Distribution of magnetic flux in the repulsion-type moving-iron instrument.

* The rods are shown much wider than is the case in an actual instrument. This is done to enable the dotted lines representing the flux passing through the rods to be shown clearly.

therefore there cannot be any lateral pressure in that region.

All the magnetic flux must be linked with the whole or part of the coil, and fig. 17.18(a) shows the approximate distribution of the flux passing through the rods when they are close together. It was mentioned in section 17.3 that magnetic flux behaves as if it were in tension; hence the flux passing through rod B tends to shorten its paths by pulling B towards the left, as shown in fig. 17.18(b). (Rod A is assumed to be fixed.) In addition to this tension effect, there is also a lateral pressure in regions D and E tending to push apart the fluxes passing through A and B, thereby helping to urge rod B towards the left. These effects combine to produce a clockwise deflecting torque in fig. 17.17.

There are two methods of controlling the deflection of the moving system, namely:

(*a*) gravity control—now practically obsolete, and

(*b*) spring control, consisting of two spiral hairsprings S (fig. 17.17), the inner ends of which are attached to spindle D. The arrangement of these springs is shown separately in fig 17.19. The outer end of spring B is fixed, whereas that of spring A is attached to one end of a lever L, pivoted at P, thereby enabling zero adjustment to be easily effected. The hairsprings are of non-magnetic alloy such as phosphor-bronze or beryllium-copper.

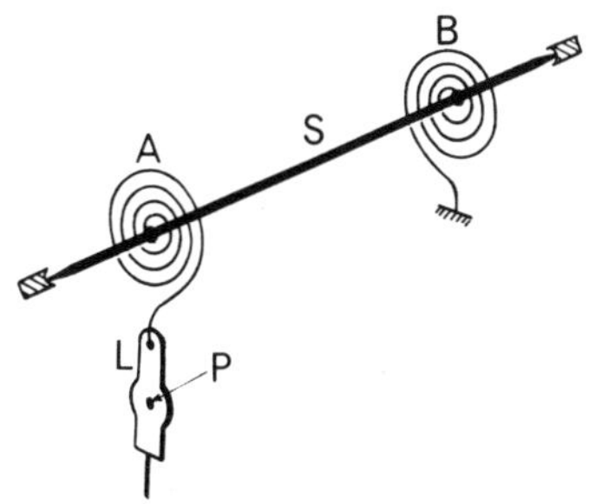


Fig. 17.19 Spring control.

With a given current through coil C (fig. 17.17), the moving system is deflected until the restoring torque exerted by the springs is equal to the deflecting torque. The value of the current is then given by the scale reading indicated by pointer P.

The tendency of the moving system to oscillate before coming to rest is damped by the thin vane V (fig. 17.17) carried by an arm attached to the spindle and moving in an air chamber.

The magnitude of the deflecting torque depends upon the product of the number of turns on coil C (fig. 17.17) and the current through the coil, i.e. it depends upon the number of *ampere turns*. For example, if full-scale deflection on a moving-iron instrument wound with 100 turns is produced by a current of 3 A,

$$\left.\begin{array}{r}\text{number of ampere turns required to}\\ \text{produce full-scale deflection}\end{array}\right\} = 3\ [\text{A}] \times 100\ [\text{t}]$$
$$= 300\ \text{A t}.$$

If a similar instrument is required to give full-scale deflection with, say, 10 A,

$$\text{number of turns} = 300\ [\text{A t}]/10\ [\text{A}] = 30.$$

It is therefore possible to arrange different instruments to have different ranges by merely winding the coils with the appropriate number of turns.

A moving-iron voltmeter is a moving-iron milliammeter connected in series with a suitable resistor, as explained in the following example.

**Example 17.1** *A moving-iron instrument requires* 400 *ampere turns to give full-scale deflection. Calculate* (a) *the number of turns required if the instrument is to be used as an ammeter reading up to* 50 *A and* (b) *the number of turns and the total resistance of the instrument if it is to be arranged as a voltmeter reading up to* 300 *V with full-scale deflection when the current is* 20 *mA.*

(*a*) Number of turns $= 400\ [\text{A t}]/50\ [\text{A}] = 8$.

(*b*) Current to give full-scale deflection $= 20\ \text{mA} = 0{\cdot}02\ \text{A}$.

$\therefore$ number of turns $= 400\ [\text{A t}]/0{\cdot}02\ [\text{A}] = 20\,000$.

Total resistance of voltmeter $= 300\ [\text{V}]/0{\cdot}02\ [\text{A}] = 15\,000\ \Omega$.

If the coil of 20 000 turns has a resistance of, say, 400 Ω,

resistance of the external series resistor $= 15000 - 400 = 14600\ \Omega$.

The advantages of moving-iron instruments are:

(i) robust construction,
(ii) relatively cheap,
(iii) can be used to measure direct and alternating currents and voltages.

The disadvantages of moving-iron instruments are:

(i) The scale divisions are not uniform, being cramped at the lower end and open at the upper end of the scale. This is due to the deflecting torque being approximately proportional to the square of the current.

(ii) Affected by stray magnetic fields. Error due to this cause is minimized by the use of a magnetic screen such as an iron casing (section 17.4).

(iii) Liable to hysteresis error when used in a direct-current circuit; i.e. for a given current, the instrument reads higher with decreasing than with increasing values of current. This error is reduced by making the iron strips of nickel-iron alloy such as Mumetal.

(iv) Moving-iron voltmeters are liable to a temperature error owing to the coil being wound with copper wire. This error is minimized by connecting in series with the coil a resistor of a material such as manganin, having a negligible temperature coefficient of resistance (section 16.7).

## 17.9 Force on a conductor carrying current across a magnetic field

In section 17.5, it was shown that a conductor carrying a current can produce a force on a magnet situated in the vicinity of the conductor. From Newton's Third Law of Motion, namely that to every force there must be an equal and opposite force, it follows that the magnet must exert an equal force on the conductor. One of the simplest methods of demonstrating this effect is to take a stiff copper wire, about 2 mm diameter, and bend it into a rectangular loop as represented by BC in fig. 17.20. The two tapered ends of the loop

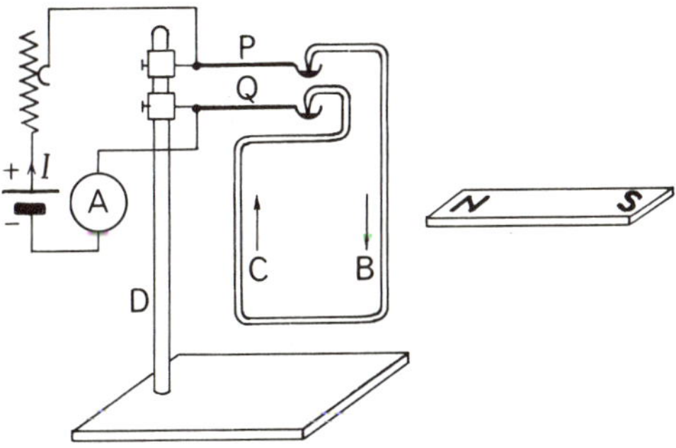


Fig. 17.20 Force on conductor carrying current across a magnetic field.

dip into mercury contained in cups, one directly above the other, the cups being attached to metal rods P and Q carried by a wooden upright rod D. A current of about 5 amperes is passed through the loop and the N pole of a permanent magnet NS is moved towards B. If the current in this wire is flowing downwards, as indicated by the arrow in fig. 17.20, it is found that the loop, when viewed from above, turns anticlockwise, as shown in plan in fig. 17.21. If the magnet is reversed and again brought up to B, the loop turns clockwise.

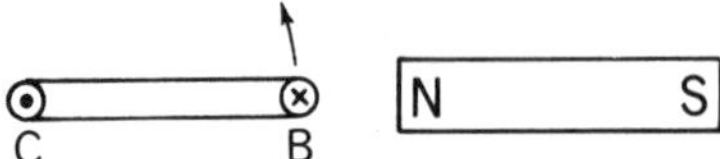


Fig. 17.21 Direction of force on conductor in Fig. 17.20.

If the magnet is placed on the other side of the loop, the latter turns clockwise when the N pole of the magnet is moved near to C, and anticlockwise when the magnet is reversed.

These effects can be explained* by the simple apparatus shown in elevation and plan in fig. 17.22. Two permanent magnets NS rest on a sheet of paper or glass G, and soft-iron pole-pieces P are added to increase the area of the magnetic field in the gap between them. Midway between the pole-pieces is a wire W passing vertically downwards through G and connected through a switch to a 6-volt battery capable of giving a very large current for a short time.

With the switch open, iron filings are sprinkled over G and the latter is gently tapped. The filings in the space between PP take up the distribution shown in fig. 17.22(b). If the switch is closed momentarily, the filings rearrange themselves as in fig. 17.22(c). It will be seen that the magnetic flux has been so distorted that it partially surrounds the wire. This distorted flux acts like a stretched elastic string bent out of the straight and therefore tries to return to the

* Many textbooks give a *left*-hand rule for deducing the direction of the force on a conductor carrying current across a magnetic field. This rule is liable to be confused with the *right*-hand rule, given in section 18.4, for determining the direction of a generated e.m.f. The latter rule is extremely useful and should be memorized. Few students can memorize both rules correctly and it is suggested that the left-hand rule should be forgotten and that the direction of the force on a current-carrying conductor in a magnetic field should be deduced from first principles by drawing separately the magnetic field due to the current in the conductor and that due to the permanent magnet or electromagnet and thus derive the distribution of the resultant magnetic field as in fig. 17.22.

shortest path between PP, thereby exerting a force $F$ urging the conductor out of the way.

It has already been shown in section 17.6 that a wire W carrying a current downwards in fig. 17.22(a) produces a magnetic field as shown in fig. 17.23. If this field is compared with that of fig. 17.22(b), it is seen that on the upper side the two fields are in the same direction, whereas on the lower side they are in opposition. Hence, the combined effect is to strengthen the magnetic field on the upper side and weaken it on the lower side, thus giving the distribution shown in fig. 17.22(c).

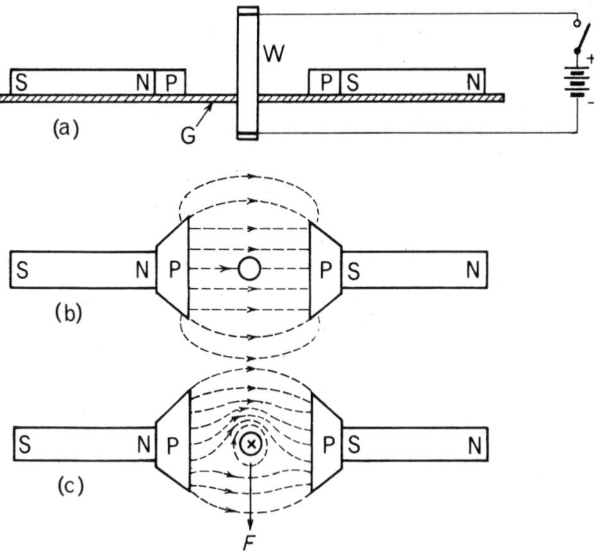


Fig. 17.22 Flux distribution with and without current.

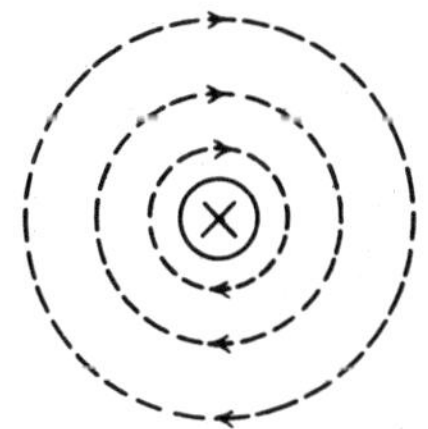

Fig. 17.23 Flux distribution due to current in a straight conductor.

By combining diagrams similar to figs. 17.22(b) and 17.23, it is easy to understand that if either the current in W or the polarity of the magnets NS is reversed, the field is strengthened on the lower side and weakened on the upper side of diagrams corresponding to fig. 17.22(b), so that the direction of the force acting on W is the reverse of that shown in fig. 17.22(c).

On the other hand, if both the current through W and the polarity of the magnets are reversed, the *distribution* of the resultant magnetic field and therefore the direction of the force on W remain unaltered.

### 17.10 Permanent-magnet moving-coil ammeters and voltmeters

The moving-coil instrument is a good example of the application of the mechanical force exerted on a conductor carrying current across a magnetic field, discussed in the preceding section.

This type of instrument consists of a rectangular coil C (fig. 17.24) of insulated copper wire wound on a light aluminium frame which is carried by steel pivots resting in jewel bearings. Current is led into and out of the coil by spiral hairsprings AA, which also provide the controlling torque. The coil is free* to move in airgaps between the soft-iron pole-pieces PP and a central soft-iron cylinder B supported by a non-magnetic plate attached to PP.

The magnetic field is provided by a permanent magnet M of modern steel alloy such as Alcomax (iron, aluminium, cobalt, nickel and copper) to which soft-iron pole-pieces PP are attached by a special process. The hardness of modern magnet materials makes machining impossible, whereas soft iron can be easily machined to give exact airgap dimensions.

The functions of the central core B are: (*a*) to intensify the magnetic field by reducing the length of airgap across which the magnetic flux has to pass and (*b*) to give a radial magnetic flux of uniform density, thereby enabling the scale to be uniformly divided.

Damping of the moving system is effected by currents induced in the aluminium frame on which coil C is wound.

An alternative arrangement that is particularly suitable for small moving-coil instruments is shown in fig. 17.25. The central cylindri-

* Students often form the impression that the moving coil is wound on the iron cylinder. It is important to realize that this cylinder is *fixed* and that the frame, on which the coil is wound, *does not touch* the cylindrical core.

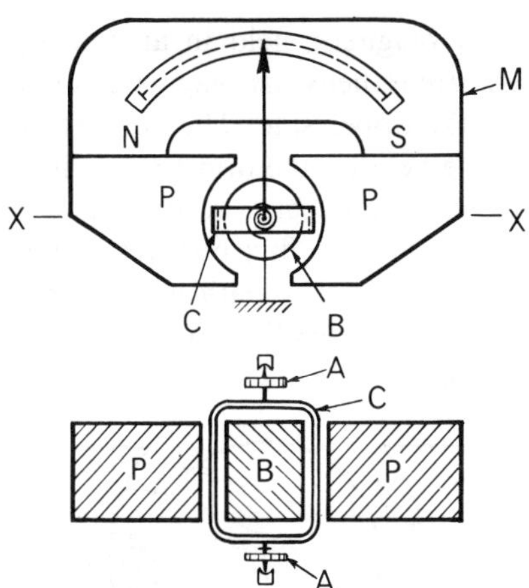


Fig. 17.24 Permanent-magnet moving-coil instrument.

cal core consists of a permanent magnet M, with soft-iron pole-pieces PP attached to M. The return path for the magnetic flux is provided by a soft-iron ring or yoke Y, concentric with the core and

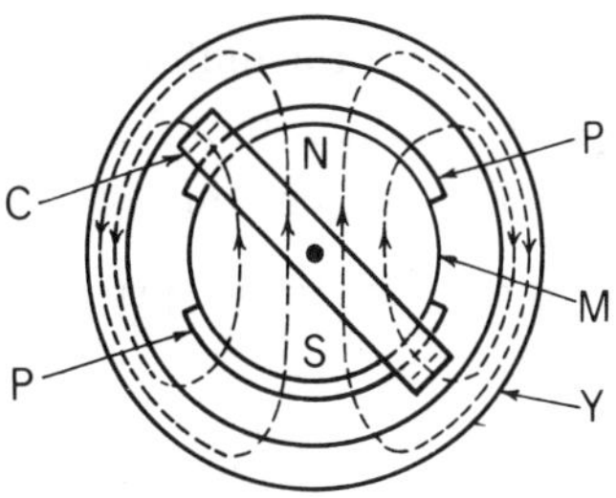


Fig. 17.25 Moving-coil instrument with centre-core magnet.

separated from it by an airgap of uniform width, the distribution of the flux being as indicated by the dotted lines in fig. 17.25. The moving coil C, wound on an aluminium frame supported by jewel bearings and controlled by spiral hairsprings, is free to move in the airgap between the pole-pieces and the yoke.

The manner in which a torque is produced when the coil is carrying a current may be understood more easily by considering a single turn PQ, as in fig. 17.26. Suppose P to carry current outwards from the paper; then Q is carrying current towards the paper. Current in

P tends to set up a magnetic field in an anticlockwise direction around P and thus strengthens the magnetic field on the lower side and weakens it on the upper side. The current in Q, on the other hand, strengthens the field on the upper side while weakening it on the lower side. Hence, the effect is to distort the magnetic flux as shown in fig. 17.26. Since this flux behaves like a stretched elastic cord, it tries to take the shortest path between poles NS, and thus exerts forces *FF* on coil PQ, tending to move it out of the magnetic field.

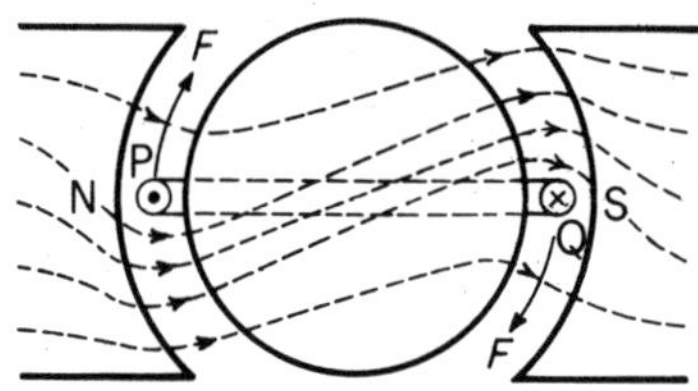


Fig. 17.26 Distribution of resultant magnetic field.

The deflecting torque $\propto$ current through coil $\times$ flux density in gap

$$= kI \text{ for uniform flux density,}$$

where $k$ = a constant for a given instrument

and $I$ = current through coil.

The controlling torque of the spiral springs $\propto$ angular deflection

$$= c\theta$$

where $c$ = a constant for given springs

and $\theta$ = angular deflection.

For a steady deflection,

$$\text{controlling torque} = \text{deflecting torque},$$

hence
$$c\theta = kI$$

$\therefore$
$$\theta = \frac{k}{c}I,$$

i.e. the deflection is proportional to the current and the scale is therefore uniformly divided.

A numerical example on the calculation of the torque on a moving coil is given in example 17.5.

Owing to the delicate nature of the moving system, this type of instrument is only suitable for measuring currents up to about 50 milliamperes directly. When a larger current has to be measured, a *shunt* S (fig. 17.27), having a low resistance, is connected in parallel with the moving coil MC, and the instrument scale may be calibrated to read directly the total current $I$. Shunts are made of a material, such as manganin (copper, manganese and nickel), having negligible temperature coefficient of resistance. A 'swamping' resistor $r$, of material having negligible temperature coefficient of resistance, is connected in series with the moving coil. The latter is wound with copper wire and the function of $r$ is to reduce the error due to the variation of resistance of the moving coil with variation of temperature.

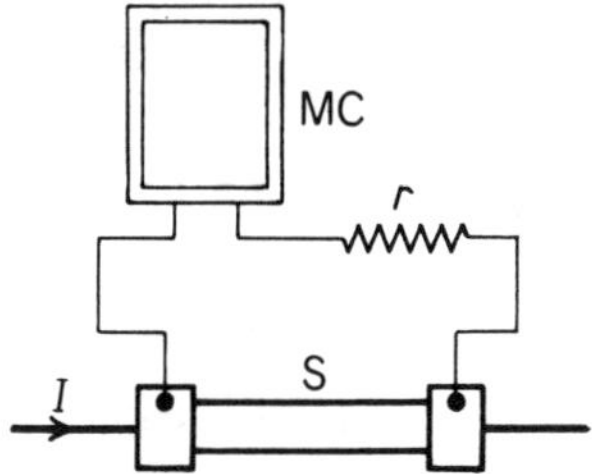


Fig. 17.27 Moving-coil instrument as an ammeter.

The shunt shown in fig. 17.27 is provided with four terminals, the milliammeter being connected across the 'potential' terminals. If the instrument were connected across the 'current' terminals, there might be considerable error due to the contact resistance at these terminals being appreciable compared with the resistance of the shunt.

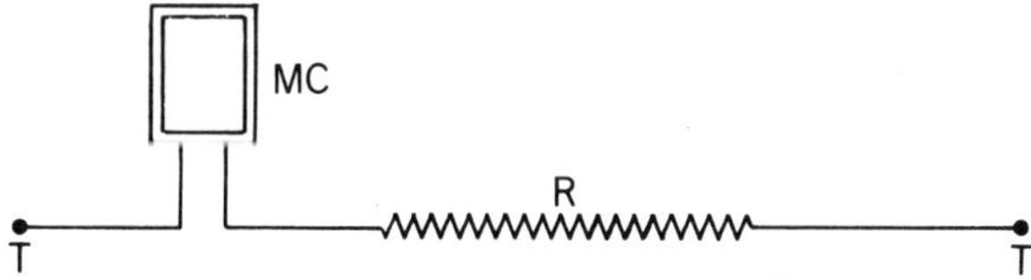


Fig. 17.28 Moving-coil instrument as a voltmeter.

The moving-coil instrument may be made into a voltmeter by connecting a resistor R of manganin or other similar material in series, as in fig. 17.28. Again the scale may be calibrated to read directly the voltage applied across terminals TT.

The main advantages of moving-coil instruments are:

(i) high sensitivity,
(ii) uniform scale,
(iii) well shielded from any stray magnetic field.

The main disadvantages are:

(i) more expensive than the moving-iron instrument,
(ii) only suitable for direct currents and voltages.

**Example 17.2** *A moving coil instrument gives full-scale deflection with* 15 *mA and has a resistance of* 5 Ω, *including that of the swamping resistor. Calculate the resistance required* (a) *in parallel to enable the instrument to read up to* 1 *A and* (b) *in series to enable it to read up to* 10*V*.

(*a*) Current for full-scale deflection = 15 mA = 0·015 A.
From Ohm's law,

$$\left.\begin{matrix}\text{current through coil}\\ \text{(fig. 17.27)}\end{matrix}\right\} = \frac{\text{p.d. across coil}}{\text{resistance of coil}}$$

$$\therefore \qquad 0{\cdot}015\ [\text{A}] = \frac{\text{p.d. across coil}}{5\ [\Omega]}$$

$$\text{so that p.d. across coil} = 0{\cdot}075\ \text{V}.$$

From fig. 17.27 it follows that:

$$\begin{aligned}\text{current through S} &= \text{total current} - \text{current through coil}\\ &= 1 - 0{\cdot}015 = 0{\cdot}985\ \text{A}.\end{aligned}$$

$$\text{Similarly, current through S} = \frac{\text{p.d. across S}}{\text{resistance of S}}$$

$$\therefore \qquad 0{\cdot}985\ [\text{A}] = \frac{0{\cdot}075\ [\text{V}]}{\text{resistance of S}}$$

$$\text{so that resistance of S} = 0{\cdot}075/0{\cdot}985 = 0{\cdot}076\ 14\ \Omega.$$

(*b*) From Ohm's law it follows that for fig. 17.28,

$$\text{current through coil} = \frac{\text{p.d. across TT}}{\text{resistance between TT}}$$

$$\therefore \qquad 0{\cdot}015\ [\text{A}] = \frac{10\ [\text{V}]}{\text{resistance between TT}}$$

so that resistance between TT $= 666{\cdot}7\ \Omega$.

Hence, resistance required in series with coil

$$= \text{total resistance between TT} - \text{resistance of coil}$$
$$= 666{\cdot}7 - 5 = 661{\cdot}7\ \Omega.$$

The moving-coil instrument can be arranged as a multi-range ammeter by making the shunt of different sections as shown in fig. 17.29, where A represents a milliammeter in series with a 'swamping' resistor $r$ of material having negligible temperature coefficient of resistance.

With the selector switch S on, say the 50-A stud, a shunt having a very low resistance is connected across the instrument, the value of its resistance being such that full-scale deflection is produced when $I = 50$ A. With S on the 10-A stud, the resistance of the two sections of the shunt is approximately five times that of the 50-A section, and full-scale deflection is obtained when $I = 10$ A. Similarly, with S on the 1-A stud, the total resistance of the three sections is such that full-scale deflection is obtained with $I = 1$ A. Such a multi-range instrument is provided with three scales so that the value of the current can be read directly.

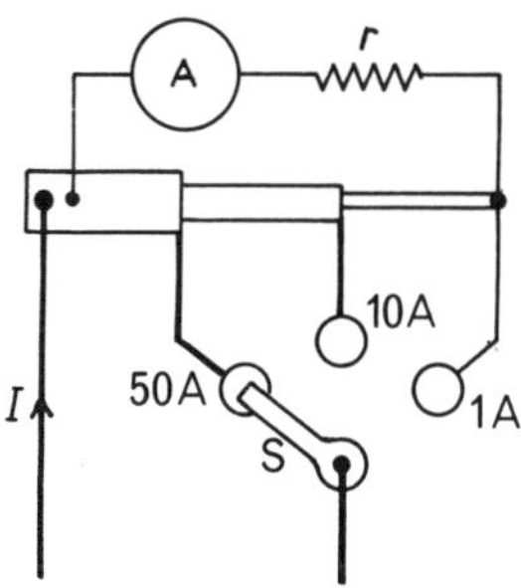


Fig. 17.29 Multi-range moving-coil ammeter.

A multi-range voltmeter is easily arranged by using a tapped resistor in series with a milliammeter A, as shown in fig. 17.30. For instance, with the data given in example 17.2, the resistance of section BC would be $661{\cdot}7\ \Omega$ for the 10-V range. If D be the tapping for, say, 100 V, the total resistance between O and D $= 100/0{\cdot}015 = 6666{\cdot}7\ \Omega$, so that the resistance of section CD $= 6666{\cdot}7 - 666{\cdot}7 = 6000\ \Omega$. Similarly, if E is to be a 500-V tapping, the p.d. across section DE must be 400 V at full-scale deflection; hence the resistance of DE = 400/0·015 = 26 667 Ω. With the aid of selector switch S, the instrument can be used on three voltage ranges, and the scales can be calibrated to enable the value of the voltage to be read directly.

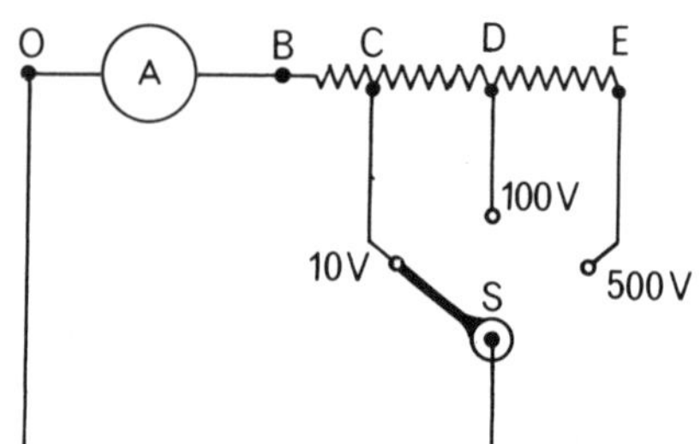


Fig. 17.30 Multi-range moving-coil voltmeter.

## 17.11 Magnitude of the force on a conductor carrying current across a magnetic field

With the apparatus of fig. 17.20, it can be shown qualitatively that the force on a conductor carrying a current at right angles to a magnetic field is increased when (*a*) the current in the conductor is increased and (*b*) when the magnetic field is made stronger by bringing the magnet nearer to the conductor. With the aid of more elaborate apparatus, the force on the conductor can be measured for various currents and various densities of the magnetic flux, and it is found that:

force on conductor ∝ current × flux density × length of conductor

If $F$ = force on conductor in newtons,

$I$ = current through conductor in amperes

and $l$ = length, in metres, of conductor at right angles to the magnetic flux,

$$F\,[\text{newtons}] \propto \text{flux density} \times l\,[\text{metres}] \times I\,[\text{amperes}].$$

The *unit of flux density* is taken as *the density of a magnetic flux such that a conductor carrying* 1 *ampere at right angles to that flux has a force of* 1 *newton per metre acting upon it*. This unit is termed a *tesla** (T). Hence, for a flux density $B$, in teslas,

* Nikola Tesla (1857–1943), a Yugoslav who emigrated to U.S.A. in 1884, was a very famous electrical inventor.

$$\text{force on conductor} = BlI \text{ newtons} \qquad (17.1)$$

For a magnetic field having a cross-sectional area $a$, in square metres, and a uniform flux density $B$, in teslas, the *total flux*, in *webers*† (Wb), is represented by the Greek *capital* letter $\Phi$ (phi), where

$$\Phi \text{ [webers]} = B \text{ [teslas]} \times a \text{ [metres}^2\text{]}$$

or

$$B \text{ [teslas]} = \frac{\Phi \text{ [webers]}}{a \text{ [metres}^2\text{]}} \qquad (17.2)$$

The weber is a large unit and either the milliweber (mWb) or the microweber (μWb) is often a more convenient unit to employ, where

$$1000 \text{ milliwebers} = 1 \text{ weber}$$

and

$$1\,000\,000 \text{ microwebers} = 1 \text{ weber.}$$

**Example 17.3** *The pole core (fig. 17.16) of an electrical machine is circular in cross-section and has a diameter of* 120 *mm. If the total flux in the core is* 16 *mWb, calculate the flux density.*

$$\text{Diameter of core} = 120 \text{ mm} = 0{\cdot}12 \text{ m}$$

$$\therefore \text{ cross-sectional area of core} = (\pi/4) \times (0{\cdot}12)^2 = 0{\cdot}011\,32 \text{ m}^2.$$

$$\text{Total magnetic flux} = 16 \text{ mWb} = 0{\cdot}016 \text{ Wb}$$

so that

$$\text{flux density} = 0{\cdot}016 \text{ [Wb]}/0{\cdot}011\,32 \text{ [m}^2\text{]} = 1{\cdot}413 \text{ T.}$$

**Example 17.4** *A conductor carries a current of* 800 *A at right angles to a magnetic field having a density of* 0·5 *tesla. Calculate the force on the conductor in newtons per metre length.*

Substituting for $B$, $l$ and $I$ in expression (17.1), we have:

$$\text{force per metre length} = 0{\cdot}5 \text{ [T]} \times 1 \text{ [m]} \times 800 \text{ [A]} = 400 \text{ N.}$$

**Example 17.5** *The coil of a moving-coil instrument (fig.* 17.24) *is wound with* $42\frac{1}{2}$ *turns. The mean width of the coil is* 25 *mm and the*

† Wilhelm Eduard Weber (1804–91), a German physicist, was the first to develop a system of absolute electrical and magnetic units.

*axial length of the magnetic field is* 20 *mm. If the flux density in the airgap is* 0·2 *T, calculate the torque for a current of* 15 *mA.*

Since the coil has $42\frac{1}{2}$ turns, one side will have 42 wires and the other side will have 43 wires.

From expression (17.1), force on the side having 42 wires

$$= 0{\cdot}2\ [\mathrm{T}] \times 0{\cdot}02\ [\mathrm{m}] \times 0{\cdot}015\ [\mathrm{A}] \times 42$$
$$= 2520 \times 10^{-6}\ \mathrm{N},$$

$\therefore$ torque on that side of coil

$$= 2520 \times 10^{-6}\ [\mathrm{N}] \times 0{\cdot}0125\ [\mathrm{m}] = 31{\cdot}5 \times 10^{-6}\ \mathrm{N\ m}.$$

Similarly, torque on side of coil having 43 wires

$$= 31{\cdot}5 \times 10^{-6} \times 43/42 = 32{\cdot}2 \times 10^{-6}\ \mathrm{N\ m},$$

$\therefore$ total torque on coil

$$= (31{\cdot}5 + 32{\cdot}2) \times 10^{-6} = 63{\cdot}7 \times 10^{-6}\ \mathrm{N\ m}$$
$$= 63{\cdot}7\ \mu\mathrm{N\ m}.$$

## 17.12 Force between two long parallel conductors carrying electric current

It was shown in section 17.5 that a current-carrying conductor is surrounded by a magnetic field and in section 17.9 that a current-carrying conductor placed across a magnetic field has a force acting upon the conductor. It therefore follows that when two current-carrying conductors are parallel to each other, there is a force acting on each of the conductors. This effect can be very easily demonstrated by means of the apparatus referred to in section 17.9. Thus, in fig. 17.31, the rectangular loop BC has its tapered ends

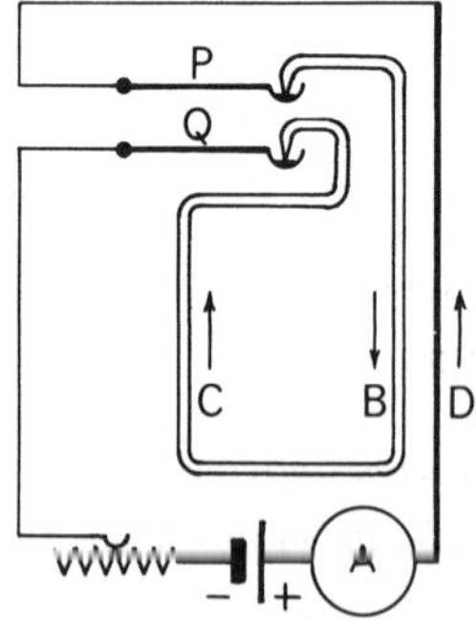


Fig. 17.31 Force between two parallel current-carrying conductors.

dipping into mercury in cups supported one directly above the other by rods P and Q, and a current of about 10 to 15 amperes is passed through the loop. Part of the electrical circuit consists of a long straight rod D which can be placed alongside B, as shown in plan in fig. 17.32. When the currents in D and B are in opposite directions, as in fig. 17.32(a), the two conductors repel each other and the loop (viewed from above) is deflected clockwise. On the other hand, if rod D is turned through 180° so that the currents in B and D are in the same direction, as in fig. 17.32(b), the conductors attract each other and the loop turns anticlockwise.

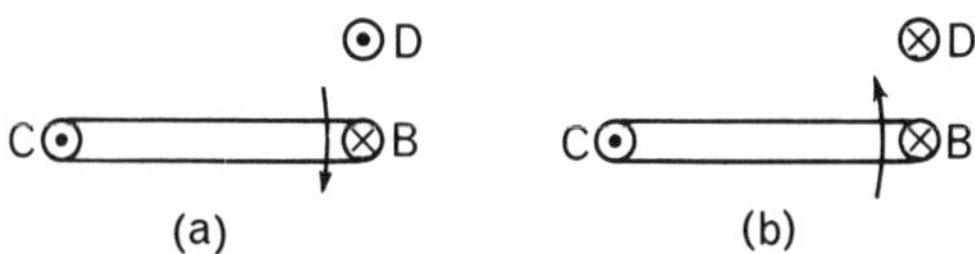


Fig. 17.32 Force between two parallel current-carrying conductors.

These effects are most easily explained by first drawing the magnetic fields produced by each conductor and then combining these fields. Thus, fig. 17.33(a) shows two conductors, A and B, each carrying current towards the paper. The magnetic flux due to the current in A alone is represented by the uniformly dotted circles in fig. 17.33(a), and that due to B alone is represented by the chain-dotted circles. It is evident that in the space between A and B the two fields tend to neutralize each other, but in the space outside A and B they assist each other. Hence the resultant distribution is somewhat as shown in fig. 17.33(b). Since magnetic flux behaves like a stretched elastic cord, the effect is to try to move conductors A and B towards each other; in other words, there is a force of attraction between A and B.

If the current in B is reversed, the magnetic fields due to A and B assist each other in the space between the conductors and the resultant distribution of the flux is shown in fig. 17.33(c). The lateral pressure in the magnetic flux exerts a force on the conductors tending to push them apart (section 17.3 (5) ).

It is this force between parallel current-carrying conductors that forms the basis for the definition of the *ampere*, adopted internationally in 1948, namely, *the constant current which, if maintained in two straight parallel conductors of infinite length, of negligible circular cross-section, and placed at a distance of* 1 *metre apart in a vacuum, would produce between these conductors a force equal to* $2 \times 10^{-7}$ *newton per metre length.*

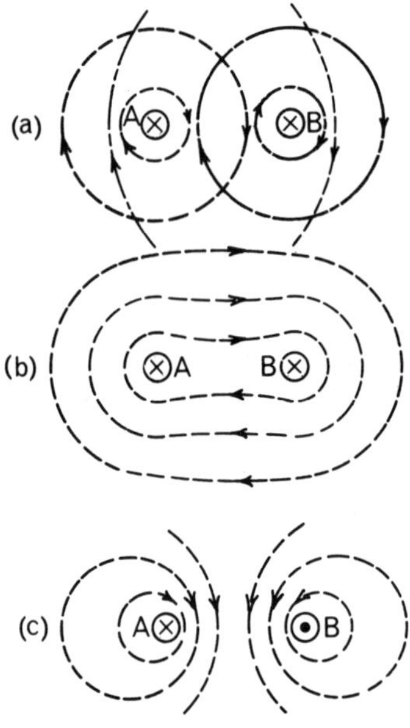


Fig. 17.33 Magnetic fields due to parallel current-carrying conductors.

## 17.13 Force between coils carrying electric current

We are now in a position to explain the effect observed in section 14.2, namely, that two co-axial coils, placed one above the other,

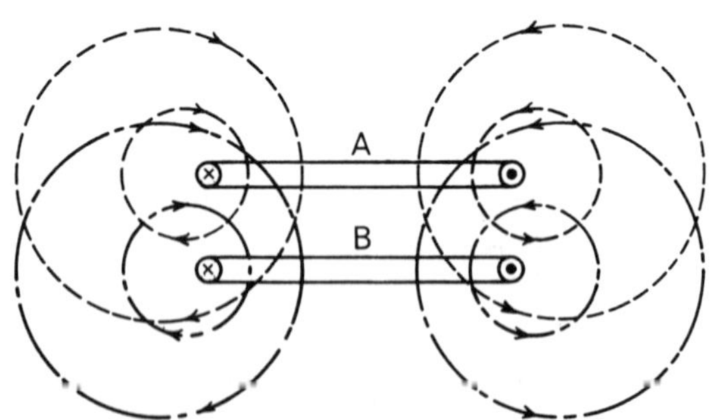


Fig. 17.34 Magnetic fluxes due to currents in A and B separately.

attract or repel each other, depending upon the relative direction of their currents. Suppose A and B in fig. 17.34 to represent the cross-section of two coils carrying currents in the directions shown by the dots and crosses. Let us first consider the distribution of the magnetic fluxes due to the coils acting independently. Thus, current through A alone gives the flux distribution represented by the uniformly dotted lines in fig. 17.34, while current in the same direction through B alone gives the distribution indicated by the chain-dotted lines. It will be seen that in the space between the coils the two fields oppose each other, while on the outside they are in the same direction. Consequently the combined effect is to give the flux

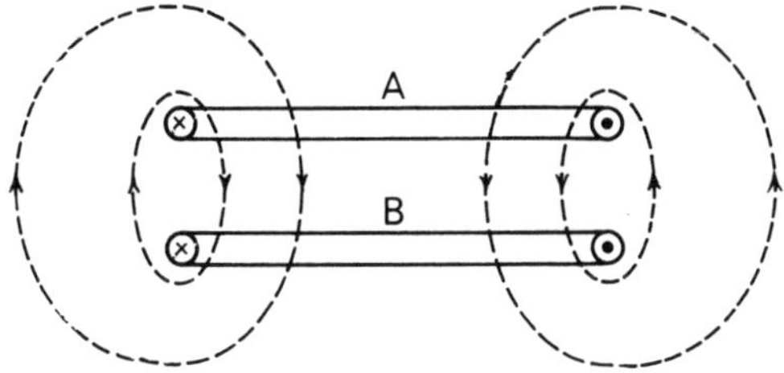


Fig. 17.35 Resultant magnetic flux when currents in A and B are in the same direction.

distribution shown in fig. 17.35. Since magnetic flux acts as if it is in tension, it tends to move coils A and B towards each other.

On the other hand, if the current through B is reversed, the direction of the arrowheads on the chain-dotted lines in fig. 17.34 is

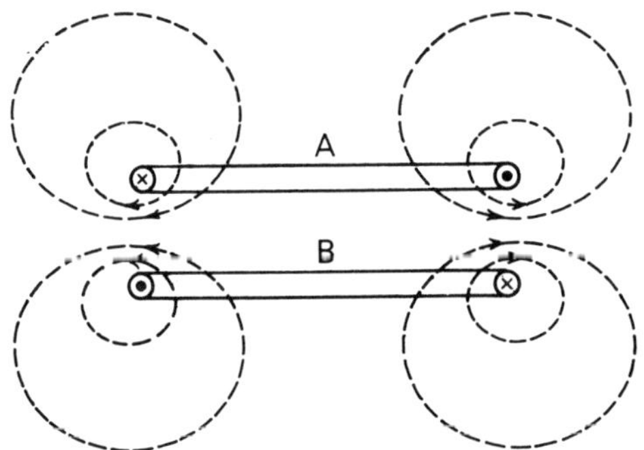


Fig. 17.36 Resultant magnetic flux when currents in A and B are in opposite directions.

reversed. Consequently, the magnetic fields of A and B are in the same direction in the space between the coils and in opposition outside the coils, so that the resultant flux distribution becomes that shown in fig. 17.36. But magnetic flux exerts a lateral pressure—just as stretched rubber cords try to swell when allowed to contract in length, and thereby exert sideways pressure on other rubber cords that are alongside. This lateral pressure in the flux is passed on to coils A and B, so that they try to move away from each other.

This attraction and repulsion between coils carrying an electric current has been applied in the current balance used at the National Physical Laboratory for determining the value of an electric current in terms of the definition given in section 17.12.

## *Summary of Chapter* 17

The characteristics of magnetic fields produced by permanent magnets and by an electric current through a straight wire and through a solenoid have been discussed, and rules are given for the direction of the magnetic field produced by an electric current.

The direction of the force exerted on a conductor carrying current across a magnetic field is deduced by deriving the distribution of the resultant flux.

Applications of electromagnetism, such as moving-iron and moving-coil instruments, are described.

For a conductor carrying current at right angles to a magnetic field,

$$\left.\begin{array}{r}\text{force on the conductor}\\ \text{[newtons]}\end{array}\right\} = B\,[\text{teslas}] \times l\,[\text{metres}] \times I\,[\text{amperes}] \qquad (17.1)$$

$$B\,[\text{teslas}] = \frac{\Phi\,[\text{webers}]}{a\,[\text{square metres}]} \qquad (17.2)$$

## EXAMPLES 17

1. Indicate, with the aid of a sketch, a method of determining the magnetic polarity of a solenoid carrying a current. (N.C.T.E.C., G2)
2. A straight conductor is situated in, and at right angles to, a uniform magnetic field. When a current is passed through the conductor there will be a force on it and the field distribution will be changed. Draw a diagram to show the direction of the current, the direction of the force and a picture of the resultant field. (U.E.I., G2)

3. Describe how you would construct an electromagnet, given a suitable soft-iron bar, insulated wire and a source of p.d. Sketch the circuit and indicate, by means of a diagram, the polarity of the magnet in relation to the direction of current flow. Explain how you would test for polarity. (E.M.E.U., G2)
4. The magnetic flux in the pole of an electric motor is 0·013 Wb. If the pole has a circular cross-section and a diameter of 120 mm, calculate the value of the flux density.
5. If the flux density inside a solenoid is 0·08 tesla and the cross-sectional area of the solenoid is 2000 mm$^2$, what is the value of the total flux in microwebers?
6. A straight conductor is carrying a current of 2500 A at right angles to a magnetic field of density 0·12 tesla. Calculate the force on the conductor in newtons per metre length.
7. A conductor, 0·3 m long, is carrying a current of 60 A at right angles to a magnetic field. The force on the conductor is 8 N. Calculate the density of the magnetic field.
8. A conductor is carrying current at right angles to a uniform magnetic field of density 1·2 teslas. The length of the conductor in the magnetic field is 150 mm. Calculate the value of the current required in order that the force on the conductor may be 20 N.
9. The coil of a moving-coil instrument is wound with $50\frac{1}{2}$ turns on a rectangular former. The axial length of the pole shoes is 23 mm and the mean width of the coil is 17 mm. If the flux density in the gap is 0·18 tesla, calculate the current required to give a torque of 30 μN m.
10. The coil of a moving-coil instrument is wound with $40\frac{1}{2}$ turns on a former having an effective length of 25 mm and an effective breadth of 20 mm. The flux density in the gap is 0·16 tesla. Calculate the torque when the current is 15 mA.
11. The armature of a certain electric motor has 900 conductors and the current per conductor is 24 A. The flux density in the airgap under the poles is 0·6 T. The armature core is 160 mm long and has a diameter of 250 mm. Assume that the core is smooth (i.e. there are no slots and the winding is on the cylindrical surface of the core) and also assume that only two-thirds of the conductors are simultaneously in the magnetic field. Calculate (*a*) the torque in newton metres and (*b*) the power developed, in kilowatts, if the speed is 700 rev/min.
    (*Note.* In the case of slotted cores, the flux density in the slots is very low, so that there is very little torque on the conductors; nearly all the torque is exerted on the teeth.)
12. A moving-coil instrument has a resistance of 5 Ω and requires a p.d. of 75 mV to give full-scale deflection.
    (*a*) Calculate the additional resistance to enable the instrument to read: (i) up to 1 A, (ii) up to 30 V.
    (*b*) Draw circuit diagrams showing how the resistances would be connected. (N.C.T.E.C., G2)
13. A milliammeter gives full-scale deflection with 5 mA and has a resistance of 12 Ω. Calculate the resistance necessary (*a*) in parallel to enable the

instrument to read up to 10 A, (*b*) in series to enable it to read up to 100 V.

14. If the shunt for Question 13(*a*) is to be made of manganin strip having a resistivity of 0·5 μΩ m, a thickness of 0·5 mm and a length of 60 mm, calculate the width of the strip.
15. A moving-iron ammeter is wound with 40 turns and gives full-scale deflection with 5 A. How many turns would be required on the same bobbin to give full-scale deflection with 20 A?

    Calculate the number of turns and the total resistance of the instrument if it is arranged as a voltmeter giving full-scale reading with a current of 25 mA and a p.d. of 250 V.

## ANSWERS TO EXAMPLES 17

4. 1·15 T
5. 160 μWb.
6. 300 N/m.
7. 0·444 T.
8. 111 A.
9. 8·43 mA.
10. 48·6 μN m.
11. 173 N m, 12·7 kW.
12. 0·0762 Ω, 1995 Ω.
13. 0·006 003 Ω, 19 988 Ω.
14. 9·995 mm.
15. 10 turns; 8000 turns, 10 000 Ω.

CHAPTER 18

# Electromagnetic induction

## 18.1 Induced e.m.f.

It was mentioned in section 17.5 that the magnetic effect of an electric current was discovered by Oersted in 1820. The knowledge of this connection between electricity and magnetism caused many scientists of the time, particularly Michael Faraday in England, to try to discover a method of obtaining an electric current from a magnetic field. Failure after failure dogged Faraday's efforts until 1831 when he made the great discovery of *electromagnetic induction* with which his name will be for ever associated.

As far as we are concerned, it is more convenient to approach this matter experimentally in a different sequence from that followed by Faraday. Let us connect a coil C (fig. 18.1) to a galvanometer G, namely a very sensitive moving-coil ammeter. If a permanent

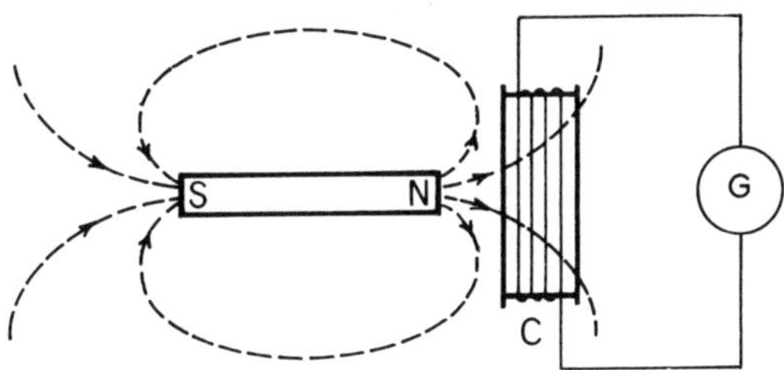


Fig. 18.1 Electromagnetic induction.

magnet NS is moved up to and along the axis of C, as shown, the moving coil of G is deflected, thereby indicating that there must be an electromotive force induced or generated in coil C. Immediately the movement of NS ceases, the moving coil of G returns to it original position. This effect proves that an e.m.f. is induced onl while magnet NS is moving relative to C.

Let us now move NS away from C. The galvanometer deflecti is found to be in the reverse direction, showing that the direction the induced e.m.f. depends upon the direction in which NS is moved relative to coil C.

If, next, we hold the magnet stationary but move the coil towards the magnet and then away from it, the deflection of the galvanometer is found to follow exactly the same sequence as it did when the magnet was moved and the coil held stationary. This result shows that the generation of an e.m.f. in C depends only upon the *relative* movement of the magnet and the coil.

If the permanent magnet is turned through 180 degrees so that its S pole is pointing towards the coil, it is found that a repetition of the movements described above is accompanied by galvanometer deflections similar to those previously obtained, except that their directions are reversed. Thus, the direction of the e.m.f. induced by bringing the S pole up to the coil is the same as that previously obtained when the N pole was moved away from the coil.

The arrowheads on the dotted lines in fig. 18.1 represent the direction of the magnetic field in their respective regions. It will be seen that as the magnet is moved towards the coil, the magnetic flux of NS also moves across the wires forming the coil; that is, the magnetic flux is said to *cut* the coil. Similarly, when the coil is moved towards the magnet, the magnetic flux is said *to be cut* by the coil. It is this relative movement of the magnetic flux and the coil that causes an e.m.f. to be induced (or generated) in the latter. Alterna-ively,* we can say that the induced e.m.f. is due to a change in the 'alue of the magnetic flux passing through the coil. The above xperiments also show that the direction of the induced e.m.f. epends upon the direction of the magnetic flux and also upon the irection in which the coil moves relative to the magnetic flux.

Let us bring the magnet up to the coil at different speeds. It is und that the greater the speed, the greater is the deflection of the lvanometer and, therefore, the greater must be the e.m.f. induced the coil.

It is immaterial whether we consider the e.m.f. as being due to change of flux ed with a coil or due to the coil cutting or being cut by magnetic flux; the result actly the same. The fact of the matter is that we do not know what is really pening; but we can calculate the effect by imagining the magnetic field in the of lines of flux, some of which expand from nothing when the field is increased llapse to nothing when the field is reduced. In so doing, the flux may be re-d as cutting the turns of the coil, or alternatively, the effect may be regarded a ng due to a change in the value of the flux passing through the coil.

**18.2 Induced e.m.f.** (*continued*)

Let us now replace magnet NS of fig. 18.1 by a coil A (fig. 18.2) connected through a switch S to a cell. At the instant when S is closed, there is a momentary deflection on G; and when S is opened, G is deflected momentarily in the reverse direction. On the other hand, if S is kept closed and coil A moved towards C, the galvanometer is deflected in the same direction as when S was closed with A

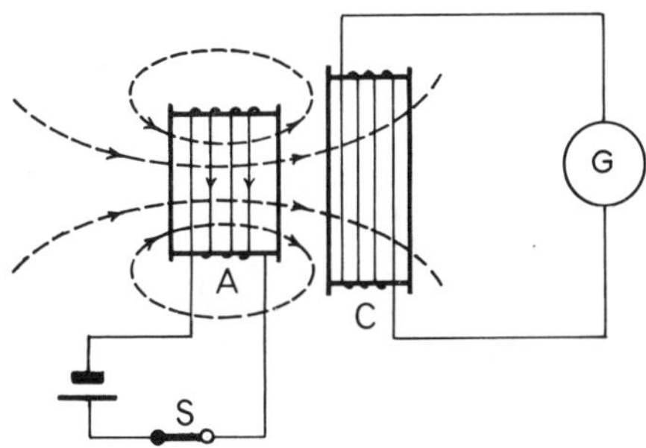


Fig. 18.2 Electromagnetic induction.

stationary. The withdrawal of A causes a deflection in the reverse direction. Deflection of G continues only while there is relative movement between the two coils, i.e. while the magnetic flux passing through coil C is changing.

The dotted lines in fig. 18.2 represent the distribution of the magnetic flux due to current in coil A. When S is opened, the current falls to zero. Consequently, the magnetic flux of A must also disappear; in other words, the magnetic flux is said to *collapse* towards A, and in so doing, the flux that passed through (or was linked with) coil C cuts the latter and induces an e.m.f. in it.

Similarly, when S is closed, the current through A causes a magnetic field to come into existence; and in this process the magnetic flux may be regarded as spreading outwards from coil A, and some of this flux will extend sufficiently to cut coil C and thereby induce an e.m.f. in it. It will be seen that as far as the e.m.f. induced in C is concerned, both the closing of S in fig. 18.2 and the moving of A towards C, with S closed, have the same effect as moving the magnet towards C in fig. 18.1.

The effects observed with the apparatus of fig. 18.2 may be accentuated by placing an iron core inside the coils, thereby increasing the magnetic flux linked with C due to a given current in A. In fact, we may go still further and wind the two coils A and C on an iron ring R, as in fig. 18.3. When S is closed, the current in A sets up magnetic flux through R, as indicated by the dotted circles. This flux, in becoming linked with coil C, induces in the latter an e.m.f. which circulates a current causing G to be deflected momentarily. So long as S remains closed, there is no further change of magnetic flux and therefore no e.m.f. induced in C. When S is opened, the magnetic flux decreases and an e.m.f. is induced in C in the reverse direction.

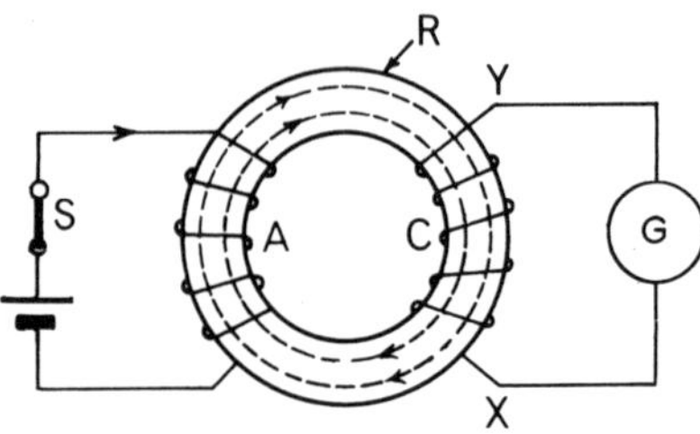


Fig. 18.3 Electromagnetic induction.

It was by means of apparatus similar to that shown in fig. 18.3 that Faraday discovered electromagnetic induction, namely that a change in the value of the magnetic flux through a coil causes an e.m.f. to be induced in that coil.

It should be pointed out that when S (fig. 18.3) is closed, the flux which becomes linked with coil C has also to grow in coil A; consequently, an e.m.f. is induced in A as well as in C. Similarly, when S is opened, the decrease of flux causes an e.m.f. to be induced in both A and C.

The results obtained from the above experiments on electromagnetic induction may now be summarized thus:

(*a*) When a conductor cuts or is cut by magnetic flux, an e.m.f. is induced in the conductor; or alternatively, when there is a change of magnetic flux passing through a circuit, an e.m.f. is induced in that circuit.

(*b*) The direction of the induced e.m.f. depends upon the direction of the magnetic flux and upon the direction in which the flux moves relative to the conductor.

(*c*) The magnitude of the e.m.f. is proportional to the rate at which the conductor cuts or is cut by the magnetic flux; or alternatively, the magnitude of the e.m.f. induced in a circuit is proportional to the rate of change of magnetic flux through the circuit. This last statement is often referred to as *Faraday's Law of Electromagnetic Induction*, although it was not stated in this form by Faraday.

## 18.3 The transformer

It is only a small step from the apparatus shown in fig. 18.3 to a transformer. The function of the latter is to change the voltage of an alternating-current supply from one value to another. An alternating voltage is maintained across coil A, termed the *primary* winding, and the alternating current through A sets up an alternating flux in the iron core. The variation of this flux causes an alternating e.m.f. to be induced in coil C (termed the *secondary* winding) as well as in coil A; and if the whole of the flux produced by the current in A passes through C, the e.m.f. induced in each turn is the same for the two coils.

Suppose the e.m.f. induced per turn to be, say, 4 V and the number of turns on the primary and secondary windings to be 50 and 500 respectively, then the e.m.f. induced in the primary winding is 200 V and that induced in the secondary winding is 2000 V. The voltage applied to the primary winding is practically equal and opposite to the e.m.f. induced in the primary and is therefore approximately 200 V. Hence such a transformer steps *up* the voltage about ten times.

Had the secondary winding been wound with only five turns, the secondary voltage would have been 20 V and the transformer would therefore step *down* the voltage to roughly a tenth of the voltage applied to the primary winding.

Since the alternating flux induces an e.m.f. in the iron core as well as in the windings, it is necessary to reduce the magnitude of the 'eddy' currents circulating in the core so as to prevent excessive loss of power in the latter. This is done by constructing the core of laminations, about 0·3 to 0·5 mm thick, insulated from one another.

## 18.4 Direction of the induced e.m.f.

The simplest method of determining the direction of the e.m.f. induced or generated in a conductor is to find the direction of the current due to that e.m.f. Thus, in fig. 18.4, AB represents a metal rod with its ends connected through a changeover switch S to a moving-coil galvanometer G. With S on side *a*, let us move AB *downwards* between the poles NS of an electromagnet and note the direction of G's deflection. Let us then move S over to *b*, so as to connect G in series with a high-resistance circuit R across a cell C. In order that G may again be deflected in the same direction, the polarity of C must be that shown in fig. 18.4; i.e., the current through the galvanometer must be in the direction indicated by the arrow alongside G. Hence, the e.m.f. generated in AB must be acting from A towards B when the rod is moved downwards through the magnetic field between poles NS.

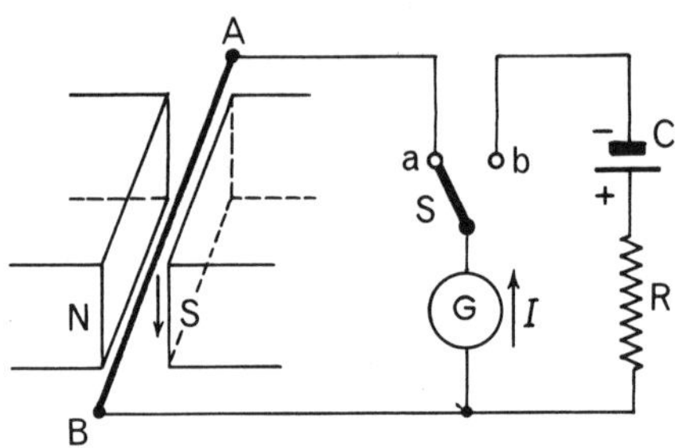


Fig. 18.4 Direction of induced e.m.f.

Now arises the problem: how can we remember this relationship in a form that can be easily applied to any other case? Two methods are available for this purpose, namely:

**(a) Fleming's* right-hand rule** *If the first finger of the right hand be pointed in the direction of the magnetic flux, as in fig.* 18.5, *and if the thumb be pointed in the direction of motion of the conductor* **relative** *to the magnetic field, then the second finger, held at right angles to both the thumb and the first finger, represents the direction of the e.m.f.* The manipulation of the thumb and fingers and their association with the correct quantity present some difficulty to many students. Easy manipulation can only be acquired by experience; and it may be helpful to associate *f*ield or *f*lux with *f*irst finger, *m*otion of the conductor relative to the field with the *m* in thu*m*b and *e*.m.f. with the *e* in s*e*cond finger. If any two of these are correctly applied, the third is correct automatically.

**(b) Lenz's law** In 1834, almost immediately after the discovery of induced currents, Heinrich Lenz, a German physicist (1804–65),

* John Ambrose Fleming (1849-1945) was Professor of Electrical Engineering at University College, London.

gave a simple rule, known as Lenz's law, which can be expressed thus: *The direction of an induced e.m.f. is always such that it tends to set up a current opposing the motion or the change of flux responsible for inducing that e.m.f.*

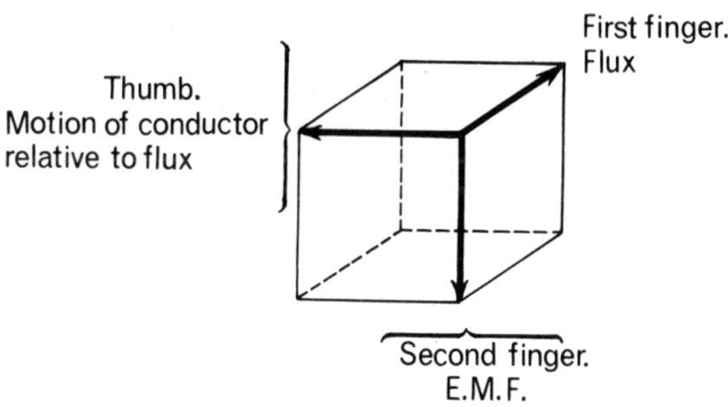


Fig. 18.5 Fleming's right-hand rule.

Let us consider the application of this law to the experiment described in connection with fig. 18.4. The current due to the e.m.f. induced in AB tends to set up an anticlockwise magnetic flux around the rod, so that the resultant flux in the vicinity of AB is distorted in the opposite direction to that shown in fig. 17.22(c). But such a distorted flux exerts an upward force upon the conductor, trying to oppose its downward movement and therefore trying to prevent that which is responsible for the generation of the e.m.f. Hence, when Lenz's law is applied to such an example, it is necessary to find the direction of the current which will distort the flux in such a direction as to try to prevent the relative movement of the conductor and the magnetic flux. The direction of such a current is also the direction of the generated e.m.f.

Let us also consider the application of Lenz's law to the ring shown in fig. 18.3. By applying either the screw or the grip rule given in section 17.7, we find that when S is closed and the cell has the polarity shown, the direction of the magnetic flux in the ring is clockwise. Consequently, the current in C must be such as to try to produce a flux in an anticlockwise direction, tending to oppose the growth of the flux due to A, namely the flux which is responsible for the e.m.f. induced in C. But an anticlockwise flux in the ring would require the current in C to be passing through the coil from X to Y (fig. 18.3). Hence, this must also be the direction of the e.m.f. induced in C.

## 18.5 Magnitude of the generated or induced e.m.f.

Fig. 18.6 represents the elevation and plan of a conductor AA situated in an airgap between poles NS. Suppose AA to be carrying a current $I$, in amperes, in direction shown. By applying either the screw or the grip rule of section 17.6, it is found that the effect of this current is to strengthen the flux on the right and weaken that on the left of A, so that there is a force of $BlI$ newtons (section 17.11) urging the conductor towards the left, where $B$ is the flux density in teslas (or webers per square metre) and $l$ is the length in metres of

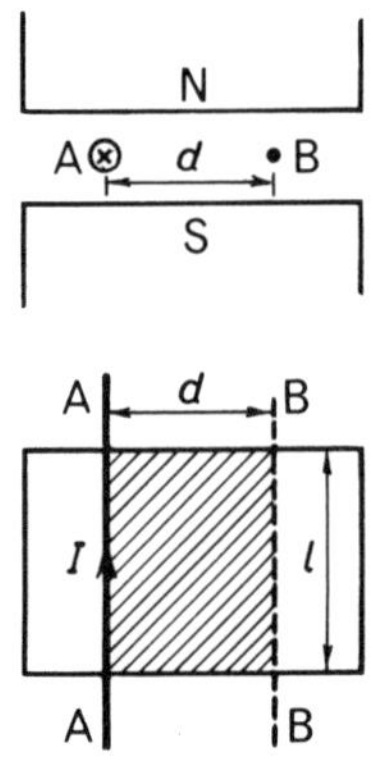


Fig. 18.6 Conductor moved across magnetic field.

the conductor in the magnetic field. Hence, a force of this magnitude has to be applied in the opposite direction to move A towards the right.

The work done in moving conductor AA through a distance $d$ metres to position BB in fig. 18.6 is $(BlI \times d)$ joules. If this movement of AA takes place at a uniform velocity in $t$ seconds, the e.m.f. induced in the conductor is constant at, say, $E$ volts. Hence the electrical power generated in AA is $IE$ watts and the electrical energy is $IEt$ watt seconds or joules. Since the mechanical energy expended in moving the conductor horizontally across the gap is all converted into electrical energy, then

$$IEt = BlId$$

$$\therefore \qquad E = \frac{Bld}{t} \text{ volts.}$$

But $Bld$ = the total magnetic flux $\Phi$, in webers, in the area shown shaded in fig. 18.6, and is therefore the flux cut by the conductor when the latter is moved from AA to BB. Hence

$$E\,[\text{volts}] = \frac{\Phi\,[\text{webers}]}{t\,[\text{seconds}]} \qquad (18.1)$$

i.e. the e.m.f., in volts, generated in a conductor is equal to the rate (in webers/second) at which the magnetic flux is cutting or being cut by the conductor; and the *weber* may therefore be defined as *that magnetic flux which, when cut at a uniform rate by a conductor in* 1 *second, generates an e.m.f. of* 1 *volt.*

**Example 18.1** *Calculate the e.m.f. generated in the axle of a car travelling at* 90 *km/h, assuming the length of the axle to be* 1·8 *m and the vertical component of the density of the earth's magnetic field to be* 40 *microteslas.*

Speed of car = 90 × 1000/3600 = 25 m/s.

Vertical component of the density of the earth's field

$= 40\ \mu\text{T} = 40\ \mu\text{Wb/m}^2$;

∴ rate at which axle cuts magnetic flux

$= 40\ [\mu\text{Wb/m}^2] \times 25\ [\text{m/s}] \times 1{\cdot}8\ [\text{m}]$

$= 1800\ \mu\text{Wb/s}$

so that e.m.f. generated in axle

$= 1800\ \mu\text{V}$.

## 18.6 Magnitude of e.m.f. induced in a coil

Suppose the magnetic flux through a coil of $N$ turns to be increased by $\Phi$ webers in $t$ seconds due to, say, the relative movement of the coil and a magnet (fig. 18.1). Since the magnetic flux cuts each turn, one turn can be regarded as a conductor cut by $\Phi$ webers in $t$ seconds; hence, from expression (18.1), the average e.m.f. induced in each turn is $\Phi/t$ volts. The current due to this e.m.f., by Lenz's law, tries to prevent the increase of flux, i.e. tends to set up an opposing flux. Thus, if the magnet NS in fig. 18.1 is moved towards coil C, the flux passing from left to right through the latter is increased. The e.m.f. induced in the coil circulates a current in the direction represented by the dot and cross in fig. 18.7, where—for simplicity—coil C is represented as one turn. The effect of this current is to distort the magnetic field as shown by the dotted lines, thereby tending to push the coil away from the magnet. By Newton's Third Law of Motion, there must be an equal force tending to oppose the movement of the magnet towards the coil.

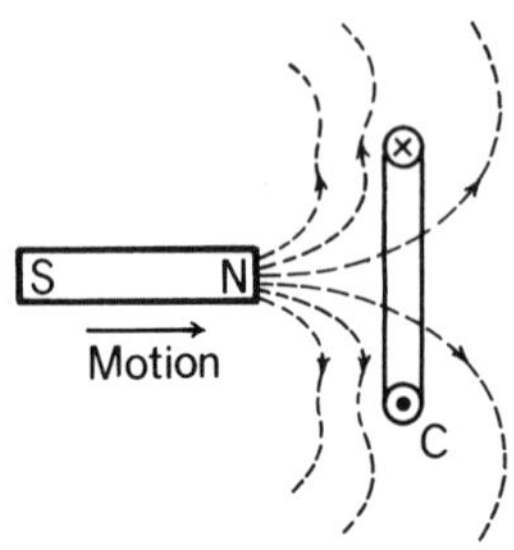


Fig. 18.7 Distortion of magnetic field caused by induced current.

Owing to the fact that the induced e.m.f. circulates a current tending to oppose the increase of flux through the coil, its direction is regarded as negative; hence

average e.m.f. induced in 1 turn $= -\Phi/t$ volts
$=$ —average rate of change of flux in webers per second

and average e.m.f. induced in coil $= -N\Phi/t$ volts (18.2)
$=$ —average rate of change of flux-linkages per second.

From expression (18.2) we can define the *weber as that magnetic flux which, linking a circuit of one turn, induces in it an e.m.f. of* 1 *volt when the flux is reduced to zero at a uniform rate in* 1 *second.* This is an alternative to the definition already given on page 305.

**Example 18.2** *A magnetic flux of* 400 $\mu$*Wb passing through a coil of* 1200 *turns is reversed in* 0·1 *s. Calculate the average value of the e.m.f. induced in the coil.*

The magnetic flux has to decrease from 400 μWb to zero and then increase to 400 μWb in the reverse direction;

hence the change of flux $= -800\ \mu\text{Wb} = -800 \times 10^{-6}$ Wb.

Substituting in expression (18.2), we have:

$$\left.\begin{array}{r}\text{average e.m.f.}\\ \text{induced in coil}\end{array}\right\} = -\frac{1200\ [\text{turns}] \times (-800 \times 10^{-6})\ [\text{Wb}]}{0{\cdot}1\ [\text{s}]}$$

$$= 9{\cdot}6\ \text{V}.$$

This e.m.f. is positive because its direction is the same as the original direction of the current, at first tending to prevent the current decreasing and then tending to prevent it increasing in the reverse direction.

## 18.7 Self inductance

Any circuit in which a change of current is accompanied by a change of flux linked with the circuit, and therefore by an e.m.f. induced in the circuit (as explained in section 18.6) is said to possess *self inductance* or merely *inductance* (symbol, $L$). The unit of inductance is termed the *henry* (symbol, H), in memory of an American physicist, Joseph Henry (1797–1878) who, quite independently, discovered electromagnetic induction within a year after it had been discovered in this country by Michael Faraday.

The *henry* is defined as *the inductance of a circuit in which an e.m.f. of* 1 *volt is induced when the current in the circuit varies at the rate of* 1 *ampere per second.* If either the inductance or the rate of change of current is doubled, the value of the induced e.m.f. is doubled. Hence we may generalize by saying that if a coil has an inductance $L$ and if the current increases from zero to $I$ in time $t$,

$$\text{average rate of change of current} = I/t\ \text{amperes/second}$$

$$\text{and average e.m.f. induced in coil} = -L \times \text{rate of change of current}$$

$$= -L \times I/t\ \text{volts} \qquad (18.3)$$

The minus sign signifies that if the current is increasing in a certain direction, the e.m.f. is induced in the *opposite* direction, thereby delaying the growth of the current, in accordance with Lenz's law.

If the coil has $N$ turns and if the increase of current from zero to $I$ has caused the flux to increase from zero to $\Phi$ in time $t$, then from expression (18.2),

$$\text{average e.m.f. induced in coil} = -\,N\Phi/t\ \text{volts.}$$

Equating expressions (18.2) and (18.3), we have:

$$-N\Phi/t = -LI/t$$

$$\therefore \quad L = N\Phi/I \text{ henrys} \quad (18.4)$$

$$= \text{flux-linkages per ampere.}$$

**Example 18.3** *The current in a coil wound with* 80 *turns is reduced from* 5 *A to zero in* 0·1 *s. The coil has an inductance of* 0·2 *H. Calculate* (a) *the average e.m.f. induced in the coil and* (b) *the value of the flux produced by* 5 *A.*

(*a*) Average rate of change of current $= -5$ [A]/0·1 [s]
$= -50$ A/s.

Substituting in expression (18.3), we have:

average e.m.f. induced in coil $= -0{\cdot}2$ [H] $\times$ $(-50)$ [A/s]
$= 10$ V.

The direction of this e.m.f. is the same as that of the current, thereby tending to delay the decrease in the value of the current.

(*b*) Substituting for *L*, *N* and *I* in expression (18.4), we have:

0·2 [H] $=$ 80 [turns] $\times$ $\Phi$/5 [A]

$\therefore$ $\Phi = 0{\cdot}0125$ Wb $= 12{\cdot}5$ mWb.

## 18.8 Mutual inductance

Let us consider the two coils, A and C, shown in fig. 18.2. When the current in coil A is varied, there is a change in the value of the flux passing through coil C. Consequently, an e.m.f. is induced in C. Coils A and C are therefore said to possess *mutual inductance* (symbol, *M*).

The unit of *mutual inductance* is the same as that of self inductance, namely the *henry* and may be defined thus: *two coils have a mutual inductance of* 1 *henry if an e.m.f. of* 1 *volt is induced in one coil when the current in the other coil varies at the rate of* 1 *ampere per second.*

If two coils have mutual inductance *M*, and if the current in one coil increases from zero to *I* in time *t*,

average e.m.f. induced in other coil $= -M \times I/t$ volts (18.5)

The minus sign signifies that the e.m.f. induced in coil C circulates a current in such a direction as to tend to set up a flux in opposition to that produced by the current in coil A, as already explained in section 18.6.

**Example 18.4** *Two coils, A and B, have a mutual inductance of* 0·4 *H. If the current in coil A increases from zero to* 5 *A in* 0·2 *s, what is the average value of the e.m.f. induced in coil B?*

Average rate of change of current in coil A

$$= 5\ [\text{A}]/0{\cdot}2\ [\text{s}] = 25\ \text{A/s}.$$

From expression (18.5), we have:

$$\text{average e.m.f. induced in coil B} = -0{\cdot}4\ [\text{H}] \times 25\ [\text{A/s}] = -10\ \text{V}.$$

## *Summary of Chapter* 18

An e.m.f. is induced in a coil when a magnet or another coil carrying a current is moved either towards or away from that coil. The same effect may also be obtained by moving the first coil towards or away from the magnet or the second coil. An e.m.f. can also be induced in a coil by switching on or off the current in an adjacent coil. In general, an e.m.f. is induced in a coil whenever the latter cuts or is cut by magnetic flux, i.e. whenever there is a change of flux through the coil.

It is shown how Fleming's right-hand rule and Lenz's law can be applied for determining the direction of the induced e.m.f.

Average e.m.f. generated in a conductor $= \Phi/t$ volts (18.1)

where $\Phi$ = flux, in webers, cut in $t$ seconds.

Average e.m.f. induced in a coil of $N$ turns $= -N\Phi/t$ volts (18.2)

where $\Phi$ = change of flux, in webers, in $t$ seconds.

For a coil having an inductance $L$,

induced e.m.f. $= -L \times$ rate of change of current (18.3)

and $L = N\Phi/I$ (18.4)

For two coils, A and B, having mutual inductance $M$,

e.m.f. induced in B $= -M \times$ rate of change of current in A (18.5)

## EXAMPLES 18

1. A conductor, 500 mm long, is moved through a magnetic field at 0·5 m/s perpendicular to the flux. What is the flux density when an e.m.f. of 0·1 V is generated? (E.M.E.U., G2)
2. A generator conductor having a length of 200 mm at radius 100 mm is moving at 573 rev/min at right angles to the magnetic field which has a flux density of 1·4 Wb/m$^2$. If the conductor current is 30 A, what is the e.m.f. induced, and what is the force on the conductor? (U.E.I., G2)
3. A conductor, 120 mm long, lies at right angles to a magnetic field having a uniform density of 0·4 T. Calculate the speed at which the conductor must be moved in a direction at right angles to its length and to the magnetic field in order that an e.m.f. of 0·8 V may be generated in the conductor.
4. The axle of a certain motor car is 1·5 m long. Calculate the e.m.f. generated in the axle when the car is travelling at 140 km/h along a level road. Assume the vertical component of the density of the earth's magnetic field to be 40 μT.
5. A wire, 200 mm long, is moved at a uniform speed of 6 m/s at right angles to its length and to a magnetic field. Calculate the density of the magnetic field if the e.m.f. generated in the wire is 0·4 V.

   If the wire forms part of a closed circuit having a resistance of 0·1 Ω, calculate the force on the wire.
6. A copper disc, 300 mm in diameter, is rotated at 200 rev/min about a horizontal axis through its centre and perpendicular to its plane. If the axis points magnetic north and south, calculate the e.m.f. between the circumference of the disc and the axis. Assume the horizontal component of the density of the earth's field to be 18 μT.
7. A coil of 1500 turns produces a magnetic flux of 2500 μWb when carrying a certain current. If this current is reversed in 0·2 s, what is the average value of the e.m.f. induced in the coil?
8. Two coils, A and B, are wound on the same iron core. There are 300 turns on A and 2800 turns on B. A current of 4 A through coil A produces a flux of 800 μWb in the core. If this current is reversed in 0·02 s, what are the values of the average e.m.f.s induced in A and B?
9. Calculate the inductance of a circuit in which an e.m.f. of 10 V is induced when the current is varying at the rate of 40 A/s.
10. A circuit has an inductance of 200 mH. At what rate is the current varying when the induced e.m.f. is 4 V?
11. A certain coil is wound with 100 turns and a current of 4 A produces a flux of 500 μWb. Calculate (*a*) the inductance of the coil, in millihenrys and (*b*) the average e.m.f. induced in the coil when the current is reversed in 0·2 s.
12. If the mutual inductance between two coils is 0·05 H, what is the average value of the e.m.f. induced in one coil when a current of 3 A in the other coil is reversed in 0·1 s?
13. If an e.m.f. of 1 V is induced in a coil when the current in an adjacent coil varies at the rate of 200 A/s, calculate the mutual inductance of the two coils, in millihenrys.

## ANSWERS TO EXAMPLES 18

1. 0·4 T.
2. 1·68 V, 8·4 N.
3. 16·67 m/s.
4. 2·334 mV.
5. 0·333 T, 0·267 N.
6. 4·23 μV.
7. 37·5 V.
8. 24 V, 224 V.
9. 0·25 H.
10. 20 A/s.
11. 12·5 mH, 0·5 V.
12. 3 V.
13. 20 mH.

CHAPTER 19

# Primary and secondary cells

## 19.1 A simple voltaic cell

It was in 1789 that a chance observation by Luigi Galvani* (1737–98) led to the idea of generating an electric current from a source of chemical energy. Galvani, a professor of anatomy at Bologna, noticed that recently-skinned frogs' legs, hung by copper wire to an iron balcony, were convulsed whenever they touched the iron. Another Italian, Alessandro Volta* (1745–1827), a professor of physics at Pavia, subsequently showed that if a rod of two dissimilar metals, such as copper and iron, was placed so that one end was in contact with a nerve on a frog's leg and the other end in contact with a muscle on the foot, muscular contraction took place. Following his investigations of this phenomenon, Volta, in 1799, constructed a simple battery—known as Volta's pile—by assembling discs of zinc, cloth soaked with brine, and copper, piled upon one another in that order. By this means, a large number of cells were obtained in series, giving a high electromotive force, but having the disadvantage of high internal resistance.

It was a comparatively short step to replace the wet cloth of Volta's pile by an electrolyte to give the simple voltaic cell shown in fig. 19.1. In this cell, plates of copper and zinc are immersed in dilute sulphuric acid contained in a glass vessel.

When a resistor R and an ammeter A are connected in series across the terminals, it is found that current flows through R from the copper electrode to the zinc electrode and that the difference of potential between the plates is about 0·8 V, the copper plate being the positive electrode.

It is also found that when current is flowing, hydrogen gas is released in the form of bubbles on the surface of the copper plate. This gas film sets up a back electromotive force and also increases the internal resistance of the cell. These effects are known as

* Galvani's name is perpetuated in such terms as 'galvanize' and 'galvanometer', and the unit of e.m.f. and of potential difference, the 'volt', has been named after Volta (see section 15.7).

*polarization.* Many types of cells were developed to reduce polarization but they are now all obsolete except the 'dry' Leclanché cell used in electric torches, etc., and the mercury cell (sections 19.2 and 19.3).

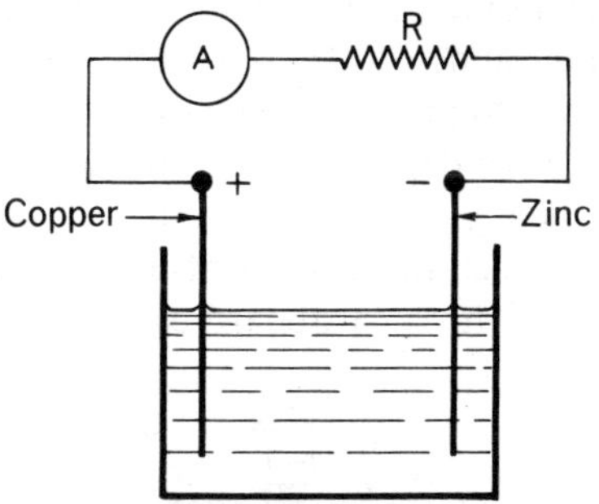


Fig. 19.1 A simple voltaic cell.

In these cells it is only possible to transform chemical energy into electrical energy, and a cell can be replenished only by renewal of the active materials. This type is usually referred to as a *primary cell.* In a *secondary cell,* the chemical action is reversible, i.e. chemical energy is converted into electrical energy when the cell is discharging, and electrical energy is converted into chemical energy when the cell is being charged, as already referred to in section 14.1.

## 19.2 Leclanché cell

The 'wet' type of Leclanché cell (now practically obsolete) consists of a carbon plate, surrounded by a mixture of manganese dioxide ($MnO_2$) and powdered carbon, in an unglazed earthenware pot. This pot and an amalgamated zinc rod are immersed in a saturated solution of salammoniac (ammonium chloride, $NH_4Cl$) in water. The carbon plate and the zinc rod form the positive and negative electrodes respectively.

The function of the manganese dioxide is to reduce polarization by combining with the hydrogen released at the carbon plate to form water and a brown oxide of manganese ($Mn_2O_3$), thus:

$$H_2 + 2MnO_2 = Mn_2O_3 + H_2O.$$

In the 'dry' type of Leclanché cell, the same ingredients are present, and fig. 19.2 is a sectional view of the construction most commonly used. A carbon rod A is surrounded by a black depolarizing paste B, consisting of manganese dioxide, powdered carbon, salammoniac, zinc chloride and water, the paste being usually contained in a bag of coarse linen. Around this depolarizer is a mixture P of flour, plaster of Paris, salammoniac and zinc chloride, with water added to form a white paste. The latter need only be thick enough to prevent the black paste B touching the zinc container Z. Above the depolarizer and the white paste is a layer S of sawdust or similar porous material, the cell being sealed with a layer of pitch T in which there is a vent tube V. The zinc container Z is usually covered with a cardboard case.

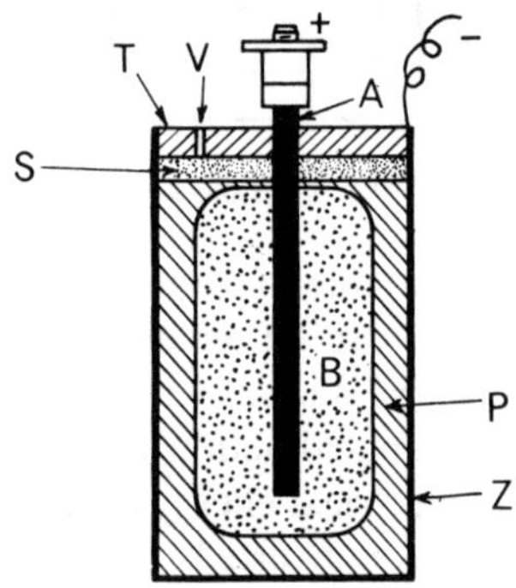


Fig. 19.2 Dry Leclanché cell

A Leclanché cell has an e.m.f. of about 1·5 V when new, but this e.m.f. falls fairly rapidly if the cell is in continuous use. This fall is due to polarization—the hydrogen film at the carbon electrode forms faster than can be dissipated by the depolarizer. However, if the cell is disconnected from the external circuit, depolarization continues and the e.m.f. recovers its normal value. Hence the Leclanché cell is suitable only for intermittent use, e.g. for electric torches, radio receivers, etc.

The life of a dry Leclanché cell is reduced by local action and even the shelf life is limited to about two years owing to the local action that goes on continuously in the cell. Also, this type of cell does not lend itself to miniaturization, since the number of ampere hours obtainable from the cell falls off rapidly as the size is reduced.

## 19.3 Mercury cell*

This type of cell was developed to meet the requirements of miniaturization, e.g. for guided missiles, medical electronics, hearing aids, etc., where it is necessary to reduce the size of the cell but at the same time obtain: (*a*) a high ratio of output energy/mass; (*b*) a constant e.m.f. over a relatively long period; and (*c*) a long shelf life, i.e. absence of local action.

A cross-section of the basic type of mercury cell is shown in fig. 19.3. The negative electrode is zinc, either as a foil or as powder compressed into a hollow cylinder. This electrode is surrounded by a layer of electrolyte consisting of a concentrated aqueous solution of potassium hydroxide (KOH) and zinc oxide (ZnO). Surrounding the electrolyte is a layer of mercuric oxide (HgO). This oxide contains a small percentage of finely powdered graphite to reduce the internal resistance of the cell.

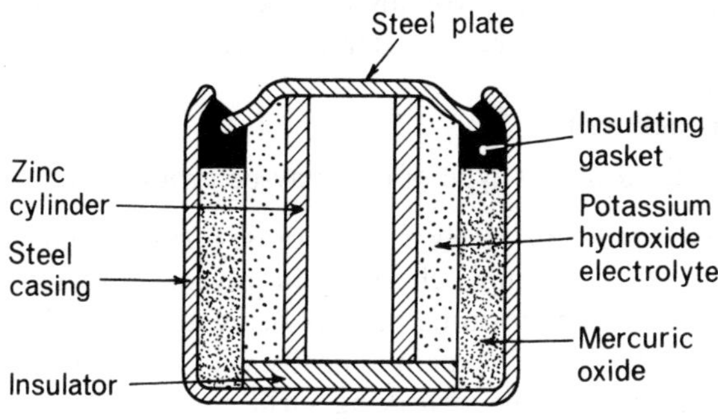


Fig. 19.3 A mercury cell.

The above constituents are assembled in a nickel-plated or stainless-steel cylinder which forms the positive electrode. The zinc cylinder and the electrolyte are supported on a disc of insulating material; and the cell is sealed by an insulating gasket between the container and a nickel-plated steel plate resting on top of the zinc cylinder.

When the cell is supplying current, no gases are evolved at either electrode, except under abnormal operating conditions. Consequently there is no polarization so that the cell is able to maintain its terminal voltage practically constant at about 1·2–1·3 V (depend-

* The authors are indebted to Mallory Batteries, Ltd, for information concerning this cell.

ing upon the value of the load current) for a relatively long time. Also, owing to local action being practically negligible, the cell can be stored for a long period at normal atmospheric temperature without appreciable loss of capacity.

## 19.4 Secondary cells

Whenever a battery is required to supply a relatively large amount of power, secondary cells must be used. Such batteries may supply power in telephone exchanges, emergency lighting for hospitals, etc. They also form portable sources of power for starting, ignition and lighting of motor vehicles and as motive power for electrically propelled vehicles.

Secondary cells may be divided into two types: (*a*) the lead–acid cell, in which lead plates covered with compounds of lead are immersed in a dilute solution of sulphuric acid in water; (*b*) the nickel–cadmium and the nickel–iron alkaline cells.

These cells will now be considered in greater detail.

## 19.5 Lead–acid cell

The plates used in this type of cell may be grouped thus:

(*a*) *Formed* or *Planté* plates, namely those formed from lead plates by charging, discharging, charging in reverse direction, etc., a number of times, the forming process being accelerated by the use of suitable chemicals. The main difficulty with this type of construction is to secure as large a working surface as possible for a given mass of plate. One method of increasing the surface area is to make the plates with deep corrugations, as in fig. 19.4, with reinforcing ribs at intervals.

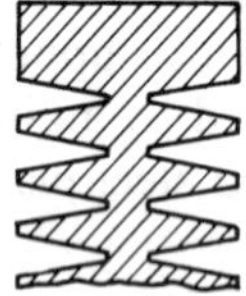

Fig. 19.4 Section of a Planté plate.

(*b*) *Pasted* or *Faure* plates, namely those in which a paste of the active material is either pressed into recesses in a lead–antimony grid or held between two finely perforated lead sheets cast with ribs and flanges so that the two sheets, when riveted together, form in effect a number of boxes which hold the paste securely in position. The paste is usually sulphuric acid mixed with red lead ($Pb_3O_4$) for the positive plates and with litharge (PbO) for the negative plates. A small percentage of a material such as powdered pumice is added to increase the porosity of the paste. For a given ampere hour capacity, the mass of a pasted plate is only about a third of that of a formed plate.

Students are advised to examine specimen plates or plates taken from disused accumulators.

When weight is of no importance it is common practice to make the positive plates of the 'formed' type and the negative plates of the 'pasted' type. The active material on the positive plates expands when it is subjected to chemical changes; consequently it is found that the greater mechanical stiffness of the 'formed' construction is an important advantage in reducing the tendency of the plates to buckle. This tendency to buckle is reduced still further by constructing the cell with an odd number of plates, as shown in fig. 19.5, the

Fig. 19.5 Arrangement of plates.

outer plates being always negative. This arrangement enables both sides of each positive plate to be actively employed, and the tendency of one side of a plate to expand and cause buckling is neutralized by a similar tendency on the other side.

The plates are assembled in glass, polystyrene, vulcanized-rubber or resin-rubber containers and separated by a special grade of paper or microporous sheets of a plastic material.

The most suitable relative density of the acid depends upon the type of cell and the state of charge of the cell. An average value, however, is about 1·21.

The terminal voltage of a lead–acid cell is about 2 V to 1·85 V when it is being discharged and about 2·1 V to 2·6 V when being charged.

## 19.6 Chemical reactions in a lead–acid cell

The chemical reactions taking place during charge and discharge are complicated, and all we can do here is to indicate the most important reactions and to account for the variation in the density of the electrolyte.

When the cell is fully charged, the active material on the positive plate is lead dioxide ($PbO_2$) and that on the negative plate is spongy or porous lead (Pb). During discharge, the lead dioxide and the spongy lead are converted into lead sulphate ($PbSO_4$). These chemical reactions are accompanied by the decomposition of some of the sulphuric acid molecules and the formation of water molecules.

During charge, the chemical reactions are reversed so that the active material is converted back to lead dioxide on the positive plates and to spongy lead on the negative plates.

The above reactions may be summarized thus:

| | *Positive plate* | *Electrolyte* | *Negative plate* | |
|---|---|---|---|---|
| Dis-charge ↓ | Lead dioxide ($PbO_2$) | Sulphuric acid ($2H_2SO_4$) | Lead (Pb) | ↑ Charge |
| | Lead sulphate ($PbSO_4$) | Water ($2H_2O$) | Lead sulphate ($PbSO_4$) | |

It will be seen that for every two molecules of sulphuric acid decomposed during discharge, two molecules of water are formed; hence the density of the electrolyte falls as the cell discharges. The reverse process occurs during charging, so that when the cell is fully charged, the density of the electrolyte is restored to its initial value.

## 19.7 Alkaline cells

In both the nickel–iron and the nickel–cadmium types, the positive plates are made of nickel hydroxide enclosed in finely perforated steel tubes or pockets, the electrical resistance being reduced by the addition of flakes of pure nickel or graphite. These tubes or pockets are assembled in nickelled-steel plates. In the nickel–iron cell the negative plate is made of iron oxide with a little mercuric oxide to reduce the resistance, the mixture being enclosed in perforated steel pockets, also assembled in nickelled-steel plates. In the nickel–cadmium cell the active material is cadmium mixed with a little iron, the purpose of the latter being to prevent the active material caking and losing its porosity.

In both types of cell, the electrolyte is a solution of potassium hydroxide (KOH) having a relative density of about 1·15–1·2, depending upon the type of cell and the conditions of service. The electrolyte does not undergo any chemical change; consequently the quantity of electrolyte can be reduced to the minimum necessitated by adequate clearance between the plates.

The plates are separated by insulating rods and assembled in sheet-steel containers, the latter being mounted in non-metallic crates to insulate the cells from one another.

The advantages of the alkaline accumulator are: (*a*) its mechanical construction enables it to withstand considerable vibration, and (*b*) it is free from 'sulphating' or any similar trouble and can therefore be left in any state of charge without damage. Its disadvantages are: (*a*) its cost is greater than that of the corresponding lead cell; (*b*) its average discharge p.d. is about 1·2 V compared with 2 V for the lead cell, so that for a given voltage the number of alkaline cells is about 67 per cent greater than that of lead cells.

Due partly to these disadvantages of the alkaline cell and partly to the improvements made in the construction of the lead–acid cell —especially the portable type—during the past twenty years, the great majority of modern batteries are of the lead–acid type.

**Example 19.1** *A constant terminal voltage of* 120 *V is to be maintained by a battery of alkaline cells. The initial and final values of the e.m.f. per cell are* 1·3 *V and* 1·15 *V respectively. The internal resistance per cell is* 0·01 Ω *and the discharge current is* 10 *A. Calculate the initial and final number of cells required in series.*

Initial terminal voltage/cell = initial e.m.f./cell − voltage drop/cell due to internal resistance

$$= 1{\cdot}3\ [\text{V}] - (10\ [\text{A}] \times 0{\cdot}01\ [\Omega])$$

$$= 1{\cdot}3 - 0{\cdot}1 = 1{\cdot}2\ \text{V}.$$

∴ initial number of cells = 120 [V]/1·2 [V] = 100.

Final terminal voltage/cell = 1·15 — 0·1 = 1·05 V,

∴ final number of cells = 120 [V]/1·05 [V] = 114.

**Example 19.2** *A battery of* 30 *lead-acid cells is to be charged at a constant current of* 8 *A from a* 110-*V d.c. supply. The terminal voltage per cell is* 1·9 *V at the commencement of charging and* 2·6 *V at the end. Calculate the maximum and minimum values of the resistor required in series with the battery.*

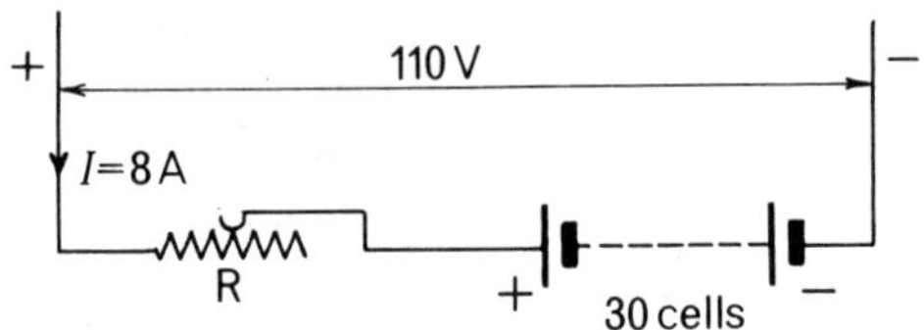


Fig. 19.6 Circuit diagram for Example 19.2.

Fig. 19.6 represents a variable resistor R connected in series with the battery of 30 cells.

At commencement of charging,

total p.d. across battery = 1·9 [V] × 30 = 57 V,

∴ corresponding p.d. across R = 110 — 57 = 53 V

and corresponding resistance of R = 53 [V]/8 [A] = 6·625 Ω.

At end of charging,

total p.d. across battery = 2·6 [V] × 30 = 78 V,

∴ corresponding p.d. across R = 110 — 78 = 32 V,

and corresponding resistance of R = 32 [V]/8 [A] = 4 Ω.

**Example 19.3** *A battery of* 80 *cells is charged through a fixed resistor from a* 240-*V d.c. supply. At the beginning of the charge, the e.m.f. per cell is* 1·9 *V and the charging current is* 5 *A. The internal resistance per cell is* 0·06 Ω. *Calculate the value of the resistor.*

At beginning of charge,

terminal voltage/cell = e.m.f./cell + voltage drop/cell due to internal resistance

= 1·9 [V] + (5 [A] × 0·06 [Ω])

= 1·9 + 0·3 = 2·2 V.

∴ total voltage across battery = 2·2 [V] × 80 = 176 V.

$$\text{Corresponding voltage across resistor} = 240 - 176 = 64\ \text{V},$$
$$\text{so that value of resistor} = 64\ [\text{V}]/5\ [\text{A}] = 12{\cdot}8\ \Omega.$$

**Example 19.4** *A fully-charged lead-acid cell was completely discharged in* 10 *h, the discharge current being constant at* 6 *A. The average terminal voltage during discharge was* 1·95 *V. A charging current of* 4 *A, maintained constant for* 17 *h, was required to restore the cell to its initial state of charge, the average terminal voltage being* 2·3 *V. Calculate* (a) *the ampere hour efficiency and* (b) *the watt hour efficiency.*

The *ampere hour efficiency* of a cell is the ratio of the number of ampere hours obtainable during discharge to that required to restore the cell to its original condition.

The *watt hour efficiency* of a cell is the ratio of the number of watt hours obtainable during discharge to that required to restore the cell to its original condition.

(*a*) $\text{Output of cell, in ampere hours} = 6\ [\text{A}] \times 10\ [\text{h}] = 60\ \text{A h}.$

$\text{Input to cell, in ampere hours} = 4\ [\text{A}] \times 17\ [\text{h}] = 68\ \text{A h}.$

$$\therefore \quad \text{ampere hour efficiency} = \frac{60\ [\text{A h}]}{68\ [\text{A h}]} = 0{\cdot}882 \text{ per unit} = 88{\cdot}2 \text{ per cent.}$$

(*b*) $\text{Output of cell, in watt hours} = 6\ [\text{A}] \times 1{\cdot}95\ [\text{V}] \times 10\ [\text{h}] = 117\ \text{W h}.$

$\text{Input to cell, in watt hours} = 4\ [\text{A}] \times 2{\cdot}3\ [\text{V}] \times 17\ [\text{h}] = 156{\cdot}4\ \text{W h}.$

$$\therefore \quad \text{watt hour efficiency} = \frac{117\ [\text{W h}]}{156{\cdot}4\ [\text{W h}]} = 0{\cdot}748 \text{ per unit} = 74{\cdot}8 \text{ per cent.}$$

## *Summary of Chapter* 19

In both primary and secondary cells, chemical energy is converted into electrical energy when the cells are discharging; but primary cells can only be replenished by renewal of active material, whereas in secondary cells the chemical action is reversible. Descriptions are given of the dry Leclanché and mercury cells and of the lead–acid and alkaline types of secondary cells.

## EXAMPLES 19

1. An accumulator is overcharged by 5 A for 10 h. If the electrochemical equivalents of hydrogen and oxygen are 0·010 45 mg/C and 0·082 95 mg/C respectively, calculate the volume of water required to be added to compensate for gassing. Assume the mass of 1 mm$^3$ of water to be 1 mg.
2. A current of 20 A is supplied by a cell having an e.m.f. of 2 V to a load with a terminal voltage of 1·8 V. What is the internal resistance of the cell? (E.M.E.U., G2)
3. If the terminal voltage of a lead–acid cell varies between 2·1 V and 1·85 V during discharge, calculate the number of cells required to give 230 V (*a*) at the beginning of discharge and (*b*) at the end of discharge.
4. Calculate the number of lead–acid cells to be connected in series to give a terminal voltage of 240 V when the battery is supplying a current of 12 A. Assume each cell to have an e.m.f. of 2 V and an internal resistance of 0·025 Ω.
5. A battery of 50 cells in series is charged through a 4-Ω resistor from a 230-V supply. If the terminal voltage per cell is 2 V and 2·7 V respectively at the beginning and the end of the charge, calculate the charging current (*a*) at the beginning and (*b*) at the end of the charge.
6. A battery of 40 cells in series is to be charged from a 240-V supply. The average terminal voltage per cell during charge is 2·2 V. Calculate the value of the resistor required in series with the battery to give an average charging current of 5 A.
7. A battery of 10 cells in series is to be charged at a constant current of 8 A from a generator having an e.m.f. of 35 V and an internal resistance of 0·2 Ω. The internal resistance of each cell is 0·08 Ω and its e.m.f. ranges from 1·85 V discharged to 2·15 V charged. Calculate the maximum and minimum values of the series resistor required in the circuit.
8. A battery of 12 cells in series is charged through a fixed resistor from a 30-V supply. When charging commences, the e.m.f. per cell is 1·85 V and the charging current is 4 A. At the end of the charge, the e.m.f. per cell has risen to 2·25 V. If each cell has an internal resistance of 0·05 Ω, calculate the value of the external resistor and the current at the end of the charge.
9. An alkaline cell is discharged at a constant current of 5 A for 10 h, the average terminal voltage being 1·2 V. A charging current of 3 A, maintained for 21 h, is required to bring the cell back to the initial state of charge, the average terminal voltage being 1·48 V. Calculate the ampere hour and watt hour efficiencies.

## ANSWERS TO EXAMPLES 19

1. 16 800 mm$^3$.
2. 0·01 Ω.
3. 110 cells, 123 cells.
4. 141 cells.
5. 32·5 A, 23·75 A.
6. 30·4 Ω.
7. 1·062 Ω, 0·687 Ω.
8. 1·35 Ω, 1·54 A.
9. 79·4 per cent, 64·4 per cent.

# CONVERSION FACTORS

| | | |
|---|---|---|
| | 1 radian | = 57·30 degrees |
| Length | 1 mile | = 1·609 km |
| | 1 yd | = 0·9144 m |
| | 1 ft | = 0·3048 m |
| | 1 in | = 25·4 mm = 2·54 cm |
| Area | 1 $yd^2$ | = 0·836 $m^2$ |
| | 1 $ft^2$ | = 0·0929 $m^2$ = 929 $cm^2$ |
| | 1 $in^2$ | = 645 $mm^2$ = 6·45 $cm^2$ |
| Volume | 1 $yd^3$ | = 0·7646 $m^3$ = 764·6 litres |
| | 1 $ft^3$ | = 28·32 litres |
| | 1 $in^3$ | = 16·39 $cm^3$ = 16 390 $mm^3$ |
| | 1 gallon | = 4·546 litres |
| | 1 pint | = 0·568 litre |
| | 1 litre | = 1000 $cm^3$ = 0·001 $m^3$ |
| Velocity | 1 mile/h | = 1·609 km/h = 0·447 m/s |
| | 1 ft/s | = 0·3048 m/s |
| | 1 km/h | = 0·2778 m/s |
| Acceleration | 1 ft/$s^2$ | = 0·3048 m/$s^2$ |
| Mass | 1 ton | = 1016 kg = 1·016 Mg |
| | 1 lb | = 0·4536 kg |
| | 1 tonne | = 1 Mg = 1000 kg |
| Density | 1 lb/$ft^3$ | = 16·02 kg/$m^3$ |
| | 1 lb/$in^3$ | = 27·68 g/$cm^3$ = 27·68 Mg/$m^3$ |
| Force | 1 tonf | = 9·964 kN |
| | 1 lbf | = 4·448 N |
| | 1 kgf | = 9·806 65 N |
| Torque | 1 lbf ft | = 1·356 N m |
| Pressure, stress | 1 lbf/$ft^2$ | = 47·88 N/$m^2$ |
| | 1 lbf/$in^2$ | = 6895 N/$m^2$ |
| | 1 in$H_2O$ | = 249 N/$m^2$ = 2·49 mbar |
| | 1 inHg | = 3386 N/$m^2$ = 33·86 mbar |
| | 1 pascal | = 1 N/$m^2$ |
| | 1 bar | = $10^5$ N/$m^2$ |
| | 1 standard atmosphere | = 14·7 lbf/$in^2$ = 101·325 kN/$m^2$ |
| | | = 1·013 25 bars |
| Energy | 1 ft lbf | = 1·356 J |
| | 1 Btu | = 1·055 kJ |
| | 1 therm | = 105·5 MJ |
| | 1 calorie | = 4·187 J |
| | 1 kW h | = 3·6 MJ = 3600 kJ |
| | 1 Btu/lb | = 2·326 kJ/kg |
| | 1 Btu/$ft^3$ | = 37·26 kJ/$m^3$ |
| Power | 1 hp | = 745·7 W |
| | 1 ft lbf/s | = 1·356 W |
| Temperature | $t$ [°C] | = $\frac{5}{9}\{t$ [°F] — 32} |
| | $t$ [°C] | = $T$ [K] — 273·15 |

# LOGARITHMS

| | 0 | 1 | 2 | 3 | 4 | 5 | 6 | 7 | 8 | 9 | 1 | 2 | 3 | 4 | 5 | 6 | 7 | 8 | 9 |
|---|---|---|---|---|---|---|---|---|---|---|---|---|---|---|---|---|---|---|---|
| **10** | 0000 | 0043 | 0086 | 0128 | 0170 | | | | | | 4 | 9 | 13 | 17 | 21 | 26 | 30 | 34 | 38 |
| | | | | | | 0212 | 0253 | 0294 | 0334 | 0374 | 4 | 8 | 12 | 16 | 20 | 24 | 28 | 32 | 37 |
| **11** | 0414 | 0453 | 0492 | 0531 | 0569 | | | | | | 4 | 8 | 12 | 15 | 19 | 23 | 27 | 31 | 35 |
| | | | | | | 0607 | 0645 | 0682 | 0719 | 0755 | 4 | 7 | 11 | 15 | 19 | 22 | 26 | 30 | 33 |
| **12** | 0792 | 0828 | 0864 | 0899 | 0934 | | | | | | 3 | 7 | 11 | 14 | 18 | 21 | 25 | 28 | 32 |
| | | | | | | 0969 | 1004 | 1038 | 1072 | 1106 | 3 | 7 | 10 | 14 | 17 | 20 | 24 | 27 | 31 |
| **13** | 1139 | 1173 | 1206 | 1239 | 1271 | | | | | | 3 | 7 | 10 | 13 | 16 | 20 | 23 | 26 | 30 |
| | | | | | | 1303 | 1335 | 1367 | 1399 | 1430 | 3 | 7 | 10 | 12 | 16 | 19 | 22 | 25 | 29 |
| **14** | 1461 | 1492 | 1523 | 1553 | 1584 | | | | | | 3 | 6 | 9 | 12 | 15 | 18 | 21 | 24 | 28 |
| | | | | | | 1614 | 1644 | 1673 | 1703 | 1732 | 3 | 6 | 9 | 12 | 15 | 17 | 20 | 23 | 26 |
| **15** | 1761 | 1790 | 1818 | 1847 | 1875 | | | | | | 3 | 6 | 9 | 11 | 14 | 17 | 20 | 23 | 26 |
| | | | | | | 1903 | 1931 | 1959 | 1987 | 2014 | 3 | 5 | 8 | 11 | 14 | 16 | 19 | 22 | 25 |
| **16** | 2041 | 2068 | 2095 | 2122 | 2148 | | | | | | 3 | 5 | 8 | 11 | 14 | 16 | 19 | 22 | 24 |
| | | | | | | 2175 | 2201 | 2227 | 2253 | 2279 | 3 | 5 | 8 | 10 | 13 | 15 | 18 | 21 | 23 |
| **17** | 2304 | 2330 | 2355 | 2380 | 2405 | | | | | | 3 | 5 | 8 | 10 | 13 | 15 | 18 | 20 | 23 |
| | | | | | | 2430 | 2455 | 2480 | 2504 | 2529 | 2 | 5 | 7 | 10 | 12 | 15 | 17 | 19 | 22 |
| **18** | 2553 | 2577 | 2601 | 2625 | 2648 | | | | | | 2 | 5 | 7 | 9 | 12 | 14 | 16 | 19 | 21 |
| | | | | | | 2672 | 2695 | 2718 | 2742 | 2765 | 2 | 5 | 7 | 9 | 11 | 14 | 16 | 18 | 21 |
| **19** | 2788 | 2810 | 2833 | 2856 | 2878 | | | | | | 2 | 4 | 7 | 9 | 11 | 13 | 16 | 18 | 20 |
| | | | | | | 2900 | 2923 | 2945 | 2967 | 2989 | 2 | 4 | 6 | 8 | 11 | 13 | 15 | 17 | 19 |
| **20** | 3010 | 3032 | 3054 | 3075 | 3096 | 3118 | 3139 | 3160 | 3181 | 3201 | 2 | 4 | 6 | 8 | 11 | 13 | 15 | 17 | 19 |
| **21** | 3222 | 3243 | 3263 | 3284 | 3304 | 3324 | 3345 | 3365 | 3385 | 3404 | 2 | 4 | 6 | 8 | 10 | 12 | 14 | 16 | 18 |
| **22** | 3424 | 3444 | 3464 | 3483 | 3502 | 3522 | 3541 | 3560 | 3579 | 3598 | 2 | 4 | 6 | 8 | 10 | 12 | 14 | 15 | 17 |
| **23** | 3617 | 3636 | 3655 | 3674 | 3692 | 3711 | 3729 | 3747 | 3766 | 3784 | 2 | 4 | 6 | 7 | 9 | 11 | 13 | 15 | 17 |
| **24** | 3802 | 3820 | 3838 | 3856 | 3874 | 3892 | 3909 | 3927 | 3945 | 3962 | 2 | 4 | 5 | 7 | 9 | 11 | 12 | 14 | 16 |
| **25** | 3979 | 3997 | 4014 | 4031 | 4048 | 4065 | 4082 | 4099 | 4116 | 4133 | 2 | 3 | 5 | 7 | 9 | 10 | 12 | 14 | 15 |
| **26** | 4150 | 4166 | 4183 | 4200 | 4216 | 4232 | 4249 | 4265 | 4281 | 4298 | 2 | 3 | 5 | 7 | 8 | 10 | 11 | 13 | 15 |
| **27** | 4314 | 4330 | 4346 | 4362 | 4378 | 4393 | 4409 | 4425 | 4440 | 4456 | 2 | 3 | 5 | 6 | 8 | 9 | 11 | 13 | 14 |
| **28** | 4472 | 4487 | 4502 | 4518 | 4533 | 4548 | 4564 | 4579 | 4594 | 4609 | 2 | 3 | 5 | 6 | 8 | 9 | 11 | 12 | 14 |
| **29** | 4624 | 4639 | 4654 | 4669 | 4683 | 4698 | 4713 | 4728 | 4742 | 4757 | 1 | 3 | 4 | 6 | 7 | 9 | 10 | 12 | 13 |
| **30** | 4771 | 4786 | 4800 | 4814 | 4829 | 4843 | 4857 | 4871 | 4886 | 4900 | 1 | 3 | 4 | 6 | 7 | 9 | 10 | 11 | 13 |
| **31** | 4914 | 4928 | 4942 | 4955 | 4969 | 4983 | 4997 | 5011 | 5024 | 5038 | 1 | 3 | 4 | 6 | 7 | 8 | 10 | 11 | 12 |
| **32** | 5051 | 5065 | 5079 | 5092 | 5105 | 5119 | 5132 | 5145 | 5159 | 5172 | 1 | 3 | 4 | 5 | 7 | 8 | 9 | 11 | 12 |
| **33** | 5185 | 5198 | 5211 | 5224 | 5237 | 5250 | 5263 | 5276 | 5289 | 5302 | 1 | 3 | 4 | 5 | 6 | 8 | 9 | 10 | 12 |
| **34** | 5315 | 5328 | 5340 | 5353 | 5366 | 5378 | 5391 | 5403 | 5416 | 5428 | 1 | 3 | 4 | 5 | 6 | 8 | 9 | 10 | 11 |
| **35** | 5441 | 5453 | 5465 | 5478 | 5490 | 5502 | 5514 | 5527 | 5539 | 5551 | 1 | 2 | 4 | 5 | 6 | 7 | 9 | 10 | 11 |
| **36** | 5563 | 5575 | 5587 | 5599 | 5611 | 5623 | 5635 | 5647 | 5658 | 5670 | 1 | 2 | 4 | 5 | 6 | 7 | 8 | 10 | 11 |
| **37** | 5682 | 5694 | 5705 | 5717 | 5729 | 5740 | 5752 | 5763 | 5775 | 5786 | 1 | 2 | 3 | 5 | 6 | 7 | 8 | 9 | 10 |
| **38** | 5798 | 5809 | 5821 | 5832 | 5843 | 5855 | 5866 | 5877 | 5888 | 5899 | 1 | 2 | 3 | 5 | 6 | 7 | 8 | 9 | 10 |
| **39** | 5911 | 5922 | 5933 | 5944 | 5955 | 5966 | 5977 | 5988 | 5999 | 6010 | 1 | 2 | 3 | 4 | 5 | 7 | 8 | 9 | 10 |
| **40** | 6021 | 6031 | 6042 | 6053 | 6064 | 6075 | 6085 | 6096 | 6107 | 6117 | 1 | 2 | 3 | 4 | 5 | 6 | 8 | 9 | 10 |
| **41** | 6128 | 6138 | 6149 | 6160 | 6170 | 6180 | 6191 | 6201 | 6212 | 6222 | 1 | 2 | 3 | 4 | 5 | 6 | 7 | 8 | 9 |
| **42** | 6232 | 6243 | 6253 | 6263 | 6274 | 6284 | 6294 | 6304 | 6314 | 6325 | 1 | 2 | 3 | 4 | 5 | 6 | 7 | 8 | 9 |
| **43** | 6335 | 6345 | 6355 | 6365 | 6375 | 6385 | 6395 | 6405 | 6415 | 6425 | 1 | 2 | 3 | 4 | 5 | 6 | 7 | 8 | 9 |
| **44** | 6435 | 6444 | 6454 | 6464 | 6474 | 6484 | 6493 | 6503 | 6513 | 6522 | 1 | 2 | 3 | 4 | 5 | 6 | 7 | 8 | 9 |
| **45** | 6532 | 6542 | 6551 | 6561 | 6571 | 6580 | 6590 | 6599 | 6609 | 6618 | 1 | 2 | 3 | 4 | 5 | 6 | 7 | 8 | 9 |
| **46** | 6628 | 6637 | 6646 | 6656 | 6665 | 6675 | 6684 | 6693 | 6702 | 6712 | 1 | 2 | 3 | 4 | 5 | 6 | 7 | 7 | 8 |
| **47** | 6721 | 6730 | 6739 | 6749 | 6758 | 6767 | 6776 | 6785 | 6794 | 6803 | 1 | 2 | 3 | 4 | 5 | 5 | 6 | 7 | 8 |
| **48** | 6812 | 6821 | 6830 | 6839 | 6848 | 6857 | 6866 | 6875 | 6884 | 6893 | 1 | 2 | 3 | 4 | 4 | 5 | 6 | 7 | 8 |
| **49** | 6902 | 6911 | 6920 | 6928 | 6937 | 6946 | 6955 | 6964 | 6972 | 6981 | 1 | 2 | 3 | 4 | 4 | 5 | 6 | 7 | 8 |
| **50** | 6990 | 6998 | 7007 | 7016 | 7024 | 7033 | 7042 | 7050 | 7059 | 7067 | 1 | 2 | 3 | 3 | 4 | 5 | 6 | 7 | 8 |

| | 0 | 1 | 2 | 3 | 4 | 5 | 6 | 7 | 8 | 9 | 1 | 2 | 3 | 4 | 5 | 6 | 7 | 8 | 9 |
|---|---|---|---|---|---|---|---|---|---|---|---|---|---|---|---|---|---|---|---|
| **51** | 7076 | 7084 | 7093 | 7101 | 7110 | 7118 | 7126 | 7135 | 7143 | 7152 | 1 | 2 | 3 | 3 | 4 | 5 | 6 | 7 | 8 |
| **52** | 7160 | 7168 | 7177 | 7185 | 7193 | 7202 | 7210 | 7218 | 7226 | 7235 | 1 | 2 | 2 | 3 | 4 | 5 | 6 | 7 | 7 |
| **53** | 7243 | 7251 | 7259 | 7267 | 7275 | 7284 | 7292 | 7300 | 7308 | 7316 | 1 | 2 | 2 | 3 | 4 | 5 | 6 | 6 | 7 |
| **54** | 7324 | 7332 | 7340 | 7348 | 7356 | 7364 | 7372 | 7380 | 7388 | 7396 | 1 | 2 | 2 | 3 | 4 | 5 | 6 | 6 | 7 |
| **55** | 7404 | 7412 | 7419 | 7427 | 7435 | 7443 | 7451 | 7459 | 7466 | 7474 | 1 | 2 | 2 | 3 | 4 | 5 | 5 | 6 | 7 |
| **56** | 7482 | 7490 | 7497 | 7505 | 7513 | 7520 | 7528 | 7536 | 7543 | 7551 | 1 | 2 | 2 | 3 | 4 | 5 | 5 | 6 | 7 |
| **57** | 7559 | 7566 | 7574 | 7582 | 7589 | 7597 | 7604 | 7612 | 7619 | 7627 | 1 | 2 | 2 | 3 | 4 | 5 | 5 | 6 | 7 |
| **58** | 7634 | 7642 | 7649 | 7657 | 7664 | 7672 | 7679 | 7686 | 7694 | 7701 | 1 | 1 | 2 | 3 | 4 | 4 | 5 | 6 | 7 |
| **59** | 7709 | 7716 | 7723 | 7731 | 7738 | 7745 | 7752 | 7760 | 7767 | 7774 | 1 | 1 | 2 | 3 | 4 | 4 | 5 | 6 | 7 |
| **60** | 7782 | 7789 | 7796 | 7803 | 7810 | 7818 | 7825 | 7832 | 7839 | 7846 | 1 | 1 | 2 | 3 | 4 | 4 | 5 | 6 | 6 |
| **61** | 7853 | 7860 | 7868 | 7875 | 7882 | 7889 | 7896 | 7903 | 7910 | 7917 | 1 | 1 | 2 | 3 | 4 | 4 | 5 | 6 | 6 |
| **62** | 7924 | 7931 | 7938 | 7945 | 7952 | 7959 | 7966 | 7973 | 7980 | 7987 | 1 | 1 | 2 | 3 | 3 | 4 | 5 | 6 | 6 |
| **63** | 7993 | 8000 | 8007 | 8014 | 8021 | 8028 | 8035 | 8041 | 8048 | 8055 | 1 | 1 | 2 | 3 | 3 | 4 | 5 | 5 | 6 |
| **64** | 8062 | 8069 | 8075 | 8082 | 8089 | 8096 | 8102 | 8109 | 8116 | 8122 | 1 | 1 | 2 | 3 | 3 | 4 | 5 | 5 | 6 |
| **65** | 8129 | 8136 | 8142 | 8149 | 8156 | 8162 | 8169 | 8176 | 8182 | 8189 | 1 | 1 | 2 | 3 | 3 | 4 | 5 | 5 | 6 |
| **66** | 8195 | 8202 | 8209 | 8215 | 8222 | 8228 | 8235 | 8241 | 8248 | 8254 | 1 | 1 | 2 | 3 | 3 | 4 | 5 | 5 | 6 |
| **67** | 8261 | 8267 | 8274 | 8280 | 8287 | 8293 | 8299 | 8306 | 8312 | 8319 | 1 | 1 | 2 | 3 | 3 | 4 | 5 | 5 | 6 |
| **68** | 8325 | 8331 | 8338 | 8344 | 8351 | 8357 | 8363 | 8370 | 8376 | 8382 | 1 | 1 | 2 | 3 | 3 | 4 | 4 | 5 | 6 |
| **69** | 8388 | 8395 | 8401 | 8407 | 8414 | 8420 | 8426 | 8432 | 8439 | 8445 | 1 | 1 | 2 | 2 | 3 | 4 | 4 | 5 | 6 |
| **70** | 8451 | 8457 | 8463 | 8470 | 8476 | 8482 | 8488 | 8494 | 8500 | 8506 | 1 | 1 | 2 | 2 | 3 | 4 | 4 | 5 | 6 |
| **71** | 8513 | 8519 | 8525 | 8531 | 8537 | 8543 | 8549 | 8555 | 8561 | 8567 | 1 | 1 | 2 | 2 | 3 | 4 | 4 | 5 | 5 |
| **72** | 8573 | 8579 | 8585 | 8591 | 8597 | 8603 | 8609 | 8615 | 8621 | 8627 | 1 | 1 | 2 | 2 | 3 | 4 | 4 | 5 | 5 |
| **73** | 8633 | 8639 | 8645 | 8651 | 8657 | 8663 | 8669 | 8675 | 8681 | 8686 | 1 | 1 | 2 | 2 | 3 | 4 | 4 | 5 | 5 |
| **74** | 8692 | 8698 | 8704 | 8710 | 8716 | 8722 | 8727 | 8733 | 8739 | 8745 | 1 | 1 | 2 | 2 | 3 | 4 | 4 | 5 | 5 |
| **75** | 8751 | 8756 | 8762 | 8768 | 8774 | 8779 | 8785 | 8791 | 8797 | 8802 | 1 | 1 | 2 | 2 | 3 | 3 | 4 | 5 | 5 |
| **76** | 8808 | 8814 | 8820 | 8825 | 8831 | 8837 | 8842 | 8848 | 8854 | 8859 | 1 | 1 | 2 | 2 | 3 | 3 | 4 | 5 | 5 |
| **77** | 8865 | 8871 | 8876 | 8882 | 8887 | 8893 | 8899 | 8904 | 8910 | 8915 | 1 | 1 | 2 | 2 | 3 | 3 | 4 | 4 | 5 |
| **78** | 8921 | 8927 | 8932 | 8938 | 8943 | 8949 | 8954 | 8960 | 8965 | 8971 | 1 | 1 | 2 | 2 | 3 | 3 | 4 | 4 | 5 |
| **79** | 8976 | 8982 | 8987 | 8993 | 8998 | 9004 | 9009 | 9015 | 9020 | 9025 | 1 | 1 | 2 | 2 | 3 | 3 | 4 | 4 | 5 |
| **80** | 9031 | 9036 | 9042 | 9047 | 9053 | 9058 | 9063 | 9069 | 9074 | 9079 | 1 | 1 | 2 | 2 | 3 | 3 | 4 | 4 | 5 |
| **81** | 9085 | 9090 | 9096 | 9101 | 9106 | 9112 | 9117 | 9122 | 9128 | 9133 | 1 | 1 | 2 | 2 | 3 | 3 | 4 | 4 | 5 |
| **82** | 9138 | 9143 | 9149 | 9154 | 9159 | 9165 | 9170 | 9175 | 9180 | 9186 | 1 | 1 | 2 | 2 | 3 | 3 | 4 | 4 | 5 |
| **83** | 9191 | 9196 | 9201 | 9206 | 9212 | 9217 | 9222 | 9227 | 9232 | 9238 | 1 | 1 | 2 | 2 | 3 | 3 | 4 | 4 | 5 |
| **84** | 9243 | 9248 | 9253 | 9258 | 9263 | 9269 | 9274 | 9279 | 9284 | 9289 | 1 | 1 | 2 | 2 | 3 | 3 | 4 | 4 | 5 |
| **85** | 9294 | 9299 | 9304 | 9309 | 9315 | 9320 | 9325 | 9330 | 9335 | 9340 | 1 | 1 | 2 | 2 | 3 | 3 | 4 | 4 | 5 |
| **86** | 9345 | 9350 | 9355 | 9360 | 9365 | 9370 | 9375 | 9380 | 9385 | 9390 | 1 | 1 | 2 | 2 | 3 | 3 | 4 | 4 | 5 |
| **87** | 9395 | 9400 | 9405 | 9410 | 9415 | 9420 | 9425 | 9430 | 9435 | 9440 | 0 | 1 | 1 | 2 | 2 | 3 | 3 | 4 | 4 |
| **88** | 9445 | 9450 | 9455 | 9460 | 9465 | 9469 | 9474 | 9479 | 9484 | 9489 | 0 | 1 | 1 | 2 | 2 | 3 | 3 | 4 | 4 |
| **89** | 9494 | 9499 | 9504 | 9509 | 9513 | 9518 | 9523 | 9528 | 9533 | 9538 | 0 | 1 | 1 | 2 | 2 | 3 | 3 | 4 | 4 |
| **90** | 9542 | 9547 | 9552 | 9557 | 9562 | 9566 | 9571 | 9576 | 9581 | 9586 | 0 | 1 | 1 | 2 | 2 | 3 | 3 | 4 | 4 |
| **91** | 9590 | 9595 | 9600 | 9605 | 9609 | 9614 | 9619 | 9624 | 9628 | 9633 | 0 | 1 | 1 | 2 | 2 | 3 | 3 | 4 | 4 |
| **92** | 9638 | 9643 | 9647 | 9652 | 9657 | 9661 | 9666 | 9671 | 9675 | 9680 | 0 | 1 | 1 | 2 | 2 | 3 | 3 | 4 | 4 |
| **93** | 9685 | 9689 | 9694 | 9699 | 9703 | 9708 | 9713 | 9717 | 9722 | 9727 | 0 | 1 | 1 | 2 | 2 | 3 | 3 | 4 | 4 |
| **94** | 9731 | 9736 | 9741 | 9745 | 9750 | 9754 | 9759 | 9763 | 9768 | 9773 | 0 | 1 | 1 | 2 | 2 | 3 | 3 | 4 | 4 |
| **95** | 9777 | 9782 | 9786 | 9791 | 9795 | 9800 | 9805 | 9809 | 9814 | 9818 | 0 | 1 | 1 | 2 | 2 | 3 | 3 | 4 | 4 |
| **96** | 9823 | 9827 | 9832 | 9836 | 9841 | 9845 | 9850 | 9854 | 9859 | 9863 | 0 | 1 | 1 | 2 | 2 | 3 | 3 | 4 | 4 |
| **97** | 9868 | 9872 | 9877 | 9881 | 9886 | 9890 | 9894 | 9899 | 9903 | 9908 | 0 | 1 | 1 | 2 | 2 | 3 | 3 | 4 | 4 |
| **98** | 9912 | 9917 | 9921 | 9926 | 9930 | 9934 | 9939 | 9943 | 9948 | 9952 | 0 | 1 | 1 | 2 | 2 | 3 | 3 | 4 | 4 |
| **99** | 9956 | 9961 | 9965 | 9969 | 9974 | 9978 | 9983 | 9987 | 9991 | 9996 | 0 | 1 | 1 | 2 | 2 | 3 | 3 | 3 | 4 |

# ANTILOGS

| | 0 | 1 | 2 | 3 | 4 | 5 | 6 | 7 | 8 | 9 | 1 | 2 | 3 | 4 | 5 | 6 | 7 | 8 | 9 |
|---|---|---|---|---|---|---|---|---|---|---|---|---|---|---|---|---|---|---|---|
| ·00 | 1000 | 1002 | 1005 | 1007 | 1009 | 1012 | 1014 | 1016 | 1019 | 1021 | 0 | 0 | 1 | 1 | 1 | 1 | 2 | 2 | 2 |
| ·01 | 1023 | 1026 | 1028 | 1030 | 1033 | 1035 | 1038 | 1040 | 1042 | 1045 | 0 | 0 | 1 | 1 | 1 | 1 | 2 | 2 | 2 |
| ·02 | 1047 | 1050 | 1052 | 1054 | 1057 | 1059 | 1062 | 1064 | 1067 | 1069 | 0 | 0 | 1 | 1 | 1 | 1 | 2 | 2 | 2 |
| ·03 | 1072 | 1074 | 1076 | 1079 | 1081 | 1084 | 1086 | 1089 | 1091 | 1094 | 0 | 0 | 1 | 1 | 1 | 1 | 2 | 2 | 2 |
| ·04 | 1096 | 1099 | 1102 | 1104 | 1107 | 1109 | 1112 | 1114 | 1117 | 1119 | 0 | 1 | 1 | 1 | 1 | 2 | 2 | 2 | 2 |
| ·05 | 1122 | 1125 | 1127 | 1130 | 1132 | 1135 | 1138 | 1140 | 1143 | 1146 | 0 | 1 | 1 | 1 | 1 | 2 | 2 | 2 | 2 |
| ·06 | 1148 | 1151 | 1153 | 1156 | 1159 | 1161 | 1164 | 1167 | 1169 | 1172 | 0 | 1 | 1 | 1 | 1 | 2 | 2 | 2 | 2 |
| ·07 | 1175 | 1178 | 1180 | 1183 | 1186 | 1189 | 1191 | 1194 | 1197 | 1199 | 0 | 1 | 1 | 1 | 1 | 2 | 2 | 2 | 2 |
| ·08 | 1202 | 1205 | 1208 | 1211 | 1213 | 1216 | 1219 | 1222 | 1225 | 1227 | 0 | 1 | 1 | 1 | 1 | 2 | 2 | 2 | 3 |
| ·09 | 1230 | 1233 | 1236 | 1239 | 1242 | 1245 | 1247 | 1250 | 1253 | 1256 | 0 | 1 | 1 | 1 | 1 | 2 | 2 | 2 | 3 |
| ·10 | 1259 | 1262 | 1265 | 1268 | 1271 | 1274 | 1276 | 1279 | 1282 | 1285 | 0 | 1 | 1 | 1 | 1 | 2 | 2 | 2 | 3 |
| ·11 | 1288 | 1291 | 1294 | 1297 | 1300 | 1303 | 1306 | 1309 | 1312 | 1315 | 0 | 1 | 1 | 1 | 2 | 2 | 2 | 2 | 3 |
| ·12 | 1318 | 1321 | 1324 | 1327 | 1330 | 1334 | 1337 | 1340 | 1343 | 1346 | 0 | 1 | 1 | 1 | 2 | 2 | 2 | 2 | 3 |
| ·13 | 1349 | 1352 | 1355 | 1358 | 1361 | 1365 | 1368 | 1371 | 1374 | 1377 | 0 | 1 | 1 | 1 | 2 | 2 | 2 | 3 | 3 |
| ·14 | 1380 | 1384 | 1387 | 1390 | 1393 | 1396 | 1400 | 1403 | 1406 | 1409 | 0 | 1 | 1 | 1 | 2 | 2 | 2 | 3 | 3 |
| ·15 | 1413 | 1416 | 1419 | 1422 | 1426 | 1429 | 1432 | 1435 | 1439 | 1442 | 0 | 1 | 1 | 1 | 2 | 2 | 2 | 3 | 3 |
| ·16 | 1445 | 1449 | 1452 | 1455 | 1459 | 1462 | 1466 | 1469 | 1472 | 1476 | 0 | 1 | 1 | 1 | 2 | 2 | 2 | 3 | 3 |
| ·17 | 1479 | 1483 | 1486 | 1489 | 1493 | 1496 | 1500 | 1503 | 1507 | 1510 | 0 | 1 | 1 | 1 | 2 | 2 | 2 | 3 | 3 |
| ·18 | 1514 | 1517 | 1521 | 1524 | 1528 | 1531 | 1535 | 1538 | 1542 | 1545 | 0 | 1 | 1 | 1 | 2 | 2 | 2 | 3 | 3 |
| ·19 | 1549 | 1552 | 1556 | 1560 | 1563 | 1567 | 1570 | 1574 | 1578 | 1581 | 0 | 1 | 1 | 1 | 2 | 2 | 3 | 3 | 3 |
| ·20 | 1585 | 1589 | 1592 | 1596 | 1600 | 1603 | 1607 | 1611 | 1614 | 1618 | 0 | 1 | 1 | 1 | 2 | 2 | 3 | 3 | 3 |
| ·21 | 1622 | 1626 | 1629 | 1633 | 1637 | 1641 | 1644 | 1648 | 1652 | 1656 | 0 | 1 | 1 | 2 | 2 | 2 | 3 | 3 | 3 |
| ·22 | 1660 | 1663 | 1667 | 1671 | 1675 | 1679 | 1683 | 1687 | 1690 | 1694 | 0 | 1 | 1 | 2 | 2 | 2 | 3 | 3 | 3 |
| ·23 | 1698 | 1702 | 1706 | 1710 | 1714 | 1718 | 1722 | 1726 | 1730 | 1734 | 0 | 1 | 1 | 2 | 2 | 2 | 3 | 3 | 4 |
| ·24 | 1738 | 1742 | 1746 | 1750 | 1754 | 1758 | 1762 | 1766 | 1770 | 1774 | 0 | 1 | 1 | 2 | 2 | 2 | 3 | 3 | 4 |
| ·25 | 1778 | 1782 | 1786 | 1791 | 1795 | 1799 | 1803 | 1807 | 1811 | 1816 | 0 | 1 | 1 | 2 | 2 | 2 | 3 | 3 | 4 |
| ·26 | 1820 | 1824 | 1828 | 1832 | 1837 | 1841 | 1845 | 1849 | 1854 | 1858 | 0 | 1 | 1 | 2 | 2 | 3 | 3 | 3 | 4 |
| ·27 | 1862 | 1866 | 1871 | 1875 | 1879 | 1884 | 1888 | 1892 | 1897 | 1901 | 0 | 1 | 1 | 2 | 2 | 3 | 3 | 3 | 4 |
| ·28 | 1905 | 1910 | 1914 | 1919 | 1923 | 1928 | 1932 | 1936 | 1941 | 1945 | 0 | 1 | 1 | 2 | 2 | 3 | 3 | 4 | 4 |
| ·29 | 1950 | 1954 | 1959 | 1963 | 1968 | 1972 | 1977 | 1982 | 1986 | 1991 | 0 | 1 | 1 | 2 | 2 | 3 | 3 | 4 | 4 |
| ·30 | 1995 | 2000 | 2004 | 2009 | 2014 | 2018 | 2023 | 2028 | 2032 | 2037 | 0 | 1 | 1 | 2 | 2 | 3 | 3 | 4 | 4 |
| ·31 | 2042 | 2046 | 2051 | 2056 | 2061 | 2065 | 2070 | 2075 | 2080 | 2084 | 0 | 1 | 1 | 2 | 2 | 3 | 3 | 4 | 4 |
| ·32 | 2089 | 2094 | 2099 | 2104 | 2109 | 2113 | 2118 | 2123 | 2128 | 2133 | 0 | 1 | 1 | 2 | 2 | 3 | 3 | 4 | 4 |
| ·33 | 2138 | 2143 | 2148 | 2153 | 2158 | 2163 | 2168 | 2173 | 2178 | 2183 | 0 | 1 | 1 | 2 | 2 | 3 | 3 | 4 | 4 |
| ·34 | 2188 | 2193 | 2198 | 2203 | 2208 | 2213 | 2218 | 2223 | 2228 | 2234 | 1 | 1 | 2 | 2 | 3 | 3 | 4 | 4 | 5 |
| ·35 | 2239 | 2244 | 2249 | 2254 | 2259 | 2265 | 2270 | 2275 | 2280 | 2286 | 1 | 1 | 2 | 2 | 3 | 3 | 4 | 4 | 5 |
| ·36 | 2291 | 2296 | 2301 | 2307 | 2312 | 2317 | 2323 | 2328 | 2334 | 2339 | 1 | 1 | 2 | 2 | 3 | 3 | 4 | 4 | 5 |
| ·37 | 2344 | 2350 | 2355 | 2360 | 2366 | 2371 | 2377 | 2382 | 2388 | 2393 | 1 | 1 | 2 | 2 | 3 | 3 | 4 | 4 | 5 |
| ·38 | 2399 | 2404 | 2410 | 2415 | 2421 | 2427 | 2432 | 2438 | 2443 | 2449 | 1 | 1 | 2 | 2 | 3 | 3 | 4 | 4 | 5 |
| ·39 | 2455 | 2460 | 2466 | 2472 | 2477 | 2483 | 2489 | 2495 | 2500 | 2506 | 1 | 1 | 2 | 2 | 3 | 3 | 4 | 5 | 5 |
| ·40 | 2512 | 2518 | 2523 | 2529 | 2535 | 2541 | 2547 | 2553 | 2559 | 2564 | 1 | 1 | 2 | 2 | 3 | 4 | 4 | 5 | 5 |
| ·41 | 2570 | 2576 | 2582 | 2588 | 2594 | 2600 | 2606 | 2612 | 2618 | 2624 | 1 | 1 | 2 | 2 | 3 | 4 | 4 | 5 | 5 |
| ·42 | 2630 | 2636 | 2642 | 2649 | 2655 | 2661 | 2667 | 2673 | 2679 | 2685 | 1 | 1 | 2 | 2 | 3 | 4 | 4 | 5 | 6 |
| ·43 | 2692 | 2698 | 2704 | 2710 | 2716 | 2723 | 2729 | 2735 | 2742 | 2748 | 1 | 1 | 2 | 3 | 3 | 4 | 4 | 5 | 6 |
| ·44 | 2754 | 2761 | 2767 | 2773 | 2780 | 2786 | 2793 | 2799 | 2805 | 2812 | 1 | 1 | 2 | 3 | 3 | 4 | 4 | 5 | 6 |
| ·45 | 2818 | 2825 | 2831 | 2838 | 2844 | 2851 | 2858 | 2864 | 2871 | 2877 | 1 | 1 | 2 | 3 | 3 | 4 | 5 | 5 | 6 |
| ·46 | 2884 | 2891 | 2897 | 2904 | 2911 | 2917 | 2924 | 2931 | 2938 | 2944 | 1 | 1 | 2 | 3 | 3 | 4 | 5 | 5 | 6 |
| ·47 | 2951 | 2958 | 2965 | 2972 | 2979 | 2985 | 2992 | 2999 | 3006 | 3013 | 1 | 1 | 2 | 3 | 3 | 4 | 5 | 5 | 6 |
| ·48 | 3020 | 3027 | 3034 | 3041 | 3048 | 3055 | 3062 | 3069 | 3076 | 3083 | 1 | 1 | 2 | 3 | 4 | 4 | 5 | 6 | 6 |
| ·49 | 3090 | 3097 | 3105 | 3112 | 3119 | 3126 | 3133 | 3141 | 3148 | 3155 | 1 | 1 | 2 | 3 | 4 | 4 | 5 | 6 | 6 |

# ANTILOGS

| | 0 | 1 | 2 | 3 | 4 | 5 | 6 | 7 | 8 | 9 | 1 | 2 | 3 | 4 | 5 | 6 | 7 | 8 | 9 |
|---|---|---|---|---|---|---|---|---|---|---|---|---|---|---|---|---|---|---|---|
| ·50 | 3162 | 3170 | 3177 | 3184 | 3192 | 3199 | 3206 | 3214 | 3221 | 3228 | 1 | 1 | 2 | 3 | 4 | 4 | 5 | 6 | 7 |
| ·51 | 3236 | 3243 | 3251 | 3258 | 3266 | 3273 | 3281 | 3289 | 3296 | 3304 | 1 | 2 | 2 | 3 | 4 | 5 | 5 | 6 | 7 |
| ·52 | 3311 | 3319 | 3327 | 3334 | 3342 | 3350 | 3357 | 3365 | 3373 | 3381 | 1 | 2 | 2 | 3 | 4 | 5 | 5 | 6 | 7 |
| ·53 | 3388 | 3396 | 3404 | 3412 | 3420 | 3428 | 3436 | 3443 | 3451 | 3459 | 1 | 2 | 2 | 3 | 4 | 5 | 6 | 6 | 7 |
| ·54 | 3467 | 3475 | 3483 | 3491 | 3499 | 3508 | 3516 | 3524 | 3532 | 3540 | 1 | 2 | 2 | 3 | 4 | 5 | 6 | 6 | 7 |
| ·55 | 3548 | 3556 | 3565 | 3573 | 3581 | 3589 | 3597 | 3606 | 3614 | 3622 | 1 | 2 | 2 | 3 | 4 | 5 | 6 | 7 | 7 |
| ·56 | 3631 | 3639 | 3648 | 3656 | 3664 | 3673 | 3681 | 3690 | 3698 | 3707 | 1 | 2 | 3 | 3 | 4 | 5 | 6 | 7 | 8 |
| ·57 | 3715 | 3724 | 3733 | 3741 | 3750 | 3758 | 3767 | 3776 | 3784 | 3793 | 1 | 2 | 3 | 3 | 4 | 5 | 6 | 7 | 8 |
| ·58 | 3802 | 3811 | 3819 | 3828 | 3837 | 3846 | 3855 | 3864 | 3873 | 3882 | 1 | 2 | 3 | 4 | 4 | 5 | 6 | 7 | 8 |
| ·59 | 3890 | 3899 | 3908 | 3917 | 3926 | 3936 | 3945 | 3954 | 3963 | 3972 | 1 | 2 | 3 | 4 | 5 | 5 | 6 | 7 | 8 |
| ·60 | 3981 | 3990 | 3999 | 4009 | 4018 | 4027 | 4036 | 4046 | 4055 | 4064 | 1 | 2 | 3 | 4 | 5 | 6 | 6 | 7 | 9 |
| ·61 | 4074 | 4083 | 4093 | 4102 | 4111 | 4121 | 4130 | 4140 | 4150 | 4159 | 1 | 2 | 3 | 4 | 5 | 6 | 7 | 8 | 9 |
| ·62 | 4169 | 4178 | 4188 | 4198 | 4207 | 4217 | 4227 | 4236 | 4246 | 4256 | 1 | 2 | 3 | 4 | 5 | 6 | 7 | 8 | 9 |
| ·63 | 4266 | 4276 | 4285 | 4295 | 4305 | 4315 | 4325 | 4335 | 4345 | 4355 | 1 | 2 | 3 | 4 | 5 | 6 | 7 | 8 | 9 |
| ·64 | 4365 | 4375 | 4385 | 4395 | 4406 | 4416 | 4426 | 4436 | 4446 | 4457 | 1 | 2 | 3 | 4 | 5 | 6 | 7 | 8 | 9 |
| ·65 | 4467 | 4477 | 4487 | 4498 | 4508 | 4519 | 4529 | 4539 | 4550 | 4560 | 1 | 2 | 3 | 4 | 5 | 6 | 7 | 8 | 9 |
| ·66 | 4571 | 4581 | 4592 | 4603 | 4613 | 4624 | 4634 | 4645 | 4656 | 4667 | 1 | 2 | 3 | 4 | 5 | 6 | 7 | 9 | 10 |
| ·67 | 4677 | 4688 | 4699 | 4710 | 4721 | 4732 | 4742 | 4753 | 4764 | 4775 | 1 | 2 | 3 | 4 | 5 | 7 | 8 | 9 | 10 |
| ·68 | 4786 | 4797 | 4808 | 4819 | 4831 | 4842 | 4853 | 4864 | 4875 | 4887 | 1 | 2 | 3 | 4 | 6 | 7 | 8 | 9 | 10 |
| ·69 | 4898 | 4909 | 4920 | 4932 | 4943 | 4955 | 4966 | 4977 | 4989 | 5000 | 1 | 2 | 3 | 5 | 6 | 7 | 8 | 9 | 10 |
| ·70 | 5012 | 5023 | 5035 | 5047 | 5058 | 5070 | 5082 | 5093 | 5105 | 5117 | 1 | 2 | 4 | 5 | 6 | 7 | 8 | 9 | 11 |
| ·71 | 5129 | 5140 | 5152 | 5164 | 5176 | 5188 | 5200 | 5212 | 5224 | 5236 | 1 | 2 | 4 | 5 | 6 | 7 | 8 | 10 | 11 |
| ·72 | 5248 | 5260 | 5272 | 5284 | 5297 | 5309 | 5321 | 5333 | 5346 | 5358 | 1 | 2 | 4 | 5 | 6 | 7 | 9 | 10 | 11 |
| ·73 | 5370 | 5383 | 5395 | 5408 | 5420 | 5433 | 5445 | 5458 | 5470 | 5483 | 1 | 3 | 4 | 5 | 6 | 8 | 9 | 10 | 11 |
| ·74 | 5495 | 5508 | 5521 | 5534 | 5546 | 5559 | 5572 | 5585 | 5598 | 5610 | 1 | 3 | 4 | 5 | 6 | 8 | 9 | 10 | 12 |
| ·75 | 5623 | 5636 | 5649 | 5662 | 5675 | 5689 | 5702 | 5715 | 5728 | 5741 | 1 | 3 | 4 | 5 | 7 | 8 | 9 | 10 | 12 |
| ·76 | 5754 | 5768 | 5781 | 5794 | 5808 | 5821 | 5834 | 5848 | 5861 | 5875 | 1 | 3 | 4 | 5 | 7 | 8 | 9 | 11 | 12 |
| ·77 | 5888 | 5902 | 5916 | 5929 | 5943 | 5957 | 5970 | 5984 | 5998 | 6012 | 1 | 3 | 4 | 5 | 7 | 8 | 10 | 11 | 12 |
| ·78 | 6026 | 6039 | 6053 | 6067 | 6081 | 6095 | 6109 | 6124 | 6138 | 6152 | 1 | 3 | 4 | 6 | 7 | 8 | 10 | 11 | 13 |
| ·79 | 6166 | 6180 | 6194 | 6209 | 6223 | 6237 | 6252 | 6266 | 6281 | 6295 | 1 | 3 | 4 | 6 | 7 | 9 | 10 | 11 | 13 |
| ·80 | 6310 | 6324 | 6339 | 6353 | 6368 | 6383 | 6397 | 6412 | 6427 | 6442 | 1 | 3 | 4 | 6 | 7 | 9 | 10 | 12 | 13 |
| ·81 | 6457 | 6471 | 6486 | 6501 | 6516 | 6531 | 6546 | 6561 | 6577 | 6592 | 2 | 3 | 5 | 6 | 8 | 9 | 11 | 12 | 14 |
| ·82 | 6607 | 6622 | 6637 | 6653 | 6668 | 6683 | 6699 | 6714 | 6730 | 6745 | 2 | 3 | 5 | 6 | 8 | 9 | 11 | 12 | 14 |
| ·83 | 6761 | 6776 | 6792 | 6808 | 6823 | 6839 | 6855 | 6871 | 6887 | 6902 | 2 | 3 | 5 | 6 | 8 | 9 | 11 | 13 | 14 |
| ·84 | 6918 | 6934 | 6950 | 6966 | 6982 | 6998 | 7015 | 7031 | 7047 | 7063 | 2 | 3 | 5 | 6 | 8 | 10 | 11 | 13 | 15 |
| ·85 | 7079 | 7096 | 7112 | 7129 | 7145 | 7161 | 7178 | 7194 | 7211 | 7228 | 2 | 3 | 5 | 7 | 8 | 10 | 12 | 13 | 15 |
| ·86 | 7244 | 7261 | 7278 | 7295 | 7311 | 7328 | 7345 | 7362 | 7379 | 7396 | 2 | 3 | 5 | 7 | 8 | 10 | 12 | 13 | 15 |
| ·87 | 7413 | 7430 | 7447 | 7464 | 7482 | 7499 | 7516 | 7534 | 7551 | 7568 | 2 | 3 | 5 | 7 | 9 | 10 | 12 | 14 | 16 |
| ·88 | 7586 | 7603 | 7621 | 7638 | 7656 | 7674 | 7691 | 7709 | 7727 | 7745 | 2 | 4 | 5 | 7 | 9 | 11 | 12 | 14 | 16 |
| ·89 | 7762 | 7780 | 7798 | 7816 | 7834 | 7852 | 7870 | 7889 | 7907 | 7925 | 2 | 4 | 5 | 7 | 9 | 11 | 13 | 14 | 16 |
| ·90 | 7943 | 7962 | 7980 | 7998 | 8017 | 8035 | 8054 | 8072 | 8091 | 8110 | 2 | 4 | 6 | 7 | 9 | 11 | 13 | 15 | 17 |
| ·91 | 8128 | 8147 | 8166 | 8185 | 8204 | 8222 | 8241 | 8260 | 8279 | 8299 | 2 | 4 | 6 | 8 | 9 | 11 | 13 | 15 | 17 |
| ·92 | 8318 | 8337 | 8356 | 8375 | 8395 | 8414 | 8433 | 8453 | 8472 | 8492 | 2 | 4 | 6 | 8 | 10 | 12 | 14 | 15 | 17 |
| ·93 | 8511 | 8531 | 8551 | 8570 | 8590 | 8610 | 8630 | 8650 | 8670 | 8690 | 2 | 4 | 6 | 8 | 10 | 12 | 14 | 16 | 18 |
| ·94 | 8710 | 8730 | 8750 | 8770 | 8790 | 8810 | 8831 | 8851 | 8872 | 8892 | 2 | 4 | 6 | 8 | 10 | 12 | 14 | 16 | 18 |
| ·95 | 8913 | 8933 | 8954 | 8974 | 8995 | 9016 | 9036 | 9057 | 9078 | 9099 | 2 | 4 | 6 | 8 | 10 | 12 | 15 | 17 | 19 |
| ·96 | 9120 | 9141 | 9162 | 9183 | 9204 | 9226 | 9247 | 9268 | 9290 | 9311 | 2 | 4 | 6 | 8 | 11 | 13 | 15 | 17 | 19 |
| ·97 | 9333 | 9354 | 9376 | 9397 | 9419 | 9441 | 9462 | 9484 | 9506 | 9528 | 2 | 4 | 7 | 9 | 11 | 13 | 15 | 17 | 20 |
| ·98 | 9550 | 9572 | 9594 | 9616 | 9638 | 9661 | 9683 | 9705 | 9727 | 9750 | 2 | 4 | 7 | 9 | 11 | 13 | 16 | 18 | 20 |
| ·99 | 9772 | 9795 | 9817 | 9840 | 9863 | 9886 | 9908 | 9931 | 9954 | 9977 | 2 | 5 | 7 | 9 | 11 | 14 | 16 | 18 | 20 |

## NATURAL SINES

| Degree | 0′ | 6′ | 12′ | 18′ | 24′ | 30′ | 36′ | 42′ | 48′ | 54′ | Mean Differences | | | | |
|---|---|---|---|---|---|---|---|---|---|---|---|---|---|---|---|
| | 0°·0 | 0°·1 | 0°·2 | 0°·3 | 0°·4 | 0°·5 | 0°·6 | 0°·7 | 0°·8 | 0°·9 | 1′ | 2′ | 3′ | 4′ | 5′ |
| **0** | ·0000 | 0017 | 0035 | 0052 | 0070 | 0087 | 0105 | 0122 | 0140 | 0157 | 3 | 6 | 9 | 12 | 15 |
| **1** | ·0175 | 0192 | 0209 | 0227 | 0244 | 0262 | 0279 | 0297 | 0314 | 0332 | 3 | 6 | 9 | 12 | 15 |
| **2** | ·0349 | 0366 | 0384 | 0401 | 0419 | 0436 | 0454 | 0471 | 0488 | 0506 | 3 | 6 | 9 | 12 | 15 |
| **3** | ·0523 | 0541 | 0558 | 0576 | 0593 | 0610 | 0628 | 0645 | 0663 | 0680 | 3 | 6 | 9 | 12 | 15 |
| **4** | ·0698 | 0715 | 0732 | 0750 | 0767 | 0785 | 0802 | 0819 | 0837 | 0854 | 3 | 6 | 9 | 12 | 14 |
| **5** | ·0872 | 0889 | 0906 | 0924 | 0941 | 0958 | 0976 | 0993 | 1011 | 1028 | 3 | 6 | 9 | 12 | 14 |
| **6** | ·1045 | 1063 | 1080 | 1097 | 1115 | 1132 | 1149 | 1167 | 1184 | 1201 | 3 | 6 | 9 | 12 | 14 |
| **7** | ·1219 | 1236 | 1253 | 1271 | 1288 | 1305 | 1323 | 1340 | 1357 | 1374 | 3 | 6 | 9 | 12 | 14 |
| **8** | ·1392 | 1409 | 1426 | 1444 | 1461 | 1478 | 1495 | 1513 | 1530 | 1547 | 3 | 6 | 9 | 12 | 14 |
| **9** | ·1564 | 1582 | 1599 | 1616 | 1633 | 1650 | 1668 | 1685 | 1702 | 1719 | 3 | 6 | 9 | 12 | 14 |
| **10** | ·1736 | 1754 | 1771 | 1788 | 1805 | 1822 | 1840 | 1857 | 1874 | 1891 | 3 | 6 | 9 | 11 | 14 |
| **11** | ·1908 | 1925 | 1942 | 1959 | 1977 | 1994 | 2011 | 2028 | 2045 | 2062 | 3 | 6 | 9 | 11 | 14 |
| **12** | ·2079 | 2096 | 2113 | 2130 | 2147 | 2164 | 2181 | 2198 | 2215 | 2233 | 3 | 6 | 9 | 11 | 14 |
| **13** | ·2250 | 2267 | 2284 | 2300 | 2317 | 2334 | 2351 | 2368 | 2385 | 2402 | 3 | 6 | 8 | 11 | 14 |
| **14** | ·2419 | 2436 | 2453 | 2470 | 2487 | 2504 | 2521 | 2538 | 2554 | 2571 | 3 | 6 | 8 | 11 | 14 |
| **15** | ·2588 | 2605 | 2622 | 2639 | 2656 | 2672 | 2689 | 2706 | 2723 | 2740 | 3 | 6 | 8 | 11 | 14 |
| **16** | ·2756 | 2773 | 2790 | 2807 | 2823 | 2840 | 2857 | 2874 | 2890 | 2907 | 3 | 6 | 8 | 11 | 14 |
| **17** | ·2924 | 2940 | 2957 | 2974 | 2990 | 3007 | 3024 | 3040 | 3057 | 3074 | 3 | 6 | 8 | 11 | 14 |
| **18** | ·3090 | 3107 | 3123 | 3140 | 3156 | 3173 | 3190 | 3206 | 3223 | 3239 | 3 | 6 | 8 | 11 | 14 |
| **19** | ·3256 | 3272 | 3289 | 3305 | 3322 | 3338 | 3355 | 3371 | 3387 | 3404 | 3 | 5 | 8 | 11 | 14 |
| **20** | ·3420 | 3437 | 3453 | 3469 | 3486 | 3502 | 3518 | 3535 | 3551 | 3567 | 3 | 5 | 8 | 11 | 14 |
| **21** | ·3584 | 3600 | 3616 | 3633 | 3649 | 3665 | 3681 | 3697 | 3714 | 3730 | 3 | 5 | 8 | 11 | 14 |
| **22** | ·3746 | 3762 | 3778 | 3795 | 3811 | 3827 | 3843 | 3859 | 3875 | 3891 | 3 | 5 | 8 | 11 | 14 |
| **23** | ·3907 | 3923 | 3939 | 3955 | 3971 | 3987 | 4003 | 4019 | 4035 | 4051 | 3 | 5 | 8 | 11 | 14 |
| **24** | ·4067 | 4083 | 4099 | 4115 | 4131 | 4147 | 4163 | 4179 | 4195 | 4210 | 3 | 5 | 8 | 11 | 13 |
| **25** | ·4226 | 4242 | 4258 | 4274 | 4289 | 4305 | 4321 | 4337 | 4352 | 4368 | 3 | 5 | 8 | 11 | 13 |
| **26** | ·4384 | 4399 | 4415 | 4431 | 4446 | 4462 | 4478 | 4493 | 4509 | 4524 | 3 | 5 | 8 | 10 | 13 |
| **27** | ·4540 | 4555 | 4571 | 4586 | 4602 | 4617 | 4633 | 4648 | 4664 | 4679 | 3 | 5 | 8 | 10 | 13 |
| **28** | ·4695 | 4710 | 4726 | 4741 | 4756 | 4772 | 4787 | 4802 | 4818 | 4833 | 3 | 5 | 8 | 10 | 13 |
| **29** | ·4848 | 4863 | 4879 | 4894 | 4909 | 4924 | 4939 | 4955 | 4970 | 4985 | 3 | 5 | 8 | 10 | 13 |
| **30** | ·5000 | 5015 | 5030 | 5045 | 5060 | 5075 | 5090 | 5105 | 5120 | 5135 | 3 | 5 | 8 | 10 | 13 |
| **31** | ·5150 | 5165 | 5180 | 5195 | 5210 | 5225 | 5240 | 5255 | 5270 | 5284 | 2 | 5 | 7 | 10 | 12 |
| **32** | ·5299 | 5314 | 5329 | 5344 | 5358 | 5373 | 5388 | 5402 | 5417 | 5432 | 2 | 5 | 7 | 10 | 12 |
| **33** | ·5446 | 5461 | 5476 | 5490 | 5505 | 5519 | 5534 | 5548 | 5563 | 5577 | 2 | 5 | 7 | 10 | 12 |
| **34** | ·5592 | 5606 | 5621 | 5635 | 5650 | 5664 | 5678 | 5693 | 5707 | 5721 | 2 | 5 | 7 | 10 | 12 |
| **35** | ·5736 | 5750 | 5764 | 5779 | 5793 | 5807 | 5821 | 5835 | 5850 | 5864 | 2 | 5 | 7 | 9 | 12 |
| **36** | ·5878 | 5892 | 5906 | 5920 | 5934 | 5948 | 5962 | 5976 | 5990 | 6004 | 2 | 5 | 7 | 9 | 12 |
| **37** | ·6018 | 6032 | 6046 | 6060 | 6074 | 6088 | 6101 | 6115 | 6129 | 6143 | 2 | 5 | 7 | 9 | 12 |
| **38** | ·6157 | 6170 | 6184 | 6198 | 6211 | 6225 | 6239 | 6252 | 6266 | 6280 | 2 | 5 | 7 | 9 | 11 |
| **39** | ·6293 | 6307 | 6320 | 6334 | 6347 | 6361 | 6374 | 6388 | 6401 | 6414 | 2 | 4 | 7 | 9 | 11 |
| **40** | ·6428 | 6441 | 6455 | 6468 | 6481 | 6494 | 6508 | 6521 | 6534 | 6547 | 2 | 4 | 7 | 9 | 11 |
| **41** | ·6561 | 6574 | 6587 | 6600 | 6613 | 6626 | 6639 | 6652 | 6665 | 6678 | 2 | 4 | 7 | 9 | 11 |
| **42** | ·6691 | 6704 | 6717 | 6730 | 6743 | 6756 | 6769 | 6782 | 6794 | 6807 | 2 | 4 | 6 | 9 | 11 |
| **43** | ·6820 | 6833 | 6845 | 6858 | 6871 | 6884 | 6896 | 6909 | 6921 | 6934 | 2 | 4 | 6 | 8 | 11 |
| **44** | ·6947 | 6959 | 6972 | 6984 | 6997 | 7009 | 7022 | 7034 | 7046 | 7059 | 2 | 4 | 6 | 8 | 10 |
| **45** | ·7071 | 7083 | 7096 | 7108 | 7120 | 7133 | 7145 | 7157 | 7169 | 7181 | 2 | 4 | 6 | 8 | 10 |

# NATURAL SINES

| Degree | 0′ | 6′ | 12′ | 18′ | 24′ | 30′ | 36′ | 42′ | 48′ | 54′ | Mean Differences | | | | |
|---|---|---|---|---|---|---|---|---|---|---|---|---|---|---|---|
| | 0°·0 | 0°·1 | 0°·2 | 0°·3 | 0°·4 | 0°·5 | 0°·6 | 0°·7 | 0°·8 | 0°·9 | 1′ | 2′ | 3′ | 4′ | 5′ |
| **45** | ·7071 | 7083 | 7096 | 7108 | 7120 | 7133 | 7145 | 7157 | 7169 | 7181 | 2 | 4 | 6 | 8 | 10 |
| **46** | ·7193 | 7206 | 7218 | 7230 | 7242 | 7254 | 7266 | 7278 | 7290 | 7302 | 2 | 4 | 6 | 8 | 10 |
| **47** | ·7314 | 7325 | 7337 | 7349 | 7361 | 7373 | 7385 | 7396 | 7408 | 7420 | 2 | 4 | 6 | 8 | 10 |
| **48** | ·7431 | 7443 | 7455 | 7466 | 7478 | 7490 | 7501 | 7513 | 7524 | 7536 | 2 | 4 | 6 | 8 | 10 |
| **49** | ·7547 | 7559 | 7570 | 7581 | 7593 | 7604 | 7615 | 7627 | 7638 | 7649 | 2 | 4 | 6 | 8 | 9 |
| **50** | ·7660 | 7672 | 7683 | 7694 | 7705 | 7716 | 7727 | 7738 | 7749 | 7760 | 2 | 4 | 6 | 7 | 9 |
| **51** | ·7771 | 7782 | 7793 | 7804 | 7815 | 7826 | 7837 | 7848 | 7859 | 7869 | 2 | 4 | 5 | 7 | 9 |
| **52** | ·7880 | 7891 | 7902 | 7912 | 7923 | 7934 | 7944 | 7955 | 7965 | 7976 | 2 | 4 | 5 | 7 | 9 |
| **53** | ·7986 | 7997 | 8007 | 8018 | 8028 | 8039 | 8049 | 8059 | 8070 | 8080 | 2 | 3 | 5 | 7 | 9 |
| **54** | ·8090 | 8100 | 8111 | 8121 | 8131 | 8141 | 8151 | 8161 | 8171 | 8181 | 2 | 3 | 5 | 7 | 8 |
| **55** | ·8192 | 8202 | 8211 | 8221 | 8231 | 8241 | 8251 | 8261 | 8271 | 8281 | 2 | 3 | 5 | 7 | 8 |
| **56** | ·8290 | 8300 | 8310 | 8320 | 8329 | 8339 | 8348 | 8358 | 8368 | 8377 | 2 | 3 | 5 | 6 | 8 |
| **57** | ·8387 | 8396 | 8406 | 8415 | 8425 | 8434 | 8443 | 8453 | 8462 | 8471 | 2 | 3 | 5 | 6 | 8 |
| **58** | ·8480 | 8490 | 8499 | 8508 | 8517 | 8526 | 8536 | 8545 | 8554 | 8563 | 2 | 3 | 5 | 6 | 8 |
| **59** | ·8572 | 8581 | 8590 | 8599 | 8607 | 8616 | 8625 | 8634 | 8643 | 8652 | 1 | 3 | 4 | 6 | 7 |
| **60** | ·8660 | 8669 | 8678 | 8686 | 8695 | 8704 | 8712 | 8721 | 8729 | 8738 | 1 | 3 | 4 | 6 | 7 |
| **61** | ·8746 | 8755 | 8763 | 8771 | 8780 | 8788 | 8796 | 8805 | 8813 | 8821 | 1 | 3 | 4 | 6 | 7 |
| **62** | ·8829 | 8838 | 8846 | 8854 | 8862 | 8870 | 8878 | 8886 | 8894 | 8902 | 1 | 3 | 4 | 5 | 7 |
| **63** | ·8910 | 8918 | 8926 | 8934 | 8942 | 8949 | 8957 | 8965 | 8973 | 8980 | 1 | 3 | 4 | 5 | 6 |
| **64** | ·8988 | 8996 | 9003 | 9011 | 9018 | 9026 | 9033 | 9041 | 9048 | 9056 | 1 | 3 | 4 | 5 | 6 |
| **65** | ·9063 | 9070 | 9078 | 9085 | 9092 | 9100 | 9107 | 9114 | 9121 | 9128 | 1 | 2 | 4 | 5 | 6 |
| **66** | ·9135 | 9143 | 9150 | 9157 | 9164 | 9171 | 9178 | 9184 | 9191 | 9198 | 1 | 2 | 3 | 5 | 6 |
| **67** | ·9205 | 9212 | 9219 | 9225 | 9232 | 9239 | 9245 | 9252 | 9259 | 9265 | 1 | 2 | 3 | 4 | 6 |
| **68** | ·9272 | 9278 | 9285 | 9291 | 9298 | 9304 | 9311 | 9317 | 9323 | 9330 | 1 | 2 | 3 | 4 | 5 |
| **69** | ·9336 | 9342 | 9348 | 9354 | 9361 | 9367 | 9373 | 9379 | 9385 | 9391 | 1 | 2 | 3 | 4 | 5 |
| **70** | ·9397 | 9403 | 9409 | 9415 | 9421 | 9426 | 9432 | 9438 | 9444 | 9449 | 1 | 2 | 3 | 4 | 5 |
| **71** | ·9455 | 9461 | 9466 | 9472 | 9478 | 9483 | 9489 | 9494 | 9500 | 9505 | 1 | 2 | 3 | 4 | 5 |
| **72** | ·9511 | 9516 | 9521 | 9527 | 9532 | 9537 | 9542 | 9548 | 9553 | 9558 | 1 | 2 | 3 | 3 | 4 |
| **73** | ·9563 | 9568 | 9573 | 9578 | 9583 | 9588 | 9593 | 9598 | 9603 | 9608 | 1 | 2 | 2 | 3 | 4 |
| **74** | ·9613 | 9617 | 9622 | 9627 | 9632 | 9636 | 9641 | 9646 | 9650 | 9655 | 1 | 2 | 2 | 3 | 4 |
| **75** | ·9659 | 9664 | 9668 | 9673 | 9677 | 9681 | 9686 | 9690 | 9694 | 9699 | 1 | 1 | 2 | 3 | 4 |
| **76** | ·9703 | 9707 | 9711 | 9715 | 9720 | 9724 | 9728 | 9732 | 9736 | 9740 | 1 | 1 | 2 | 3 | 3 |
| **77** | ·9744 | 9748 | 9751 | 9755 | 9759 | 9763 | 9767 | 9770 | 9774 | 9778 | 1 | 1 | 2 | 3 | 3 |
| **78** | ·9781 | 9785 | 9789 | 9792 | 9796 | 9799 | 9803 | 9806 | 9810 | 9813 | 1 | 1 | 2 | 2 | 3 |
| **79** | ·9816 | 9820 | 9823 | 9826 | 9829 | 9833 | 9836 | 9839 | 9842 | 9845 | 1 | 1 | 2 | 2 | 3 |
| **80** | ·9848 | 9851 | 9854 | 9857 | 9860 | 9863 | 9866 | 9869 | 9871 | 9874 | 0 | 1 | 1 | 2 | 2 |
| **81** | ·9877 | 9880 | 9882 | 9885 | 9888 | 9890 | 9893 | 9895 | 9898 | 9900 | 0 | 1 | 1 | 2 | 2 |
| **82** | ·9903 | 9905 | 9907 | 9910 | 9912 | 9914 | 9917 | 9919 | 9921 | 9923 | 0 | 1 | 1 | 2 | 2 |
| **83** | ·9925 | 9928 | 9930 | 9932 | 9934 | 9936 | 9938 | 9940 | 9942 | 9943 | 0 | 1 | 1 | 1 | 2 |
| **84** | ·9945 | 9947 | 9949 | 9951 | 9952 | 9954 | 9956 | 9957 | 9959 | 9960 | 0 | 1 | 1 | 1 | 2 |
| **85** | ·9962 | 9963 | 9965 | 9966 | 9968 | 9969 | 9971 | 9972 | 9973 | 9974 | 0 | 0 | 1 | 1 | 1 |
| **86** | ·9976 | 9977 | 9978 | 9979 | 9980 | 9981 | 9982 | 9983 | 9984 | 9985 | 0 | 0 | 1 | 1 | 1 |
| **87** | ·9986 | 9987 | 9988 | 9989 | 9990 | 9990 | 9991 | 9992 | 9993 | 9993 | 0 | 0 | 0 | 1 | 1 |
| **88** | ·9994 | 9995 | 9995 | 9996 | 9996 | 9997 | 9997 | 9997 | 9998 | 9998 | 0 | 0 | 0 | 0 | 0 |
| **89** | ·9998 | 9999 | 9999 | 9999 | 9999 | 1·000 | 1·000 | 1·000 | 1·000 | 1·000 | 0 | 0 | 0 | 0 | 0 |
| **90** | 1·000 | | | | | | | | | | | | | | |

## NATURAL COSINES

| Degree | 0′ | 6′ | 12′ | 18′ | 24′ | 30′ | 36′ | 42′ | 48′ | 54′ | Mean Differences | | | | |
|---|---|---|---|---|---|---|---|---|---|---|---|---|---|---|---|
| | 0°·0 | 0°·1 | 0°·2 | 0°·3 | 0°·4 | 0°·5 | 0°·6 | 0°·7 | 0°·8 | 0°·9 | 1′ | 2′ | 3′ | 4′ | 5′ |
| **0** | 1·000 | 1·000 | 1·000 | 1·000 | 1·000 | 1·000 | ·9999 | 9999 | 9999 | 9999 | 0 | 0 | 0 | 0 | 0 |
| **1** | ·9998 | 9998 | 9998 | 9997 | 9997 | 9997 | 9996 | 9996 | 9995 | 9995 | 0 | 0 | 0 | 0 | 0 |
| **2** | ·9994 | 9993 | 9993 | 9992 | 9991 | 9990 | 9990 | 9989 | 9988 | 9987 | 0 | 0 | 0 | 1 | 1 |
| **3** | ·9986 | 9985 | 9984 | 9983 | 9982 | 9981 | 9980 | 9979 | 9978 | 9977 | 0 | 0 | 1 | 1 | 1 |
| **4** | ·9976 | 9974 | 9973 | 9972 | 9971 | 9969 | 9968 | 9966 | 9965 | 9963 | 0 | 0 | 1 | 1 | 1 |
| **5** | ·9962 | 9960 | 9959 | 9957 | 9956 | 9954 | 9952 | 9951 | 9949 | 9947 | 0 | 1 | 1 | 1 | 2 |
| **6** | ·9945 | 9943 | 9942 | 9940 | 9938 | 9936 | 9934 | 9932 | 9930 | 9928 | 0 | 1 | 1 | 1 | 2 |
| **7** | ·9925 | 9923 | 9921 | 9919 | 9917 | 9914 | 9912 | 9910 | 9907 | 9905 | 0 | 1 | 1 | 2 | 2 |
| **8** | ·9903 | 9900 | 9898 | 9895 | 9893 | 9890 | 9888 | 9885 | 9882 | 9880 | 0 | 1 | 1 | 2 | 2 |
| **9** | ·9877 | 9874 | 9871 | 9869 | 9866 | 9863 | 9860 | 9857 | 9854 | 9851 | 0 | 1 | 1 | 2 | 2 |
| **10** | ·9848 | 9845 | 9842 | 9839 | 9836 | 9833 | 9829 | 9826 | 9823 | 9820 | 1 | 1 | 2 | 2 | 3 |
| **11** | ·9816 | 9813 | 9810 | 9806 | 9803 | 9799 | 9796 | 9792 | 9789 | 9785 | 1 | 1 | 2 | 2 | 3 |
| **12** | ·9781 | 9778 | 9774 | 9770 | 9767 | 9763 | 9759 | 9755 | 9751 | 9748 | 1 | 1 | 2 | 3 | 3 |
| **13** | ·9744 | 9740 | 9736 | 9732 | 9728 | 9724 | 9720 | 9715 | 9711 | 9707 | 1 | 1 | 2 | 3 | 3 |
| **14** | ·9703 | 9699 | 9694 | 9690 | 9686 | 9681 | 9677 | 9673 | 9668 | 9664 | 1 | 1 | 2 | 3 | 4 |
| **15** | ·9659 | 9655 | 9650 | 9646 | 9641 | 9636 | 9632 | 9627 | 9622 | 9617 | 1 | 2 | 2 | 3 | 4 |
| **16** | ·9613 | 9608 | 9603 | 9598 | 9593 | 9588 | 9583 | 9578 | 9573 | 9568 | 1 | 2 | 2 | 3 | 4 |
| **17** | ·9563 | 9558 | 9553 | 9548 | 9542 | 9537 | 9532 | 9527 | 9521 | 9516 | 1 | 2 | 3 | 3 | 4 |
| **18** | ·9511 | 9505 | 9500 | 9494 | 9489 | 9483 | 9478 | 9472 | 9466 | 9461 | 1 | 2 | 3 | 4 | 5 |
| **19** | ·9455 | 9449 | 9444 | 9438 | 9432 | 9426 | 9421 | 9415 | 9409 | 9403 | 1 | 2 | 3 | 4 | 5 |
| **20** | ·9397 | 9391 | 9385 | 9379 | 9373 | 9367 | 9361 | 9354 | 9348 | 9342 | 1 | 2 | 3 | 4 | 5 |
| **21** | ·9336 | 9330 | 9323 | 9317 | 9311 | 9304 | 9298 | 9291 | 9285 | 9278 | 1 | 2 | 3 | 4 | 5 |
| **22** | ·9272 | 9265 | 9259 | 9252 | 9245 | 9239 | 9232 | 9225 | 9219 | 9212 | 1 | 2 | 3 | 4 | 6 |
| **23** | ·9205 | 9198 | 9191 | 9184 | 9178 | 9171 | 9164 | 9157 | 9150 | 9143 | 1 | 2 | 3 | 5 | 6 |
| **24** | ·9135 | 9128 | 9121 | 9114 | 9107 | 9100 | 9092 | 9085 | 9078 | 9070 | 1 | 2 | 4 | 5 | 6 |
| **25** | ·9063 | 9056 | 9048 | 9041 | 9033 | 9026 | 9018 | 9011 | 9003 | 8996 | 1 | 3 | 4 | 5 | 6 |
| **26** | ·8988 | 8980 | 8973 | 8965 | 8957 | 8949 | 8942 | 8934 | 8926 | 8918 | 1 | 3 | 4 | 5 | 6 |
| **27** | ·8910 | 8902 | 8894 | 8886 | 8878 | 8870 | 8862 | 8854 | 8846 | 8838 | 1 | 3 | 4 | 5 | 7 |
| **28** | ·8829 | 8821 | 8813 | 8805 | 8796 | 8788 | 8780 | 8771 | 8763 | 8755 | 1 | 3 | 4 | 6 | 7 |
| **29** | ·8746 | 8738 | 8729 | 8721 | 8712 | 8704 | 8695 | 8686 | 8678 | 8669 | 1 | 3 | 4 | 6 | 7 |
| **30** | ·8660 | 8652 | 8643 | 8634 | 8625 | 8616 | 8607 | 8599 | 8590 | 8581 | 1 | 3 | 4 | 6 | 7 |
| **31** | ·8572 | 8563 | 8554 | 8545 | 8536 | 8526 | 8517 | 8508 | 8499 | 8490 | 2 | 3 | 5 | 6 | 8 |
| **32** | ·8480 | 8471 | 8462 | 8453 | 8443 | 8434 | 8425 | 8415 | 8406 | 8396 | 2 | 3 | 5 | 6 | 8 |
| **33** | ·8387 | 8377 | 8368 | 8358 | 8348 | 8339 | 8329 | 8320 | 8310 | 8300 | 2 | 3 | 5 | 6 | 8 |
| **34** | ·8290 | 8281 | 8271 | 8261 | 8251 | 8241 | 8231 | 8221 | 8211 | 8202 | 2 | 3 | 5 | 7 | 8 |
| **35** | ·8192 | 8181 | 8171 | 8161 | 8151 | 8141 | 8131 | 8121 | 8111 | 8100 | 2 | 3 | 5 | 7 | 8 |
| **36** | ·8090 | 8080 | 8070 | 8059 | 8049 | 8039 | 8028 | 8018 | 8007 | 7997 | 2 | 3 | 5 | 7 | 9 |
| **37** | ·7986 | 7976 | 7965 | 7955 | 7944 | 7934 | 7923 | 7912 | 7902 | 7891 | 2 | 4 | 5 | 7 | 9 |
| **38** | ·7880 | 7869 | 7859 | 7848 | 7837 | 7826 | 7815 | 7804 | 7793 | 7782 | 2 | 4 | 5 | 7 | 9 |
| **39** | ·7771 | 7760 | 7749 | 7738 | 7727 | 7716 | 7705 | 7694 | 7683 | 7672 | 2 | 4 | 6 | 7 | 9 |
| **40** | ·7660 | 7649 | 7638 | 7627 | 7615 | 7604 | 7593 | 7581 | 7570 | 7559 | 2 | 4 | 6 | 8 | 9 |
| **41** | ·7547 | 7536 | 7524 | 7513 | 7501 | 7490 | 7478 | 7466 | 7455 | 7443 | 2 | 4 | 6 | 8 | 10 |
| **42** | ·7431 | 7420 | 7408 | 7396 | 7385 | 7373 | 7361 | 7349 | 7337 | 7325 | 2 | 4 | 6 | 8 | 10 |
| **43** | ·7314 | 7302 | 7290 | 7278 | 7266 | 7254 | 7242 | 7230 | 7218 | 7206 | 2 | 4 | 6 | 8 | 10 |
| **44** | ·7193 | 7181 | 7169 | 7157 | 7145 | 7133 | 7120 | 7108 | 7096 | 7083 | 2 | 4 | 6 | 8 | 10 |
| **45** | ·7071 | 7059 | 7046 | 7034 | 7022 | 7009 | 6997 | 6984 | 6972 | 6959 | 2 | 4 | 6 | 8 | 10 |

# NATURAL COSINES

| Degree | 0′ 0°·0 | 6′ 0°·1 | 12′ 0°·2 | 18′ 0°·3 | 24′ 0°·4 | 30′ 0°·5 | 36′ 0°·6 | 42′ 0°·7 | 48′ 0°·8 | 54′ 0°·9 | Mean Differences 1′ | 2′ | 3′ | 4′ | 5′ |
|---|---|---|---|---|---|---|---|---|---|---|---|---|---|---|---|
| **45** | ·7071 | 7059 | 7046 | 7034 | 7022 | 7009 | 6997 | 6984 | 6972 | 6959 | 2 | 4 | 6 | 8 | 10 |
| **46** | ·6947 | 6934 | 6921 | 6909 | 6896 | 6884 | 6871 | 6858 | 6845 | 6833 | 2 | 4 | 6 | 8 | 11 |
| **47** | ·6820 | 6807 | 6794 | 6782 | 6769 | 6756 | 6743 | 6730 | 6717 | 6704 | 2 | 4 | 6 | 9 | 11 |
| **48** | ·6691 | 6678 | 6665 | 6652 | 6639 | 6626 | 6613 | 6600 | 6587 | 6574 | 2 | 4 | 7 | 9 | 11 |
| **49** | ·6561 | 6547 | 6534 | 6521 | 6508 | 6494 | 6481 | 6468 | 6455 | 6441 | 2 | 4 | 7 | 9 | 11 |
| **50** | ·6428 | 6414 | 6401 | 6388 | 6374 | 6361 | 6347 | 6334 | 6320 | 6307 | 2 | 4 | 7 | 9 | 11 |
| **51** | ·6293 | 6280 | 6266 | 6252 | 6239 | 6225 | 6211 | 6198 | 6184 | 6170 | 2 | 5 | 7 | 9 | 11 |
| **52** | ·6157 | 6143 | 6129 | 6115 | 6101 | 6088 | 6074 | 6060 | 6046 | 6032 | 2 | 5 | 7 | 9 | 12 |
| **53** | ·6018 | 6004 | 5990 | 5976 | 5962 | 5948 | 5934 | 5920 | 5906 | 5892 | 2 | 5 | 7 | 9 | 12 |
| **54** | ·5878 | 5864 | 5850 | 5835 | 5821 | 5807 | 5793 | 5779 | 5764 | 5750 | 2 | 5 | 7 | 9 | 12 |
| **55** | ·5736 | 5721 | 5707 | 5693 | 5678 | 5664 | 5650 | 5635 | 5621 | 5606 | 2 | 5 | 7 | 10 | 12 |
| **56** | ·5592 | 5577 | 5563 | 5548 | 5534 | 5519 | 5505 | 5490 | 5476 | 5461 | 2 | 5 | 7 | 10 | 12 |
| **57** | ·5446 | 5432 | 5417 | 5402 | 5388 | 5373 | 5358 | 5344 | 5329 | 5314 | 2 | 5 | 7 | 10 | 12 |
| **58** | ·5299 | 5284 | 5270 | 5255 | 5240 | 5225 | 5210 | 5195 | 5180 | 5165 | 2 | 5 | 7 | 10 | 12 |
| **59** | ·5150 | 5135 | 5120 | 5105 | 5090 | 5075 | 5060 | 5045 | 5030 | 5015 | 3 | 5 | 8 | 10 | 13 |
| **60** | ·5000 | 4985 | 4970 | 4955 | 4939 | 4924 | 4909 | 4894 | 4879 | 4863 | 3 | 5 | 8 | 10 | 13 |
| **61** | ·4848 | 4833 | 4818 | 4802 | 4787 | 4772 | 4756 | 4741 | 4726 | 4710 | 3 | 5 | 8 | 10 | 13 |
| **62** | ·4695 | 4679 | 4664 | 4648 | 4633 | 4617 | 4602 | 4586 | 4571 | 4555 | 3 | 5 | 8 | 10 | 13 |
| **63** | ·4540 | 4524 | 4509 | 4493 | 4478 | 4462 | 4446 | 4431 | 4415 | 4399 | 3 | 5 | 8 | 10 | 13 |
| **64** | ·4384 | 4368 | 4352 | 4337 | 4321 | 4305 | 4289 | 4274 | 4258 | 4242 | 3 | 5 | 8 | 11 | 13 |
| **65** | ·4226 | 4210 | 4195 | 4179 | 4163 | 4147 | 4131 | 4115 | 4099 | 4083 | 3 | 5 | 8 | 11 | 13 |
| **66** | ·4067 | 4051 | 4035 | 4019 | 4003 | 3987 | 3971 | 3955 | 3939 | 3923 | 3 | 5 | 8 | 11 | 14 |
| **67** | ·3907 | 3891 | 3875 | 3859 | 3843 | 3827 | 3811 | 3795 | 3778 | 3762 | 3 | 5 | 8 | 11 | 14 |
| **68** | ·3746 | 3730 | 3714 | 3697 | 3681 | 3665 | 3649 | 3633 | 3616 | 3600 | 3 | 5 | 8 | 11 | 14 |
| **69** | ·3584 | 3567 | 3551 | 3535 | 3518 | 3502 | 3486 | 3469 | 3453 | 3437 | 3 | 5 | 8 | 11 | 14 |
| **70** | ·3420 | 3404 | 3387 | 3371 | 3355 | 3338 | 3322 | 3305 | 3289 | 3272 | 3 | 5 | 8 | 11 | 14 |
| **71** | ·3256 | 3239 | 3223 | 3206 | 3190 | 3173 | 3156 | 3140 | 3123 | 3107 | 3 | 6 | 8 | 11 | 14 |
| **72** | ·3090 | 3074 | 3057 | 3040 | 3024 | 3007 | 2990 | 2974 | 2957 | 2940 | 3 | 6 | 8 | 11 | 14 |
| **73** | ·2924 | 2907 | 2890 | 2874 | 2857 | 2840 | 2823 | 2807 | 2790 | 2773 | 3 | 6 | 8 | 11 | 14 |
| **74** | ·2756 | 2740 | 2723 | 2706 | 2689 | 2672 | 2656 | 2639 | 2622 | 2605 | 3 | 6 | 8 | 11 | 14 |
| **75** | ·2588 | 2571 | 2554 | 2538 | 2521 | 2504 | 2487 | 2470 | 2453 | 2436 | 3 | 6 | 8 | 11 | 14 |
| **76** | ·2419 | 2402 | 2385 | 2368 | 2351 | 2334 | 2317 | 2300 | 2284 | 2267 | 3 | 6 | 8 | 11 | 14 |
| **77** | ·2250 | 2233 | 2215 | 2198 | 2181 | 2164 | 2147 | 2130 | 2113 | 2096 | 3 | 6 | 9 | 11 | 14 |
| **78** | ·2079 | 2062 | 2045 | 2028 | 2011 | 1994 | 1977 | 1959 | 1942 | 1925 | 3 | 6 | 9 | 11 | 14 |
| **79** | ·1908 | 1891 | 1874 | 1857 | 1840 | 1822 | 1805 | 1788 | 1771 | 1754 | 3 | 6 | 9 | 11 | 14 |
| **80** | ·1736 | 1719 | 1702 | 1685 | 1668 | 1650 | 1633 | 1616 | 1599 | 1582 | 3 | 6 | 9 | 12 | 14 |
| **81** | ·1564 | 1547 | 1530 | 1513 | 1495 | 1478 | 1461 | 1444 | 1426 | 1409 | 3 | 6 | 9 | 12 | 14 |
| **82** | ·1392 | 1374 | 1357 | 1340 | 1323 | 1305 | 1288 | 1271 | 1253 | 1236 | 3 | 6 | 9 | 12 | 14 |
| **83** | ·1219 | 1201 | 1184 | 1167 | 1149 | 1132 | 1115 | 1097 | 1080 | 1063 | 3 | 6 | 9 | 12 | 14 |
| **84** | ·1045 | 1028 | 1011 | 0993 | 0976 | 0958 | 0941 | 0924 | 0906 | 0889 | 3 | 6 | 9 | 12 | 14 |
| **85** | ·0872 | 0854 | 0837 | 0819 | 0802 | 0785 | 0767 | 0750 | 0732 | 0715 | 3 | 6 | 9 | 12 | 14 |
| **86** | ·0698 | 0680 | 0663 | 0645 | 0628 | 0610 | 0593 | 0576 | 0558 | 0541 | 3 | 6 | 9 | 12 | 15 |
| **87** | ·0523 | 0506 | 0488 | 0471 | 0454 | 0436 | 0419 | 0401 | 0384 | 0366 | 3 | 6 | 9 | 12 | 15 |
| **88** | ·0349 | 0332 | 0314 | 0297 | 0279 | 0262 | 0244 | 0227 | 0209 | 0192 | 3 | 6 | 9 | 12 | 15 |
| **89** | ·0175 | 0157 | 0140 | 0122 | 0105 | 0087 | 0070 | 0052 | 0035 | 0017 | 3 | 6 | 9 | 12 | 15 |
| **90** | ·0000 | | | | | | | | | | | | | | |

# NATURAL TANGENTS

| Degree | 0′ | 6′ | 12′ | 18′ | 24′ | 30′ | 36′ | 42′ | 48′ | 54′ | Mean Differences | | | | |
|---|---|---|---|---|---|---|---|---|---|---|---|---|---|---|---|
| | 0°·0 | 0°·1 | 0°·2 | 0°·3 | 0°·4 | 0°·5 | 0°·6 | 0°·7 | 0°·8 | 0°·9 | 1′ | 2′ | 3′ | 4′ | 5′ |
| **0** | ·0000 | 0017 | 0035 | 0052 | 0070 | 0087 | 0105 | 0122 | 0140 | 0157 | 3 | 6 | 9 | 12 | 15 |
| **1** | ·0175 | 0192 | 0209 | 0227 | 0244 | 0262 | 0279 | 0297 | 0314 | 0332 | 3 | 6 | 9 | 12 | 15 |
| **2** | ·0349 | 0367 | 0384 | 0402 | 0419 | 0437 | 0454 | 0472 | 0489 | 0507 | 3 | 6 | 9 | 12 | 15 |
| **3** | ·0524 | 0542 | 0559 | 0577 | 0594 | 0612 | 0629 | 0647 | 0664 | 0682 | 3 | 6 | 9 | 12 | 15 |
| **4** | ·0699 | 0717 | 0734 | 0752 | 0769 | 0787 | 0805 | 0822 | 0840 | 0857 | 3 | 6 | 9 | 12 | 15 |
| **5** | ·0875 | 0892 | 0910 | 0928 | 0945 | 0963 | 0981 | 0998 | 1016 | 1033 | 3 | 6 | 9 | 12 | 15 |
| **6** | ·1051 | 1069 | 1086 | 1104 | 1122 | 1139 | 1157 | 1175 | 1192 | 1210 | 3 | 6 | 9 | 12 | 15 |
| **7** | ·1228 | 1246 | 1263 | 1281 | 1299 | 1317 | 1334 | 1352 | 1370 | 1388 | 3 | 6 | 9 | 12 | 15 |
| **8** | ·1405 | 1423 | 1441 | 1459 | 1477 | 1495 | 1512 | 1530 | 1548 | 1566 | 3 | 6 | 9 | 12 | 15 |
| **9** | ·1584 | 1602 | 1620 | 1638 | 1655 | 1673 | 1691 | 1709 | 1727 | 1745 | 3 | 6 | 9 | 12 | 15 |
| **10** | ·1763 | 1781 | 1799 | 1817 | 1835 | 1853 | 1871 | 1890 | 1908 | 1926 | 3 | 6 | 9 | 12 | 15 |
| **11** | ·1944 | 1962 | 1980 | 1998 | 2016 | 2035 | 2053 | 2071 | 2089 | 2107 | 3 | 6 | 9 | 12 | 15 |
| **12** | ·2126 | 2144 | 2162 | 2180 | 2199 | 2217 | 2235 | 2254 | 2272 | 2290 | 3 | 6 | 9 | 12 | 15 |
| **13** | ·2309 | 2327 | 2345 | 2364 | 2382 | 2401 | 2419 | 2438 | 2456 | 2475 | 3 | 6 | 9 | 12 | 15 |
| **14** | ·2493 | 2512 | 2530 | 2549 | 2568 | 2586 | 2605 | 2623 | 2642 | 2661 | 3 | 6 | 9 | 12 | 16 |
| **15** | ·2679 | 2698 | 2717 | 2736 | 2754 | 2773 | 2792 | 2811 | 2830 | 2849 | 3 | 6 | 9 | 13 | 16 |
| **16** | ·2867 | 2886 | 2905 | 2924 | 2943 | 2962 | 2981 | 3000 | 3019 | 3038 | 3 | 6 | 9 | 13 | 16 |
| **17** | ·3057 | 3076 | 3096 | 3115 | 3134 | 3153 | 3172 | 3191 | 3211 | 3230 | 3 | 6 | 10 | 13 | 16 |
| **18** | ·3249 | 3269 | 3288 | 3307 | 3327 | 3346 | 3365 | 3385 | 3404 | 3424 | 3 | 6 | 10 | 13 | 16 |
| **19** | ·3443 | 3463 | 3482 | 3502 | 3522 | 3541 | 3561 | 3581 | 3600 | 3620 | 3 | 7 | 10 | 13 | 16 |
| **20** | ·3640 | 3659 | 3679 | 3699 | 3719 | 3739 | 3759 | 3779 | 3799 | 3819 | 3 | 7 | 10 | 13 | 17 |
| **21** | ·3839 | 3859 | 3879 | 3899 | 3919 | 3939 | 3959 | 3979 | 4000 | 4020 | 3 | 7 | 10 | 13 | 17 |
| **22** | ·4040 | 4061 | 4081 | 4101 | 4122 | 4142 | 4163 | 4183 | 4204 | 4224 | 3 | 7 | 10 | 14 | 17 |
| **23** | ·4245 | 4265 | 4286 | 4307 | 4327 | 4348 | 4369 | 4390 | 4411 | 4431 | 3 | 7 | 10 | 14 | 17 |
| **24** | ·4452 | 4473 | 4494 | 4515 | 4536 | 4557 | 4578 | 4599 | 4621 | 4642 | 4 | 7 | 11 | 14 | 18 |
| **25** | ·4663 | 4684 | 4706 | 4727 | 4748 | 4770 | 4791 | 4813 | 4834 | 4856 | 4 | 7 | 11 | 14 | 18 |
| **26** | ·4877 | 4899 | 4921 | 4942 | 4964 | 4986 | 5008 | 5029 | 5051 | 5073 | 4 | 7 | 11 | 15 | 18 |
| **27** | ·5095 | 5117 | 5139 | 5161 | 5184 | 5206 | 5228 | 5250 | 5272 | 5295 | 4 | 7 | 11 | 15 | 18 |
| **28** | ·5317 | 5340 | 5362 | 5384 | 5407 | 5430 | 5452 | 5475 | 5498 | 5520 | 4 | 8 | 11 | 15 | 19 |
| **29** | ·5543 | 5566 | 5589 | 5612 | 5635 | 5658 | 5681 | 5704 | 5727 | 5750 | 4 | 8 | 12 | 15 | 19 |
| **30** | ·5774 | 5797 | 5820 | 5844 | 5867 | 5890 | 5914 | 5938 | 5961 | 5985 | 4 | 8 | 12 | 16 | 20 |
| **31** | ·6009 | 6032 | 6056 | 6080 | 6104 | 6128 | 6152 | 6176 | 6200 | 6224 | 4 | 8 | 12 | 16 | 20 |
| **32** | ·6249 | 6273 | 6297 | 6322 | 6346 | 6371 | 6395 | 6420 | 6445 | 6469 | 4 | 8 | 12 | 16 | 20 |
| **33** | ·6494 | 6519 | 6544 | 6569 | 6594 | 6619 | 6644 | 6669 | 6694 | 6720 | 4 | 8 | 13 | 17 | 21 |
| **34** | ·6745 | 6771 | 6796 | 6822 | 6847 | 6873 | 6899 | 6924 | 6950 | 6976 | 4 | 9 | 13 | 17 | 21 |
| **35** | ·7002 | 7028 | 7054 | 7080 | 7107 | 7133 | 7159 | 7186 | 7212 | 7239 | 4 | 9 | 13 | 18 | 22 |
| **36** | ·7265 | 7292 | 7319 | 7346 | 7373 | 7400 | 7427 | 7454 | 7481 | 7508 | 5 | 9 | 14 | 18 | 23 |
| **37** | ·7536 | 7563 | 7590 | 7618 | 7646 | 7673 | 7701 | 7729 | 7757 | 7785 | 5 | 9 | 14 | 18 | 23 |
| **38** | ·7813 | 7841 | 7869 | 7898 | 7926 | 7954 | 7983 | 8012 | 8040 | 8069 | 5 | 9 | 14 | 19 | 24 |
| **39** | ·8098 | 8127 | 8156 | 8185 | 8214 | 8243 | 8273 | 8302 | 8332 | 8361 | 5 | 10 | 15 | 20 | 24 |
| **40** | ·8391 | 8421 | 8451 | 8481 | 8511 | 8541 | 8571 | 8601 | 8632 | 8662 | 5 | 10 | 15 | 20 | 25 |
| **41** | ·8693 | 8724 | 8754 | 8785 | 8816 | 8847 | 8878 | 8910 | 8941 | 8972 | 5 | 10 | 16 | 21 | 26 |
| **42** | ·9004 | 9036 | 9067 | 9099 | 9131 | 9163 | 9195 | 9228 | 9260 | 9293 | 5 | 11 | 16 | 21 | 27 |
| **43** | ·9325 | 9358 | 9391 | 9424 | 9457 | 9490 | 9523 | 9556 | 9590 | 9623 | 6 | 11 | 17 | 22 | 28 |
| **44** | ·9657 | 9691 | 9725 | 9759 | 9793 | 9827 | 9861 | 9896 | 9930 | 9965 | 6 | 11 | 17 | 23 | 29 |
| **45** | 1·0000 | 0035 | 0070 | 0105 | 0141 | 0176 | 0212 | 0247 | 0283 | 0319 | 6 | 12 | 18 | 24 | 30 |

# NATURAL TANGENTS

| Degree | 0′ | 6′ | 12′ | 18′ | 24′ | 30′ | 36′ | 42′ | 48′ | 54′ | Mean Differences | | | | |
|---|---|---|---|---|---|---|---|---|---|---|---|---|---|---|---|
| | 0°·0 | 0°·1 | 0°·2 | 0°·3 | 0°·4 | 0°·5 | 0°·6 | 0°·7 | 0°·8 | 0°·9 | 1′ | 2′ | 3′ | 4′ | 5′ |
| **45** | 1·0000 | 0035 | 0070 | 0105 | 0141 | 0176 | 0212 | 0247 | 0283 | 0319 | 6 | 12 | 18 | 24 | 30 |
| **46** | 1·0355 | 0392 | 0428 | 0464 | 0501 | 0538 | 0575 | 0612 | 0649 | 0686 | 6 | 12 | 18 | 25 | 31 |
| **47** | 1·0724 | 0761 | 0799 | 0837 | 0875 | 0913 | 0951 | 0990 | 1028 | 1067 | 6 | 13 | 19 | 25 | 32 |
| **48** | 1·1106 | 1145 | 1184 | 1224 | 1263 | 1303 | 1343 | 1383 | 1423 | 1463 | 7 | 13 | 20 | 27 | 33 |
| **49** | 1·1504 | 1544 | 1585 | 1626 | 1667 | 1708 | 1750 | 1792 | 1833 | 1875 | 7 | 14 | 21 | 28 | 34 |
| **50** | 1·1918 | 1960 | 2002 | 2045 | 2088 | 2131 | 2174 | 2218 | 2261 | 2305 | 7 | 14 | 22 | 29 | 36 |
| **51** | 1·2349 | 2393 | 2437 | 2482 | 2527 | 2572 | 2617 | 2662 | 2708 | 2753 | 8 | 15 | 23 | 30 | 38 |
| **52** | 1·2799 | 2846 | 2892 | 2938 | 2985 | 3032 | 3079 | 3127 | 3175 | 3222 | 8 | 16 | 24 | 31 | 39 |
| **53** | 1·3270 | 3319 | 3367 | 3416 | 3465 | 3514 | 3564 | 3613 | 3663 | 3713 | 8 | 16 | 25 | 33 | 41 |
| **54** | 1·3764 | 3814 | 3865 | 3916 | 3968 | 4019 | 4071 | 4124 | 4176 | 4229 | 9 | 17 | 26 | 34 | 43 |
| **55** | 1·4281 | 4335 | 4388 | 4442 | 4496 | 4550 | 4605 | 4659 | 4715 | 4770 | 9 | 18 | 27 | 36 | 45 |
| **56** | 1·4826 | 4882 | 4938 | 4994 | 5051 | 5108 | 5166 | 5224 | 5282 | 5340 | 10 | 19 | 29 | 38 | 48 |
| **57** | 1·5399 | 5458 | 5517 | 5577 | 5637 | 5697 | 5757 | 5818 | 5880 | 5941 | 10 | 20 | 30 | 40 | 50 |
| **58** | 1·6003 | 6066 | 6128 | 6191 | 6255 | 6319 | 6383 | 6447 | 6512 | 6577 | 11 | 21 | 32 | 43 | 53 |
| **59** | 1·6643 | 6709 | 6775 | 6842 | 6909 | 6977 | 7045 | 7113 | 7182 | 7251 | 11 | 23 | 34 | 45 | 56 |
| **60** | 1·7321 | 7391 | 7461 | 7532 | 7603 | 7675 | 7747 | 7820 | 7893 | 7966 | 12 | 24 | 36 | 48 | 60 |
| **61** | 1·8040 | 8115 | 8190 | 8265 | 8341 | 8418 | 8495 | 8572 | 8650 | 8728 | 13 | 26 | 38 | 51 | 64 |
| **62** | 1·8807 | 8887 | 8967 | 9047 | 9128 | 9210 | 9292 | 9375 | 9458 | 9542 | 14 | 27 | 41 | 55 | 68 |
| **63** | 1·9626 | 9711 | 9797 | 9883 | 9970 | $\bar{0}057$ | $\bar{0}145$ | $\bar{0}233$ | $\bar{0}323$ | $\bar{0}413$ | 15 | 29 | 44 | 58 | 73 |
| **64** | 2·0503 | 0594 | 0686 | 0778 | 0872 | 0965 | 1060 | 1155 | 1251 | 1348 | 16 | 31 | 47 | 63 | 78 |
| **65** | 2·1445 | 1543 | 1642 | 1742 | 1842 | 1943 | 2045 | 2148 | 2251 | 2355 | 17 | 34 | 51 | 68 | 85 |
| **66** | 2·2460 | 2566 | 2673 | 2781 | 2889 | 2998 | 3109 | 3220 | 3332 | 3445 | 18 | 37 | 55 | 73 | 92 |
| **67** | 2·3559 | 3673 | 3789 | 3906 | 4023 | 4142 | 4262 | 4383 | 4504 | 4627 | 20 | 40 | 60 | 79 | 99 |
| **68** | 2·4751 | 4876 | 5002 | 5129 | 5257 | 5386 | 5517 | 5649 | 5782 | 5916 | 22 | 43 | 65 | 87 | 108 |
| **69** | 2·6051 | 6187 | 6325 | 6464 | 6605 | 6746 | 6889 | 7034 | 7179 | 7326 | 24 | 47 | 71 | 95 | 119 |
| **70** | 2·7475 | 7625 | 7776 | 7929 | 8083 | 8239 | 8397 | 8556 | 8716 | 8878 | 26 | 52 | 78 | 104 | 131 |
| **71** | 2·9042 | 9208 | 9375 | 9544 | 9714 | 9887 | $\bar{0}061$ | $\bar{0}237$ | $\bar{0}415$ | $\bar{0}595$ | 29 | 58 | 87 | 116 | 145 |
| **72** | 3·0777 | 0961 | 1146 | 1334 | 1524 | 1716 | 1910 | 2106 | 2305 | 2506 | 32 | 64 | 96 | 129 | 161 |
| **73** | 3·2709 | 2914 | 3122 | 3332 | 3544 | 3759 | 3977 | 4197 | 4420 | 4646 | 36 | 72 | 108 | 144 | 180 |
| **74** | 3·4874 | 5105 | 5339 | 5576 | 5816 | 6059 | 6305 | 6554 | 6806 | 7062 | 41 | 81 | 122 | 163 | 204 |
| **75** | 3·7321 | 7583 | 7848 | 8118 | 8391 | 8667 | 8947 | 9232 | 9520 | 9812 | 46 | 93 | 139 | 186 | 232 |
| **76** | 4·0108 | 0408 | 0713 | 1022 | 1335 | 1653 | 1976 | 2303 | 2635 | 2972 | | | | | |
| **77** | 4·3315 | 3662 | 4015 | 4374 | 4737 | 5107 | 5483 | 5864 | 6252 | 6646 | | | | | |
| **78** | 4·7046 | 7453 | 7867 | 8288 | 8716 | 9152 | 9594 | $\bar{0}045$ | $\bar{0}504$ | $\bar{0}970$ | | | | | |
| **79** | 5·1446 | 1929 | 2422 | 2924 | 3435 | 3955 | 4486 | 5026 | 5578 | 6140 | | | | | |
| **80** | 5·6713 | 7297 | 7894 | 8502 | 9124 | 9758 | $\bar{0}405$ | $\bar{1}066$ | $\bar{1}742$ | $\bar{2}432$ | | | | | |
| **81** | 6·3138 | 3859 | 4596 | 5350 | 6122 | 6912 | 7720 | 8548 | 9395 | $\bar{0}264$ | Interpolation is no longer sufficiently accurate. | | | | |
| **82** | 7·1154 | 2066 | 3002 | 3962 | 4947 | 5958 | 6996 | 8062 | 9158 | $\bar{0}285$ | | | | | |
| **83** | 8·1443 | 2636 | 3863 | 5126 | 6427 | 7769 | 9152 | 0579 | 2052 | 3572 | | | | | |
| **84** | 9·514 | 9·677 | 9·845 | 10·02 | 10·20 | 10·39 | 10·58 | 10·78 | 10·99 | 11·20 | | | | | |
| **85** | 11·43 | 11·66 | 11·91 | 12·16 | 12·43 | 12·71 | 13·00 | 13·30 | 13·62 | 13·95 | | | | | |
| **86** | 14·30 | 14·67 | 15·06 | 15·46 | 15·89 | 16·35 | 16·83 | 17·34 | 17·89 | 18·46 | | | | | |
| **87** | 19·08 | 19·74 | 20·45 | 21·20 | 22·02 | 22·90 | 23·86 | 24·90 | 26·03 | 27·27 | | | | | |
| **88** | 28·64 | 30·14 | 31·82 | 33·69 | 35·80 | 38·19 | 40·92 | 44·07 | 47·74 | 52·08 | | | | | |
| **89** | 57·29 | 63·66 | 71·62 | 81·85 | 95·49 | 114·6 | 143·2 | 191·0 | 286·5 | 573·0 | | | | | |
| **90** | $\infty$ | | | | | | | | | | | | | | |

# Index